Aquaculture Microbiology and Biotechnology

Volume 1

Aquaculture Microbiology and Biotechnology

Volume 1

Editors

Didier Montet

Centre International de Recherche en Agronomie
pour le Développement (CIRAD)
Montpellier
France

Ramesh C. Ray

Central Tuber Crops Research Institute
Bhubaneswar
India

CRC Press
Taylor & Francis Group
Boca Raton London New York

CRC Press is an imprint of the
Taylor & Francis Group, an **informa** business
A SCIENCE PUBLISHERS BOOK

CRC Press
Taylor & Francis Group
6000 Broken Sound Parkway NW, Suite 300
Boca Raton, FL 33487-2742

First issued in paperback 2017

© 2009 by Taylor & Francis Group, LLC
CRC Press is an imprint of Taylor & Francis Group, an Informa business

General enquiries : *info@scipub.net*
Editorial enquiries : *editor@scipub.net*
Sales enquiries   : *sales@scipub.net*

ISBN 13: 978-1-138-11522-4 (pbk)
ISBN 13: 978-1-57808-574-3 (hbk)

Library of Congress Cataloging-in-Publication Data
Aquaculture microbiology and biotechnology / editors, Didier Montet,
Ramesh C. Ray.
    p. cm.
  Includes bibliographical references and index.
  ISBN 978-1-57808-574-3 (hardcover)
1. Fishes--Diseases. 2. Fishes--Genetics. 3. Microbial biotechnology.
4. Aquaculture. I. Montet, Didier. II. Ray, Ramesh C.
  SH171.A72 2009
  639.3--dc22

                                                    2009003772

PREFACE

Aquaculture is currently one of the fastest growing production sectors in the world. It now accounts for nearly half (45%) of the world's food fish (which refers to production of aquatic animals: fish, crustaceans, molluscs, amphibians) and this increase is expected to reach 50% in 2015. Genetic engineering and biotechnology have contributed vastly to this field. The most commonly used methods in fish biotechnology are chromosome manipulation and hormone treatment which can be used to produce triploid, tetraploid, haploid, gynogenetic and androgenetic fish. In Chapter 1— Transgenic Fish— Arvanitoyannis and Tserkezou have described various aspects of genetic engineering in development of transgenic fish.

Montet and his colleagues in Chapter 2 have discussed the application of PCR-DGGE method in tracing the geographical origin of fish using native bacterial flora as a biological marker. This technique is quicker (less than 24 h) than all the classical microbial techniques and can be considered as a provider of a unique biological bar code.

The various bacterial fish diseases, and both, bacterial and viral, shrimp diseases and their molecular diagnostic methods have been addressed in Chapters 3 and 4, respectively. The benefits of using molecular tools are their high sensitivity and specificity. The disadvantages are that they detect nucleic acid in general and not necessarily a viable pathogen.

Intensive use of antibiotics in aquaculture has been associated with the increase of bacterial resistance in the exposed microbial environment (water, sediment, fish bacteria). Sarter and Guichard in Chapter 5 have addressed the detrimental effects of antibiotic resistance in aquaculture such as (i) once acquired, resistance genes can be maintained even in the absence of the corresponding antibiotic, (ii) farming practices impact extends beyond the individual farm environment, and (iii) in response to the antibiotic pressure, bacteria optimizes its resistance system towards multiple drugs to survive leading to multi-resistance patterns. Consequently, the contamination of the environment with bacterial pathogens resistant to antimicrobial agents is a real threat not only as a source of disease but also as a source from which resistance genes can easily spread to other pathogens of diverse origins, which has severe implications on both animal and human health.

Deoxyribonucleic acid (DNA) vaccination is considered as a promising strategy to combat various bacterial and viral diseases in aquaculture. Chapter 6 by Gillund and his colleagues describe some of the prospects and constraints of DNA vaccination in aquaculture. There is a need for further investigation especially concerning immunological reactions following DNA vaccine injection, the fate of the DNA vaccines after injection and environmental release of the DNA vaccine.

Probiotics, commonly refers to the lactic acid bacteria (*Bifidobacterium*, *Lactobacillus*, *Streptococcus*, etc.) and yeasts (*Saccharomyces cerevisiae*), are culture products or live microbial feed supplements, which beneficially affects the host by improving the intestinal (microbial) balance of humans and animals and should be capable of commercialization. Use of probiotics in aquaculture is comparatively new and in Chapter 7, Austin and Brunt have discussed the recent progress in the application of probiotics in freshwater aquaculture, mode of action of probiotics and future development.

Aquaculture practices in India and other Asian countries are based mainly on organic inputs and hence, offer great scope for recycling of a variety of animal wastes like cow dung, cattle shed refuse, lignocellulosic wastes such as crop residues and aquatic macrophytes. In Chapter 8, Barik and Ayappan have addressed various lignocellulosic wastes as resources for freshwater aquaculture, their bioconversion and utilization for fish culture.

The last chapter of this volume (Chapter 9) by Arvanitoyannis and Kassaveti describes subjects pertaining to ethics, legislation, risk assessment concerning genetic engineering in aquaculture, particularly to the release of transgenic fish into the environment. The possible impact on human health and related issues have been also addressed. They predict that the new technology (transgenic fish) may not be totally risk-free but the benefits may vastly overweigh the risks.

Didier Montet
Ramesh C. Ray

CONTENTS

LIST OF CONTRIBUTORS

Aikaterini Kassaveti

School of Agricultural Sciences, Department of Agriculture Ichthyology & Aquatic Environment, University of Thessaly, Fytokou str., 38446, Neo Ionia Magnesia, Volos Hellas (Greece), Tel: +30 24210 93104, Fax: +30 24210 93137

Ana Roque

IRTA-Sant Carles de la Rápita, Carr al Poblenou SN km 5.5 Sant Carles de la Rapita. Spain

E-mail: Ana.Roque@irta.es

Anne Ingeborg Myhr

Genøk – Center for Biosafety, The Science Park in Breivika, Po Box 6418, 9294 Tromsø, Norway

E-mail: anne.myhr@genok.org

B. Austin

School of Life Sciences, Heriot-Watt University, Riccarton, Edinburgh, Scotland, EH14 4AS, U.K.

E-mail: b.austin@hw.ac.uk

Benjamin Guichard

AFSSA-ANMV, BP 90203, 35302 Fougères, France

Bruno Gomez-Gil

CIAD, A.C. Mazatlan Unit for Aquaculture and Environmental Management, A.P. 711 Mazatlan, Sinaloa 82010, Mexico

Didier Montet

Centre de Coopération Internationale en Recherche Agronomique pour le Développement, CIRAD, UMR Qualisud, TA 95B/16, 34398 Montpellier Cedex 5, France, Tel: 33 467615728, Fax: 33 467614444

E-mail: didier.montet@cirad.fr

Frøydis Gillund

Genøk – Center for Biosafety, The Science Park in Breivika, Po Box 6418, 9294 Tromsø, Norway

E-mail: froydis.gillund@genok.org

Gérard Loiseau

Centre de Coopération Internationale en Recherche Agronomique pour le Développement, CIRAD, UMR Qualisud, TA 95B/16, 34398 Montpellier Cedex 5, France

Ioannis S. Arvanitoyannis

School of Agricultural Sciences, Department of Agriculture Ichthyology & Aquatic Environment, University of Thessaly, Fytokou str., 38446, Neo Ionia Magnesia, Volos Hellas (Greece), Tel.: +30 24210 93104, Fax: +30 24210 93137

E-mail: parmenion@uth.gr

J.W. Brunt

School of Life Sciences, Heriot-Watt University, Riccarton, Edinburgh, Scotland, EH14 4AS, U.K.

Present address: Institute of Food Research, Norwich Research Park, Colney, Norwich NR4 7UA, England

Le Nguyen Doan Duy

Centre de Coopération Internationale en Recherche Agronomique pour le Développement, CIRAD, UMR Qualisud, TA 95B/16, 34398 Montpellier Cedex 5, France

Present address: Cantho University, College of Agriculture and Applied Biology Rue 3-2, Campus 2, Ninh Kieu District, Cantho Province, Vietnam

Persephoni Tserkezou

School of Agricultural Sciences, Department of Agriculture Ichthyology & Aquatic Environment, University of Thessaly, Fytokou str., 38446, Neo Ionia Magnesia, Volos Hellas (Greece), Tel: +30 24210 93104, Fax: +30 24210 93137

Ratanaporn Leesing

Centre de Coopération Internationale en Recherche Agronomique pour le Développement, CIRAD, UMR Qualisud, TA 95B/16, 34398 Montpellier Cedex 5, France

Present address: Department of Microbiology, Faculty of Science, Khon Kaen University, Khon Kaen 40002, Thailand

Roy Dalmo

Department of Marine Biotechnology, Norwegian College of Fishery Science, University of Tromsø, N-9037 Tromsø, Norway

E-mail: royd@nfh.uit.no

S. Ayyappan

Indian Council of Agricultural Research, Krishi Anusandhan Bhawan - II, Pusa, New Delhi, 110 012, India, Tel: 011-25846738, Fax: 011-25841955

E-mail: ayyapans@yahoo.co.uk

S.K. Barik

The Patent Office, Intellectual Property Building, CP-2, Sector–V, Salt Lake City, Kolkata, 700 091, West Bengal, India, Tel: 033-23671987, Fax: 033-23671988

E-mail: skb_ipindia@yahoo.co.in

Sarter Samira

CIRAD, UMR QUALISUD, Montpellier, F-34398 France

E-mail: samira.sarter@cirad.fr

Sonia A. Soto-Rodríguez

CIAD, A.C. Mazatlan Unit for Aquaculture and Environmental Management, A.P. 711 Mazatlan, Sinaloa 82010, Mexico, Tel: +52 669 989 8700, Fax: +52 669 989 8701

E-mail: ssoto@ciad.mx

Thierry Goli

Centre de Coopération Internationale en Recherche Agronomique pour le Développement, CIRAD, UMR Qualisud, TA 95B/16, 34398 Montpellier Cedex 5, France

Tom Christian Tonheim

Department of Marine Biotechnology, Norwegian College of Fishery Science, University of Tromsø, N-9037 Tromsø, Norway

E-mail: tom-cto@fom-as.no

1

Transgenic Fish: Issues and Applications

Ioannis S. Arvanitoyannis* and Persephoni Tserkezou

INTRODUCTION

In the last few years, genetics has contributed greatly to fish culture through the application of the more recent techniques developed in biotechnology and in genetic engineering. At present, the most commonly used methods in fish biotechnology are chromosome manipulation and hormonal treatments, which can be used to produce triploid, tetraploid, haploid, gynogenetic and androgenetic fish. These result in the production of individuals and lineages of sterile, monosex or highly endogamic fish (Foresti, 2000).

Fish have a number of characteristics that allow biotechnologists to apply techniques for their genetic improvement (Lutz, 2003). Osteichthyes are the most favourable team within Chordata for experimental application of biotechnological techniques. Fish have an important reproductive advantage. They produce a large number of eggs and certain species can reproduce more than once a year (Maisse et al., 1998).

Biotechnology is used in several different ways in aquaculture. Researchers use transgenics to introduce desirable genetic traits into the fish, thereby creating hardier stock. Transgenics involve the transfer of genes from one species into another, in this case, fish. Other ways in which biotechnology is applied to aquaculture include: improvement of growth rates and control of reproductive cycles through hormone therapy, production of new vaccines and development of disease resistance in fish.

School of Agricultural Sciences, Department of Agriculture Ichthyology & Aquatic Environment, University of Thessaly, Fytokou str., 38446, Neo Ionia Magnesia, Volos Hellas (Greece)
Tel.: +30 24210 93104; Fax: +30 24210 93137
*Corresponding author: E-mail: parmenion@uth.gr

By using different transgenic techniques, researchers are seeking to improve the genetic traits of the fish used in aquaculture. They are trying to develop fish which are: larger and grow faster, more efficient in converting their feed into muscle, resistant to disease, tolerant of low oxygen levels in the water, and tolerant to freezing temperatures (http://www.agwest.sk.ca/publications/infosource/inf_feb98.pdf).

Despite predictions of a growing aquaculture industry, stagnant world capture fisheries and increased populations are projected to lead to a global shortage of fish and fish products in the years to come (Delgado et al., 2002). Use of biotechnology in aquaculture has the potential to alleviate these predicted fish shortages and price increases by enhancing production efficiency, minimizing costs, and reducing disease. However, the incorporation of transgenic organisms into the food chain has met with massive criticism from both the environmental and human health sectors (Rasmussen and Morrissey, 2007).

TRANSGENIC FISH

Since 1985, the field of transgenics has experienced a number of technological advances. Many genetically modified fish species have been established along with various methods for foreign gene insertion (such as microinjection, electroporation, infection with pantropic defective retroviral vectors, particle gun bombardment, and sperm- and testis-mediated gene transfer methods) and detection (such as polymerase chain reaction [PCR]-based assay and Southern blot analysis) (Collares et al., 2005; Smith and Spadafora, 2005). Transgenic fish are appealing to some because attainment of desired traits is generally more effective, direct and selective than traditional breeding, and could prove to be an economic benefit for improvement of production efficiency in aquaculture worldwide (Ramirez and Morrissey, 2003).

Transgenic fish of various species of salmon (*Salmo* sp., *Oncorhynchus* sp.), tilapia (*Oreochromis* sp.), channel catfish (*Ictalurus punctatus*) and others are being actively investigated worldwide as possible new food-producing varieties. Technology developed for using transgenic fish as laboratory models to study developmental biology is being applied to food fish species with the aim of adding agronomically important traits, like improved growth rates and disease resistance (http://www.fda.gov/cvm/transgen.htm).

To create a transgenic fish, a DNA construct containing genes for the desired trait(s) along with a promoter sequence is generally introduced into the pronuclei of fertilized eggs. This is followed by *in vitro* or *in vivo* (implanted into the uterus of a pseudopregnant female) incubation of the injected embryos and subsequent maturation into a fully developed transgenic organism (Chen et al., 1996). Once transgenes have become integrated into a host organism's DNA, they can be passed on to future generations, with the possibility of 100% transmission using stable isogenic transgenic lines (Nam et al., 2002).

Gene transfer research with fish began in the mid 1980's utilizing microinjection (Zhu et al., 1985; Chourrout et al., 1986). Microinjection is a tedious and slow procedure (Powers et al., 1992) and can result in high egg mortality (Dunham et al., 1987). After the initial development of microinjection, new techniques such as electroporation, retroviral integration, liposomal-reverse-phase-evaporation, sperm-mediated transfer and high velocity microprojectile bombardment were developed (Chen and Powers, 1990) that sometimes can more efficiently produce large quantities of transgenic individuals in a shorter time period.

Efficiency of gene transfer is determined by several factors including: hatching percentage, gene integration frequency, the number of eggs, which can be manipulated in a given amount of time and the quantity of effort, required to manipulate the embryos. In this regard electroporation is a powerful technique for mass production of transgenic fish (Yee et al., 1994). The knockout approach is the ultimate method for gene inactivation because it will eliminate the gene products completely. Another technique for post-transcriptional gene silencing is utilization of RNA antisense constructs. Both double stranded RNA and antisense RNA were effective in disrupting the expression of green fluorescence protein (GFP) in transgenic zebrafish (Gong et al., 2002). Various antisense technologies appear feasible (Dunham, 2004). Regardless of the method of transfer, the foreign DNA introduced to the developing embryo, appears to initially replicate and amplify rapidly in the cytoplasm of the developing embryo, and then disappears as development proceeds (Houdebine and Chourrout, 1991).

Transgenic fish have been produced for numerous species of fish including non-commercial model species such as the loach, *Misgurnus anguillicaudatus* (Maclean et al., 1987), medaka, *Oryzias latipes* (Ozato et al., 1986), topminnows and zebrafish, although Gong et al. (2002) have developed transgenic rainbow zebrafish for the ornamental fish industry. Several experiments have evaluated transgenic farmed fish species including goldfish (Zhu et al., 1985), common carp, silver carp, mud loach, rainbow trout (Chourrout et al., 1986), Atlantic salmon, coho salmon, chinook salmon, channel catfish (Dunham et al., 1987) and Nile tilapia (Brem et al., 1988). Additionally, gene transfer has been accomplished in a game fish, northern pike (Gross et al., 1992). Transgenic fishes which are tested for use in aquaculture are summarized in Table 1.1.

Table 1.1 Transgenic fish being tested for use in aquaculture

Species	Foreign gene	Desired effect	Results	References
Goldfish (*Carassius auratus*)	Human GH	Increased growth	Success	Zhu et al., 1985
Loach (*Misgurnus anguillicaudatus*)	Human GH	3- to 4.6-fold increase in growth over 135 d	Success	Zhu et al., 1986
Carp (*Cyprinus carpio*)	Salmon and human GH Rainbow trout GH	20-150% growth improvement Improved disease resistance Tolerance of low oxygen level	Success	Zhang et al., 1990; Chen et al., 1992; Fu et al., 1998
Channel catfish (*Ictalurus punctatus*)	GH and rous sarcoma virus promoter	Some increased growth rates (23-26%)	Some normal growth rates	Dunham et al., 1992; Maclean and Laight, 2000
Pike, northern (*Esox lucius*)	Bovine GH or Chinook salmon GH	Growth enhancement	Success	Gross et al., 1992
Atlantic salmon (*Salmo salar*)	Arctic flatfish AFP	Cold tolerance	Trying to achieve freeze resistance	Fletcher et al., 1992; Hew et al., 1995
Atlantic salmon (*Salmo salar*)	Rainbow trout lysozyme gene and ocean pout AFP promoter	Bacterial resistance is desired	No results have been reported	Hew et al., 1995
Coho salmon (*Oncorhynchus kisutch*)	Arctic flatfish AFP and Chinook salmon GH	10- to 30-fold growth increase after 1 year	Success	Devlin et al., 1995
Goldfish (*Carassius auratus*)	Ocean pout type III AFP	Increased cold tolerance	Success	Wang et al., 1995
Tilapia (*Oreochromis* sp.)	Tilapia GH and Human CMV	Increase growth and feed conversion efficiency (290%)	Stable inheritance	Martinez et al., 1996; Kapuscinski, 2005
Red sea bream (*Pagrosomus major*)	Chinook salmon GH and ocean pout AFP gene promoter	9.3% increase in length 21% increase in weight after 7 mon	Success	Zhang et al., 1998

Table 1.1 contd...

Species	Gene/construct	Effect	Result/status	References
Tilapia (*Oreochromis niloticus*)	Chinook salmon GH with oceanpout type III AFP promoter	Increased growth (2.5-4x) and 20% greater feed conversion efficiency / Germline transmission / No organ growth abnormalities	Seeking regulatory approval	Rahman and Maclean, 1998; Maclean and Laight, 2000; Rahman et al., 2001; Caelers et al., 2005
Arctic charr (*Salvelinus alpinus* L.)	Sockeye salmon GH and human CMV (also tried sockeye piscine metallothionein B and histone 3 promoters)	14-fold weight increase after 10 mon	CMV promoter resulted in greatest weight increase, with no difference in muscle composition compared to controls	Krasnov et al., 1999; Pitkanen et al., 1999b
Rainbow trout (*Oncorhynchus mykiss*)	Human glucose transporter and rat hexinose type II with viral or sockeye salmon promoters	Improved carbohydrate metabolism	Success	Pitkanen et al., 1999a
Atlantic salmon (*Salmo salar*)	Antifreeze protein gene (wflAFP-6)	Cold resistance	Success	Hew et al., 1999
Carp (*Cyprinus carpio*)	Grass carp GH and common carp β-actin promoter	No harmful effects to mice fed transgenic fish	Increased growth in only 8.7% of transgenic fish	Zhang et al., 2000; Wang et al., 2001
Atlantic salmon (*Salmo salar*)	Chinook salmon GH	Greater feed efficiency 2- to 13-fold growth increase	Inheritance through 6 generations	Cook et al., 2000
Striped bass (*Morone saxatilis*)	Insect genes	Disease resistance	Success	FAO, 2000
Indian major carp (*Labeo rohita*)	Human GH	Increased growth	Success	FAO, 2000
Channel catfish (*Ictalurus punctatus*)	GH	33% growth improvement in culture conditions	Success	FAO, 2000

Table 1.1 contd...

Species	Gene/construct	Effect	Result	Reference
Goldfish (Carassius auratus)	GH and Arctic flatfish AFP	Increased growth	Success	FAO, 2000
Chinook salmon (Oncorhynchus tschawytscha)	Arctic flatfish AFP and salmon GH	Enhanced growth and feed efficiency	Success	FAO, 2000
Cutthroat trout (Oncorhynchus clarki)	Arctic flatfish AFP and Chinook salmon GH	Increased growth	Success	FAO, 2000
Rainbow trout (Oncorhynchus mykiss)	Arctic AFP and salmon GH	Increased growth and feed efficiency	Success	FAO, 2000
Crayfish (Procambarus clarkii)	Replication defective pantropic retroviral vector	Successful transformation of immature gonads	Germline transmission	Sarmasik et al., 2001
Eastern oyster (Crassostrea virginica)	Aminoglycoside phosphotransferase II (neo^r)	Increased survival rates when exposed to antibiotics	Success	Buchanan et al., 2001
Grass carp (Ctenopharyngodon idellus)	hLF and common carp β-actin promoter	Increased disease resistance to grass carp haemorrhage virus	Success	Zhong et al., 2002
Channel catfish (Ictalurus punctatus)	Silk moth (Hyalophora cecropia) cecropin genes	Bactericidal activity	Increased survival when exposed to pathogens	Dunham et al., 2002; Dunham, 2005
Japanese medaka (Oryzias latipes)	Insect cecropin or pig cecropin-like peptide genes and CMV	Enhanced bactericidal activity against common fish pathogens	Germline transmission	Sarmasik et al., 2002
Mud loach (Misgurnus mizolepis)	Mud loach GH and mud loach and mouse promoter genes	Increased growth and feed conversion efficiency 2- to 30-fold growth enhancement	100% germline transmission	Nam et al., 2002; Kapuscinski, 2005
Silver sea bream (Sparus sarba)	Rainbow trout GH and common carp β-actin promoter	Increased growth	Successful use of sperm- and testis-mediated gene transfer techniques	Lu et al., 2002

Table 1.1 contd...

Species	Gene/construct	Trait	Result	Reference
Atlantic salmon (*Salmo salar*)	Mx genes	Potential resistance to pathogens following treatment with poly I:C	Success	Jensen et al., 2002
Japanese medaka (*Oryzias latipes*)	*Aspergillus niger* phytase gene and human CMV or sockeye salmon histone type III promoter	Ability to digest phytate, the major form of phosphorus in plants	Increased survival on high phytate diet	Hostetler et al., 2003
Coho salmon (*Oncorhynchus kisutch*)	Growth hormone (GH)	Resistance to the bacterial pathogen *Vibrio anguillarum* Stress response	Increased disease resistance	Jhingan et al., 2003
Grass carp (*Ctenopharyngodon idellus*)	hLF	Increased disease resistance to bacterial pathogen	Success	Mao et al., 2004
Nile tilapia (*Oreochromis niloticus*) and redbelly tilapia (*Tilapia zillii*)	Shark (*Squalus acanthias* L.) IgM genes	Enhanced immune response and growth	Abnormal gonad development at high doses	El-Zaeem and Assem, 2004; Assem and El-Zaeem, 2005
Indian major carp (*Labeo rohita*)	Carp GH and CMV promoter and internal robisimal entry sites element	4- to 5-fold growth increases compared to controls	99% mortality by 10th wk	Pandian and Venugopal, 2005
Zebrafish (*Danio rerio*)	Masu salmon n-6-desaturase-like gene and medaka β-actin promoter	Increased ability to convert ALA to DHA and EPA	Success	Alimuddin et al., 2005
Shrimp (*Litopenaeus vannamei*)	Antisense TSV-CP and shrimp β-actin promoter	Stable expression No biological abnormalities	Increased survival (83% vs 44% in controls) when challenged with TSV	Lu and Sun, 2005
Black tiger shrimp (*Penaeus monodon*)	Kuruma prawn EF-1α promoter and GFP gene or CAT gene	Ubiquitous expression of GFP/CAT	Success	Yazawa et al., 2005

Table 1.1 contd...

Species	Gene/construct	Result/Use	Status	Reference
Zebrafish (*Danio rerio*)	4 Japanese flounder (*Paralichthys olivaceus*) promoters (complement component C3, gelatinase B, keratin and tumour necrosis factor) linked to GFP	Tissue-specific GFP gene expression: complement component C3 in liver gelatinase B in pectoral fin and gills keratin in skin and liver tumour necrosis factor in pharynx and heart	Ready for use	Yazawa et al., 2005
Clams (*Mulina lateralis*)	Retroviral insertion	Successful gene transfer method for use as a model species	Retroviral insertion method is patented	Kapuscinski, 2005
Carp (*Cyprinus carpio*)	Antisense-GnRH mRNA	Sterility	30% of injected fish did not develop gonads	Hu et al., 2006
Shrimp (*Litopenaeus vannamei*)	IHHNV promoters	Functional for use in gene transfer Potential use in expression vectors	Success	Dhar et al., 2006
Zebrafish (*Danio rerio*)	Antisense salmon GnRH and common carp β-actin promoter	Sterility	30% of injected eggs	Hu et al., 2006
Zebrafish (*Danio rerio*)	Japanese flounder keratin promoter linked to hen egg white (HEW) lysozyme gene and green fluorescence protein (GFP) gene	Resistance against bacterial disease	Success	Yazawa et al., 2006
Zebrafish (*Danio rerio*)	Cre recombinase driven by T7 promoter and fluorescent protein flanked by two loxP sites crossed with T7 RNA polymerase (gonad-specific)	Gonad-specific excision of foreign (fluorescent protein) gene	Success	Hu et al., 2006

Table 1.1 contd...

Species	Gene	Effect	Result	Reference
Turbot (*Scophthalmus maximus* L.)	*Synechocystis* sp. PCC6803 containing the *Paralichthys olivaceus* GH gene	Growth ratio Feed intake and feed efficiency ratio	Increased growth rate	Liu et al., 2007
Atlantic salmon (*Salom salar*)	Growth hormone	Increased rates of protein synthesis and lipid mobilization Affecting feed conversion efficiency, metabolic rate, body composition, head and body morphometrics, osmoregulation and age at maturity	Success	Hallerman et al., 2007
Coho salmon (*Oncorhynchus kisutch*)	Growth hormone	Increased rates of protein synthesis and lipid mobilization	Success	Hallerman et al., 2007
Coho salmon (*Oncorhynchus kisutch*)	Growth hormone	Increased tissue glutathione Increased hepatic glutathione reductase activity Increased intestinal activity of the glutathione catabolic enzyme γ-glutamyltranspeptidase	- Decreased hepatic activity of the glutathione synthesis enzyme γ-glutamyl-cysteine synthetase	Leggatt et al., 2007
Tilapia (*Oreochromis* sp.)	Growth hormone	Affecting feed conversion efficiency, metabolic rate, body composition, head and body morphometrics, osmoregulation and age at maturity	Success	Hallerman et al., 2007
Carp (*Cyprinus carpio*)	Growth hormone	Increased rates of protein synthesis and lipid mobilization Affecting feed conversion efficiency, metabolic rate, body composition	Success	Hallerman et al., 2007
Zebrafish (*Danio rerio*)	IGF-I and GHR gene	Growth hormone resistance	Success	Figueiredo et al., 2007

Disease Resistance

Disease resistance is a strong issue in aquaculture (Chevassus and Dorson, 1990) and a lot of biotechnological techniques are applied to approach this matter. A major limitation facing the aquaculture industry is outbreak of disease, as farmed fish are generally cultured at high densities and under stress, putting them at elevated risk for bacterial infection. The potential of using rainbow trout lysozyme gene as a bacterial inhibitor is being tested in the Atlantic salmon (Hew et al., 1995). Fish stockfarming for disease resistance provides an ideal modelling plan for application of marker-assisted selection since broodstock resistance in diseases can be created that have never been exposed before in pathogens (Exadactylos and Arvanitoyannis, 2006).

A large number of biologically active peptides, many of which have counterparts in neural and intestinal tissues of mammals, are present in frog skin (Bevin and Zasloff, 1990). The first antibacterial peptides identified from frog skin were named magainins 1 and 2. These peptides have broad antibacterial and antiparasitic activities. In a study using the flat oyster *Ostrea edulis* (Ostreidae), Morvan et al. (1994) demonstrated that on exposure to magainin 1 *in vitro*, the viability of the protozoam *Bonamia ostreae*, which infects the oyster, was reduced by 94%. Such peptide genes may be used for the development of disease-resistant broodstock by transgenesis. Sin (1997) was testing a gene construct containing a chinook salmon metallothionein promoter and magainin cDNA from *Xenopus laevis*. Another approach is to take advantage of the infection mechanism of viruses. For viruses to successfully infect a host, they must produce a glycoprotein for the binding of the virus to the receptor on the host cell surface. The transfer of a gene coding for the synthesis of viral glycoprotein may allow the host cells to saturate the receptors and thus inhibit the initial contact of the viruses with these receptors (Chevassis and Dorson, 1990). Another possibility is the transfer of genes coding for the antisense RNA which is complementary to the viral mRNAs for proteins that are essential for viral infection (Leong et al., 1995).

Antibiotics can help provide disease resistance, but only a limited number have been approved for use in aquaculture (Chapter 5, in this volume). Use of DNA vaccines (Chapter 6, in this volume) is often labour-intensive and can cause high stress to the fish owing to excessive handling (Sarmasik et al., 2002). Cecropins are a group of small, antibacterial peptides first identified in the silk moth *Hyalophora cecropia* (Dunham et al., 2002). Lysozyme is a non-specific antibacterial enzyme present in the blood, mucus, kidney, and lymphomyeloid tissues in fish. Rainbow trout contain elevated levels of lysozyme (10- to 20-fold higher than in the Atlantic salmon) and a rainbow trout lysozyme cDNA construct with an ocean pout antifreeze protein (AFP) promoter has been created (Hew et al., 1995). Interestingly, rainbow trout were recently reported to have two distinct types of lysozymes, with only type II having significant bactericidal activity (Mitra et al., 2003). The gene for type II lysozyme was amplified and sequenced for future use in transgenic immune system enhancement of farmed fish. Lysozyme transgenes are also being tested in agricultural products such as rice and have been found to increase disease resistance (Huang et al., 2002). Although the lysozyme gene may prove to heighten disease resistance in transgenic fish, research is still in the initial phases with no published results to date (Fletcher et al., 2004).

Human lactoferrin (hLF) is a nonspecific antimicrobial and immunomodulatory iron-binding protein that has been used widely in agriculture for production of disease-resistant transgenic crops, including potatoes and tobacco (Mao et al., 2004). One use of hLF in fish is to increase resistance against the grass carp hemorrhage virus. This virus induces deadly hemorrhaging in grass carp and is a major setback to the successful farming of these fish in China. To induce resistance against the virus, a DNA construct containing the *hLF* gene linked to a common carp β-actin promoter was electroporated into the sperm of grass carp (*Ctenopharyngodon idellus*) (Zhong et al., 2002).

An additional biotechnological application in the aquaculture industry is the treatment of fish with poly I:C, a potent inducer of type I interferons (IFNs). Type I IFNs are known to stimulate expression of myxovirus resistance (Mx) proteins, which are GTPases that inhibit the replication of single-stranded RNA viruses such as infectious salmon anaemia virus (ISAV), one of the most economically harmful pathogens in the Atlantic salmon industry. When challenged with ISAV, the Atlantic salmon treated with poly I:C experienced increased levels of Mx proteins and reduced mortality as compared to untreated controls (Jensen et al., 2002).

F2 transgenic medaka from different families and controls were challenged with *Pseudomonas fluorescens* and *Vibrio anguillarum* killing about 40% of the control fish by both pathogens, but only 0-10% of the F2 transgenic fish were killed by *P. fluorescens* and about 10-30% killed by *V. anguillarum*. When challenged with *P. fluorescens*, zero mortality was found in one transgenic fish family carrying preprocecropin B and two families with porcine cecropin P1, 0-10% cumulative mortality for five transgenic families with procecropin B and two families with cecropin B. When challenged with *V. anguillarum*, the cumulative mortality was 40% for non-transgenic control medaka, 20% in one transgenic family carrying preprocecropin B, between 20 and 30% in three transgenic families with procecropin B and 10% in one family with porcine cecropin P1 (Sarmasik et al., 2002).

Diploid and triploid coho salmon *Oncorhynchus kisutch* transgenic for growth hormone (GH) and control coho salmon were compared for differences in disease resistance and stress response. Resistance to the bacterial pathogen *V. anguillarum* was not affected in transgenic fish relative to their non-transgenic counterparts when they were infected at the fry stage, but was lower in transgenic fish when infected near smolting. Vaccination against vibriosis provided equal protection to both transgenic and non-transgenic fish. Triploid fish showed a lower resistance to vibriosis than their diploid counterparts. Diploid transgenic fish and non-transgenic fish appeared to show similar physiological and cellular stress responses to a heat shock (Jhingan et al., 2003).

Pleiotropic effects — Fast growing transgenic common carp and channel catfish containing rainbow trout growth hormone gene had lower feed conversion efficiency than controls (Chatakondi et al., 1995; Dunham and Liu, 2002). Various transgenic common carp families had increased, decreased or there was no change in food consumption. Transgenic Nile tilapia also had a 20% improvement in feed conversion efficiency and were better utilizers of protein and energy compared to controls (Rahman et al., 2001). Transgenic tilapia expressing the tilapia GH cDNA under the control of human cytomegalovirus regulatory sequences exhibited about 3.6 times less food consumption than non-transgenic controls, and food conversion efficiency was 290% better for the transgenic tilapia (Martinez et al., 2000). Efficiency of growth, synthesis retention, anabolic stimulation, and average protein synthesis were higher in transgenic than control tilapia. Martinez et al. (2000) observed differences in hepatic glucose, and in the level of enzymatic activities in target organs in the transgenic and control tilapia. GH transgenic Atlantic and coho salmon had intestinal surface area 2.2 times that of control salmon and the growth rate was about twice that of controls (Stevens and Devlin, 2000). The relative intestinal length was the same in transgenic and control salmon, but the surface area was greater for transgenics as a result of increased number of folds. This increase intestinal surface area was found in both Atlantic and coho salmon.

Yazawa and co-workers (2006) established a transgenic zebrafish strain expressing chicken lysozyme gene under the control of the Japanese flounder keratin gene promoter, and investigated its resistance to a pathogenic bacterial infection. To generate the lysozyme transgenic construct, Japanese flounder keratin promoter was linked to both the hen egg white (HEW) lysozyme gene and green fluorescence protein (GFP) gene used as a selection marker for the transgenic strains, in a recombinant plasmid. The recombinant plasmid was microinjected into fertilized zebrafish eggs. In F2 transgenic zebrafish, GFP expression was strong in the epithelial tissues, liver and gill from the embryonic stage to the adult stage. The expressions of HEW lysozyme and GFP mRNA were confirmed in the liver and skin by RT-PCR. Western blot analysis showed that both HEW lysozyme and GFP were present in protein extracts from the liver of transgenic zebrafish, but not in protein extracts from the muscle (Udvadia and Linney, 2003). The lytic activity of protein extracts from the liver (assessed by a lysoplate assay using *Micrococcus lysodeikticus* as a substrate) was 1.75 times higher in F2 transgenic zebrafish than in the wild type. In a challenge experiment, 65% of the F2 transgenic fish survived an infection of *Flavobacterium columnare* and 60% survived an infection of *Edwardsiella tarda*, whereas 100% of the control fish were killed by both pathogens. However, the survival rates of the transgenic fish were not significantly higher when higher concentrations of bacteria were used (Yazawa et al., 2006).

Growth Hormone (GH)

Positive biological effects have been obtained by transferring transgenes to fish in some, but not all cases, and the greatest amount of work has focussed on transfer of GH genes. Due to the lack of available piscine gene sequences, transgenic fish research in the mid 1980s employed existing mammalian GH gene constructs, and growth enhancement was reported for some fish species examined (Enikolopov et al., 1989; Zhu, 1992). GH is a polypeptide that is excreted from the pituitary gland, binds specific cell receptors, and induces synthesis and secretion of insulin-like growth factors (IGF-I and IGF-II), resulting in promotion of somatic growth through improved appetite, feeding efficiency, and growth rate (Hsih et al., 1997; de la Fuente et al., 1998). In fish, the central nervous system (CNS) normally controls GH excretion levels, which are highly variable, occurring seasonally and in bursts. However, the AFP (anti-freeze protein) gene of ocean pout (*Macrozoarces americanus*) is expressed year round in the liver. As a way of bypassing CNS control on GH expression, transgenic research typically involves linking the GH gene to the AFP gene promoter (Fletcher et al., 2004). Increased growth has been the most thoroughly researched of the possible transgenically induced fish traits, and it is predicted that this technology will soon be applied to commercial aquaculture production (Wu et al., 2003).

Since the initial transgenic introduction of human GH gene into goldfish and mud loach (Zhu et al., 1985; Zhu et al., 1986), extensive research has been performed on the use of GH in a wide variety of aquatic species. New, improved techniques have been developed for the introduction of transgenes into host genomes, and there has been a focus on the use of GH genes originating from fish rather than humans, with the hope of increasing consumer acceptability (Levy et al., 2000). Negative perceptions associated with the use of viral promoters to express transgenes have also driven many researchers to replace them with fish-based promoters for the creation of "all-fish" transgene constructs (Galli, 2002). These all-fish GH-transgenic strains have been developed in a number of species, including the common carp (*Cyprinus carpio*), silver sea bream (*Sparus sarba*), red sea bream (*Pagrosomus major*), tilapia (*Oreochromis niloticus*), and Atlantic salmon (*Salmo salar*).

Subsequent experiments demonstrated that growth can be enhanced through transgenesis from 10% up to an incredible 30-fold. Du et al. (1992) used an all-fish GH gene construct to make transgenic Atlantic salmon, and report 2- to 6-fold increase of the transgenic fish growth rate. *Oreochromis niloticus* possessing one copy of an eel (ocean) pout promoter-chinook salmon growth hormone fusion grew 2.5-4 folds faster and converted feed 20% better than non-transgenic siblings (Rahman and Maclean, 1999). However, F1 Nile tilapia transgenic for a construct consisting of a sockeye salmon metallothionein promoter spliced to a sockeye salmon growth hormone gene exhibited no growth enhancement (Rahman et al., 1998), although salmon transgenic for this construct show greatly enhanced growth.

An all-fish GH construct has been successfully introduced into the common carp, resulting in increased growth rate and more efficient feed conversion as compared to farmed fish controls. Middle-scale trials of these all-fish GH-transgenic common carp have shown high potential for successful commercial application in aquaculture (Wu et al., 2003). Transgenic lines of silver sea bream, an economically important cultivated species in Asia, were developed using a construct containing rainbow trout (*Oncorhynchus mykiss*) GH complementary DNA (cDNA) with a common carp promoter (Lu et al., 2002).

Insertion of a GH transgene in coho salmon results in accelerated growth, and increased feeding and metabolic rates. Whether other physiological systems within the fish are adjusted to this accelerated growth has not been well explored. Leggatt and co-workers (2007) examined the effects of a GH transgene and feeding level on the antioxidant glutathione and its associated enzymes in various tissues of coho salmon. When transgenic and control salmon were fed to satiation, transgenic fish had increased tissue glutathione, increased hepatic glutathione reductase activity, decreased hepatic activity of the glutathione synthesis enzyme γ-glutamylcysteine synthetase, and increased intestinal activity of the glutathione catabolic enzyme γ-glutamyltranspeptidase. However, these differences were mostly abolished by ration restriction and fasting, indicating that upregulation of the glutathione antioxidant system was due to accelerated growth, and not to intrinsic effects of the transgene.

Transgenic tilapia containing the GH gene driven by the human cytomegalovirus (CMV) was compared to nontransgenic siblings on a number of metabolic and physiological parameters (Martinez et al., 2000). The results showed several significant differences, with transgenic tilapia consuming 3.6-fold less food and having 290% greater feed conversion efficiency. In addition, growth efficiency, average protein synthesis, anabolic stimulation, and synthesis retention were higher for the transgenic tilapia.

Studies into the use of CMV as a promoter of the GH gene have also taken place with fish such as Arctic char (*Salvelinus alpinus* L.) and the Indian major carp *Labeo rohita* (Pitkanen et al., 1999b; Pandian and Venugopal, 2005). A further investigation into the use of CMV as a GH gene promoter showed no difference in the muscle composition of transgenic fish as compared to non-transgenic siblings (Krasnov et al., 1999). However, the transgenic fish did show some metabolic features that are often observed in farmed salmonids, such as an increased metabolic rate and faster utilization of dietary lipids, particularly in the case of triglycerides. In transgenic studies involving the Indian major carp *L. rohita*, an element known as internal ribosomal entry sites (IRES) was included in an expression vector containing the CMV promoter and Indian major carp GH (Pandian and Venugopal, 2005).

Insertion of other GH constructs into tilapia has also yielded positive results, but not as dramatic as those with the salmon GH constructs. Two possible explanations for the difference in results are the type of construct and the type of tilapia studied was different. Introduction of a CMV/tilapia GH construct into a hybrid *Oreochromis hornorum* resulted in 60-80% growth acceleration (Estrada et al., 1999) depending on the culture conditions. When introduced into coho salmon, cutthroat trout, *Oncorhynchus clarki*, rainbow trout, and Chinook salmon, GH gene constructs using either an ocean pout antifreeze promoter driving a Chinook salmon GH cDNA, or a sockeye salmon metallothionein promoter driving the full-length sockeye GH1 gene elevated circulating GH levels by as much as 40 folds (Devlin, 1997), resulting in up to 5- to 30-fold increase in weight after one year of growth (Devlin et al., 2001), and allowing precocious development of physiological capabilities necessary for marine survival (smoltification). The largest of these P1 transgenics were mated and produced offspring with extraordinary growth. As was seen with three transgenic common carp and channel catfish, the effect of GH gene insertion was variable among families, and multiple insertion sites and multiple copies of the gene were observed.

Results with the Atlantic salmon are not quite as impressive as with coho salmon. Transgenic Atlantic salmon also ingested more, exhibiting 2.14- to 2.62-fold greater daily feed consumption, with lower body protein, dry matter, ash, lipid, and energy and higher moisture than nontransgenic controls (Cook et al., 2000). The results of a recent study into the genetic expression and interactions of GH, the insulin-like growth factor I (IGF-I), and their receptors indicate involvement of the hormones in the areas of vertebral growth and bone density (Wargelius et al., 2005).

Most biological actions of GH are mediated by IGF-I that is produced after the interaction of the hormone with a specific cell surface receptor, the GH receptor (GHR). Even though the GH excess on fish metabolism is poorly known, several species have been genetically engineered for this hormone in order to improve growth for aquaculture. In some GH-transgenic fish growth has been dramatically increased, while in others high levels of transgene expression have shown inhibition of the growth response. Figueiredo et al. (2007) used for the first time different genotypes (hemizygous and homozygous) of a GH-transgenic zebrafish (*Danio rerio*) lineage as a model for studying the GH resistance induced by different GH transgene expression levels. The results obtained here demonstrated that homozygous fish did not grow as expected and have a lower condition factor, which implies a catabolic state. These findings are explained by a decreased IGF-I and GHR gene expression as a consequence of GH resistance.

Insertion of sockeye MT-B-sockeyeGHcDNA1 (Devlin, 1997) produced a similar result, 5-fold growth enhancement. Varying results among species and families might be related to different gene constructs, coding regions, chromosome positions and copy numbers. Magnification effects can explain some of the growth differences between transgenic and control salmon, however, specific growth rates of the transgenic coho were approximately 2.7-fold higher than older nontransgenic animals of similar size, and 1.7-fold higher than their nontransgenic siblings (Devlin et al., 1999) indicating that the transgenic salmon are growing at faster rates at numerous sizes and life stages. GH levels were increased dramatically (19.3- to 32.1-fold) relative to size control salmon, but IGF-I levels were only modestly affected, being slightly enhanced in one experiment and slightly reduced in another.

In comparison, GH transgenic channel catfish derived from domesticated and selectively-bred strains exhibit only a moderate growth enhancement (41%). However, additional data on transgenic rainbow trout (Devlin et al., 2001) refutes this hypothesis on the effect of wild and domestic genetic backgrounds on response to GH transgene insertion. When OnMTGH1 was transferred to another wild rainbow trout strain, F77, growth was enhanced 7-fold which was almost 4-fold greater growth than that observed in a non-transgenic domestic rainbow trout. In this case, the wild transgenic is actually superior to the domestic selected strain indicating that genetic engineering can have a greater, rather than equivalent effect to the domestication and selection. When F77 was crossbred with a domestic strain, growth of the crossbreed was intermediate to the parent strains, a typical result (Dunham and Devlin, 1998). The combined effects of transgenesis and crossbreeding had a much greater growth enhancement than crossbreeding or transgenesis alone. A transgenic with 50% of its heritage from domestic sources was much larger than a wild transgenic, so a good response from some domestic genotypes is possible.

A recent study investigated the effect of supplementing feed with transgenic *Synechocystis* sp. PCC6803 containing the *Paralichthys olivaceus* growth hormone (GH) gene on growth, feed intake and feed efficiency ratio, muscle composition, haematology and histology of turbot (*Scophthalmus maximus* L.). At the end of the 40-day feeding trial, the specific growth rate of fish fed the supplemented feed with 1% transgenic *Synechocystis* sp. PCC6803 was 21.67% higher than that of control fish. Haematological parameters, including red blood cell, white blood cell, haemoglobin, and serum biochemical indices, such as enzyme activities of alanine aminotransferase, aspartate aminotransferase, alkaline phosphatase, lactate dehydrogenase, concentrations of total protein, glucose, blood urea nitrogen, creatinine, triglyceride and cholesterol and ion levels of K, Na, Cl, P were not influenced by supplementing the transgenic *Synechocystis* sp. PCC6803. Furthermore, no histopathological alterations were induced by transgenic alga treatment in the stomach, intestine, liver, spleen and kidney of the experimental fish (Liu et al., 2007).

Hallerman et al. (2007) reviewed the impacts of GH transgene expression on the behaviour and welfare of cultured fishes, focusing on the Atlantic salmon, coho salmon, tilapia and common carp lines which have been posed for commercial production. Elevation of GH increases rates of protein synthesis and lipid mobilization, affecting not only growth, but also feed conversion efficiency, metabolic rate, body composition, head and body morphometrics, osmoregulation, and age at maturity, with other modifications particular to species and transgenic lines. Because of heightened feeding motivation, transgenic fishes often prove more active, aggressive, and willing to risk exposure to predation. Swimming ability of some transgenic lines is reduced. While limited in scope, studies suggest that fish of some transgenic lines show lessened reproductive behaviour. Welfare issues posed by morphological, physiological, and behavioural alternation in GH-transgenic fishes have not been well characterized. The use of weaker promoters in expression vectors and by selection of transgenic lines with physiologically appropriate levels of GH expression may reduce welfare issues that do arise. Also, markers of welfare for GH-transgenic fish under production conditions can be monitored to determine how culture systems and practices can be modified to promote fish welfare.

Cold and Freeze Tolerance

Most efforts in transgenic fish have been devoted to growth enhancement, although there are reports of improvement in cold resistance (Fletcher and Davies, 1991; Shears et al., 1991). Early research also involved the transfer of the antifreeze protein gene of the winter flounder (Fletcher et al., 1988). The primary purpose of this research was to produce salmon that could be farmed under arctic conditions, but expression levels obtained have been inadequate for increasing cold tolerance of salmon. However, preliminary results with goldfish show some promise for increasing survival within the normal cold temperature range.

Some Arctic fish species are known to resist freezing by producing an antifreeze protein which prevents ice crystal formation in the blood, thus lowering the freezing point of the blood plasma (Davies et al., 1989). The idea of an "antifreeze" system was first described in marine fish inhabiting the coast of Northern Labrador whose body fluids had the same freezing point of seawater (–1.7°C to –2°C) rather than freshwater (0°C) (Scholander et al., 1957; Gordon et al., 1962). This phenomenon was eventually attributed to a set of peptides and glycopeptides termed AFPs and antifreeze glycoproteins (AFGPs), respectively. These proteins are synthesized primarily in the liver and secreted into the blood and extracellular space, where they bind and modify the structure of microice crystals, thereby inhibiting ice crystal growth and lowering the freezing point of body fluids (Davies and Hew, 1990; Raymond, 1991).

Fletcher et al. (1988) transferred the antifreeze protein (AFP) gene from the winter flounder (*Pseudopleuronectes americanus*, Pleuronectidae) into Atlantic salmon embryos. The transgene was found to be integrated in 7% of the founder fish. Low levels of proAFP, an AFP precursor, were found in the transgenic founder fish and in the F1 and F2 progeny (Shears et al., 1991; Fletcher et al., 1992). However, the level of AFP was about one one-hundredth of that in winter flounder. Thus, for the production of salmon with low-temperature tolerance, improvement on the gene construct to give higher efficiency of expression would be required (Sin, 1997).

AFPs have diverse structures and are divided into three categories (types I, II, and III) depending on their protein sequences. Also, the number of copies and type of AFP genes varies with the fish species: for example, winter flounder (*Pleuronectes americanus*) have 30 to 40 copies of type I; sea raven (*Hemitripterus americanus*) have 12 to 15 copies of type II; and Newfoundland ocean pout have 150 copies of type III (Hew et al., 1995). The ocean pout type III AFP transgene has been successfully transferred and expressed in goldfish (Wang et al., 1995). The gene was microinjected into the goldfish oocytes and was inherited through two generations. The transgenic goldfish showed significantly higher cold tolerance as compared to controls, suggesting possible use of the transgene for promoting cold resistance in fish.

AFP transgenic technology could be highly beneficial to the aquaculture industry in countries with freezing and sub-zero coastline conditions. For example, winter water temperatures along the Atlantic coast of Canada can range from 0°C to –1.8°C; these conditions restrict the cultivation of salmonids and other commercially important fish to a few select areas at the southern edge of eastern Canada (Fletcher et al., 2004). Therefore, research is currently under way to develop strains of Atlantic salmon that could be cultivated over a wider geographic range. This could be accomplished by (i) introduction of a set of AFP transgenes that allow the fish to survive lower water temperatures, or (ii) introduction of GH transgenes to produce a rapidly growing strain that does not require over-wintering (Hew et al., 1995). Although AFP transgenes have been successfully introduced into, expressed in, and inherited through germlines of Atlantic salmon, the cold-tolerant transgenic salmon do not produce AFP in sufficient quantities to achieve freeze resistance (Hew and Fletcher 1992; Fletcher et al., 2004). This may be due to the need for higher expression levels and/or the fact that the gene actually codes for an AFP precursor that must first be changed into its fully functional form (Maclean and Laight, 2000).

Hew et al. (1999) have analyzed the inheritance and expression of a line of transgenic salmon harbouring the antifreeze protein gene from the winter flounder. The genomic clone 2A-7 coding for a major liver-type antifreeze protein gene (wflAFP-6) was integrated into the salmon genome. From a transgenic founder, an F3 generation was produced. In this study, southern blot analysis showed that only one copy of the antifreeze protein transgene was integrated into a unique site in F3 transgenic fish. The integration site was cloned and characterized. Northern analysis indicated that the antifreeze protein mRNA was only expressed in the liver and showed seasonal variation. All of the F3 offspring contained similar levels of the antifreeze protein precursor in the sera and the sera of these offspring showed a characteristic hexagonal ice crystal pattern indicating the presence of antifreeze activity.

One copy of the AFP transgene was integrated into the genome of the transgenic salmon, and stable expression of the transgene (approximately 250 µg/mL) up to the F3 generation has been obtained. Expression of proAFP was liver-specific and showed seasonal variations not identical but very similar to those in winter flounder. Although in Atlantic salmon the proAFP precursor may not be processed into mature protein due to the lack of the required processing enzymes, the antifreeze activity was detected in the sera of F3 offspring. However, the primary problem with AFP gene transfer remains a low level of production of AFP (Zbikowska, 2003). The concentration of AFP in the serum of winter flounder is normally approximately 10-20 mg/mL while in all fish transgenic for AFP the expression level was in the μg/mL range. An increase in the copy number of the transgene or the use of constructs with other AFPs that have a higher antifreeze activity might be helpful for successful enhancement of freeze-resistance in farm fish. The intracellular, skin-type family of AFPs that are present in the external tissues of cold-water marine fish and that were recognized as the first line of defence against freezing is currently under consideration for gene transfer (Low et al., 1998).

Transgenic Sterilization

Commercial production of transgenic fish will depend on the risk to wild aquatic species. Fish are typically raised in the sea in netted pens, and escapes are quite common. If the escaped fish breed with their wild counterparts the consequences of spreading the modified genes into environment are unpredictable. For safety, genetically modified fish for human consumption should be made sterile. Heat-shock or pressure-shock-treatment of the freshly fertilized fish eggs or treatment of females with male sex hormones is a common practice to produce sterile fish, but the methods are not 100% reliable (Razak et al., 1999). An alternative method could involve induced sterility in transgenic lines by blocking of the gonadotropin releasing hormone (GnRH) with antisense RNA. The idea came from experiments with hypogonadal mice (Mason et al., 1986).

Although use of transgenic fish in aquaculture has the potential to increase food availability and reduce production costs, there is much concern over the possibility of escapement of transgenic fish and contamination of wild populations through interbreeding. Therefore, it is of great interest to develop techniques for preventing introduction of transgenes into wild stocks. Two concepts currently being researched include induction of sterility and gonad-specific transgene excision (Hu et al., 2006). Although a method has been developed for production of sterile fish through chromosome manipulation, sterility is not achieved 100% of the time and the fish often have stunted growth (Wang et al., 2003; Dunham, 2004).

Recently, sterility was reported using a transgenic method to inhibit expression of the gene coding for the gonadotropin-releasing hormone (GnRH), an important component in gonad development and reproductive function (Hu et al., 2006). In a series of pilot studies, reversible fertility was achieved in common carp by inserting a gene coding for an antisense RNA sequence that inhibits expression of the

GnRH gene. When the antisense-GnRH messenger RNA (mRNA) was transferred into common carp eggs, around 30% of the offspring did not develop gonads; however, the fish could be made fertile by exogenous administration of hormones (Rasmussen and Morrissey, 2007).

A deletion in the GAP (GnRH-associated peptide) region of the *GnRH* gene decreased the level of gonadotropin in this mouse line resulting in complete sterility. In fish, as in other vertebrate, GnRH is thought to play an important role both in sexual maturation and in reproductive behaviour. Two forms of GnRH peptide, salmon GnRH (sGnRH) and chicken type II GnRH (cGnRH-II), has been identified in salmonids. They are present in the brain and the gonads and are separately regulated in both tissues (von Schalburg et al., 1999). Salmon GnRH is encoded by two different genes, *sGnRH* gene-1 and *sGnRH* gene-2 (Ashihara et al., 1995).

Generating transgenic fish with desirable traits (e.g., rapid growth, larger size, etc.) for commercial use has been hampered by concerns for biosafety and competition if these fish are released into the environment. These obstacles may be overcome by producing transgenic fish that are sterile, possibly by inhibiting hormones related to reproduction. In vertebrates, synthesis and release of gonadotropin (GtH) and other reproductive hormones is mediated by GnRH. Recently two cDNA sequences encoding salmon-type GnRH (sGnRH) decapeptides were cloned from the common carp (*Cyprinus carpio*). Transcripts of both genes were detected in ovary and testis in mature and regressed, but not in juvenile carp. To evaluate the effects of sGnRH inhibition, the recombinant gene CAsGnRHpc-antisense, expressing antisense sGnRH RNA driven by a carp beta-actin promoter, was constructed. Blocking sGnRH expression using antisense sGnRH significantly decreased GtH in the blood of male transgenic carp. Furthermore, some antisense transgenic fish had no gonadal development and were completely sterile (Hu et al., 2007).

Carp beta actin-tilapia salmon type GnRH antisense construct was injected into Nile tilapia. Transgenic females were crossed with wild type males. A reduction in fertility of about half that of non-transgenic control females was observed. Fertility was much more greatly reduced in transgenic males crossed to control females. In some cases, 0% fertility was obtained with an average of about an 80% reduction in fertility. Limited data on transgenic females crossed with transgenic males indicated near zero fertility. Tilapia beta actin-tilapia seabream GnRH antisense construct was injected into Nile tilapia, but no reduction in fertility of heterozygous transgenic males and females was observed (Dunham, 2004).

Limited data on transgenic females crossed with transgenic males indicated no reduction in fertility. Reciprocal crosses between seabream and salmon GnRH antisense transgenics gave hatch rates that appeared to be dictated by the salmon GnRH antisense parent. Transgenic rainbow trout containing salmon type antisense GnRH from the Atlantic salmon, *Salmo salar,* driven by either the GnRH or histone 3 promoter had reduced levels of GnRH and appear to be sterile (Uzbekova et al., 2000). The designed construct contained an antisense DNA complementary to the Atlantic salmon sGnRH cDNA that is driven by a specific promoter Pab derived from the corresponding *sGnRH* gene. However, although the transgene was integrated into the genome, transmitted through the germline and the antisense-*sGnRH* was expressed specifically in the brain of transgenic offspring, the presence of the antisense-*sGnRH* did not result in sterility of transgenic fish.

Preliminary data indicated that spermiation of transgenic males was only obtained after prolonged treatment with salmon pituitary extract, whereas control males spermiated naturally. Data is still needed for the females. Another strategy, introduction of "Sterile Feral" constructs, disrupts embryonic development, thus sterilizing brood stock. Preliminary results show promise for this approach (Thresher et al., 2001). To achieve gonad-specific transgene excision, two types of transgenic zebrafish were created (Hu et al., 2006). One line of zebrafish contained the gene coding for Cre recombinase driven by the T7 promoter along with the desired transgene (in this case fluorescent protein) flanked by 2 loxP sites. When expressed, the Cre recombinase excises the transgene in between the 2 loxP sites. The other line of zebrafish contained the T7 RNA polymerase driven by a protamine promoter specifically expressed in the gonad. When these 2 lines of zebrafish are crossed, the offspring specifically express the T7 RNA polymerase in the gonads, resulting in the expression of the Cre recombinase and excision of the foreign gene. This concept would allow for the creation of transgenic individuals lacking the ability to pass on a foreign gene.

Modification of Metabolism

A distinct approach, currently not successful, has been to use gene transfer for the modification of metabolic pathways to permit better utilization of food (Krasnov et al., 1999). First, the potential introduction of the missing L-ascorbic acid biosynthesis (L-AAB) pathway in rainbow trout by delivery of the rat L-gulono-γ-lactone oxidase (GLO) cDNA was tested. Secondly, an attempt to enhance the utilization of glucose by generation of trout transgenic for human glucose transporter (hGLUT1) or rat hexokinase (rHKII) was undertaken.

Glucose transporter and hexokinase have a rate-limiting role in glucose transport and phosphorylation, respectively, while GLO is an enzyme that catalyzes the terminal reaction of the L-AAB pathway. In fish the activities of GLO, GLUT1 and HK (hexokinase) were found to be very low or not present. The Cytomegavirus (CMV) or the sockeye salmon metallothionein promoters were used for the targeted modification of these metabolic pathways (Zbikowska, 2003). However, in the transgenic trout the GLO protein was not detectable and the interpretation of glucose utilization results was complicated by mosaicism. Further development of these biotechnology applications will strongly depend on the better understanding of metabolic pathways in fish and their regulation on the molecular level.

The omega-3-polyunsaturated fatty acids (n-3 PUFAs) eicosapentaenoic acid (EPA) and docosahexaenoic acid (DHA) are known to provide a number of benefits to human health, including promotion of visual and neurodevelopment, alleviation of diseases such as arthritis and hypertension, and reduction of cardiovascular problems (Bao et al., 1998; Horrocks and Yeo, 1999; Grimm et al., 2002; Leaf et al., 2003). Fish obtain high levels of these fatty acids through the aquatic food chain and are thus a major dietary source of EPA and DHA for humans. However, farmed fish are often fed diets rich in plant oils, which contain the fatty acid α-linolenic acid (ALA), a precursor to EPA and DHA. The rate-limiting step in the conversion of ALA to EPA and DHA involves the enzyme n-6-desaturase. Recently, a gene construct containing the n-6-desaturase-like gene from the masu salmon (*Oncorhynchus masuo*) linked to a β-actin gene promoter from the Japanese medaka was microinjected into the 1-cell embryos of zebrafish (Alimuddin et al., 2005).

The expression of the n-6-desaturase-like gene in the transgenic zebrafish resulted in a 1.4-fold increase in EPA content, a 2.1-fold increase in DHA content, and a corresponding decrease in ALA content, as compared to the non-transgenic controls. Despite these changes in fatty acid composition in the transgenic fish, total lipid content remained constant. This technology has potential in the aquaculture industry as a means to increase levels of n-3 fatty acids in farmed fish, thereby providing consumers with a healthier product. It could also reduce the reliance of the aquaculture industry on ocean-caught feed sources and allow for widespread use of cheaper, plant-based diets rich in ALA. Alternatively, advances in plant transgenics have recently allowed for the production of EPA (up to 15% of total fatty acids) and DHA (up to 3.3% of total fatty acids) in oilseed crops (Robert, 2006). Although these transgenic plants produce variable and limited levels of n-3 PUFAs, they do show potential for use in aquaculture feed.

The use of terrestrial plant based diets in aquaculture has a number of advantages over more traditional marine-derived diets (Naylor et al., 2000). Plant products such as soybean meal and vegetable oils can supply high levels of protein and energy at a lower cost than marine products such as fish meal and fish oil. Also, some argue that use of plants helps to conserve marine ecosystems by reducing the need for the wild-caught small pelagic fish often used to produce fish feed. However, since plant-based diets differ in composition from the traditional marine diets, it is important to ensure that farmed organisms are able to maintain appropriate levels of nutrients and other beneficial ingredients.

In another effort to facilitate the use of plant-based diets in aquaculture, the gene coding for the enzyme phytase was recently expressed in the Japanese medaka (Hostetler et al., 2003). Phytase breaks down phytate, the major form of phosphorus in plants. Attempts to add the phytase enzyme to the feed for reduction of phytate waste have shown limited success due to degradation of the enzyme at the high temperatures required for feed processing. Hostetler et al. (2003) reported successful expression of the *Aspergillus niger* phytase gene, driven by either the human CMV promoter or the sockeye salmon histone type III promoter, in the Japanese medaka. The transgenic fish were compared to their nontransgenic siblings on three different diets all rich in phytate.

POLYPLOIDY

One of the main reasons for handling polyploidy is for producing fish with faster growth rates, without being influenced by genital maturation or reproduction. The benefits of polyploidy include better food convertibility, higher output with regard to production systems. In fish and shellfish, gametes can produce triploid, viable offspring with suitable handling (Blanc et al., 2005). Triploid fish and shellfish are viable and tend to be sterile due to a lack of gonadal development. This sterility allows for reproductive energy to be diverted toward somatic growth, resulting in higher growth rates for some triploid individuals. Although triploidy has been highly effective for enhancing growth in shellfish, results thus far in fish have shown variable, conflicting results, with reports of triploid fish growing slower, at the same rate, or faster than diploids (Dunham, 2004).

Polyploidy has been studied in China in close to 30 shellfish species, including Pacific oysters, Chinese scallop (*Chlamys farreri*), pearl oyster, and abalone (*Haliotis discus hannai*) (Beaumont and Fairbrother, 1991). The benefits of triploids vary with species, from larger adductor muscles in scallops to increased survival in the Chinese pearl oyster (*Pinctada martensi*) (Allen, 1998). In aquaculture, research has been focused mainly in the development of techniques for producing triploid populations (Exadactylos and Arvanitoyannis, 2006). Use of triploids in aquaculture can be advantageous when reproductive efforts negatively affect growth, survival, or product quality. For example, during the spawning season, the majority of an oyster's energy is directed toward reproduction, and flesh palatability is reduced as gonad replaces stored glycogen (Shatkin et al., 1997). Processes that are used commonly can induce sterility in the next generation. Polyploidy in fish is because of individuals with more than two chromosome sequences (Leggatt and Iwama, 2003).

Besides advantages such as increased growth rates, use of sterile triploids in aquaculture can help protect the genetic diversity of native populations and prevent establishment of populations of escaped organisms. This could be particularly relevant in addressing concerns over use of transgenic organisms in the aquaculture industry and their possible escapement and subsequent mating with wild broodstocks. To prevent unauthorized breeding of farmed shrimp, research into shrimp polyploidy is currently under way, with successful production of all-female, sterile triploids with comparable growth to diploids (Sellars et al., 2006). Although most triploids are sterile, they still exhibit sexual behaviour, participating in mating rituals and thereby disrupting natural spawning processes within populations (Dunham, 2004).

A potential alternative to the frequent problematic and expensive production of triploids by mechanical induction is the crossing of tetraploids with diploids. Tetraploids are induced in a similar way as triploids, but during a more advanced stage of embryonic development. A tetraploid/diploid cross would be expected to produce all-triploid progeny that might be more viable than mechanically induced triploids, as triploid embryos would not have to undergo the same stress and damage that occurs during mechanical induction (Rasmussen and Morrissey, 2007).

POSSIBLE EFFECTS OF TRANSGENIC FISH IN THE ENVIRONMENT AND FOR HUMAN HEALTH

Transgenic fish in aquaculture have the potential to advance the industry and help supply a growing global food demand, there is concern over the potentially negative effects of introducing genetically altered fish into the food chain. Appropriate risk analyses must be carried out in order to evaluate possible detrimental effects on both the environment and human health (Kapuscinski, 2005; Rasmussen and Morrissey, 2007).

Production of transgenic animals has raised concern regarding their potential ecological impact should they escape or be released to the natural environment. This concern has arisen mainly from research on laboratory-reared animals and theoretical modelling exercises. In this study, biocontained naturalized stream environments and conventional hatchery environments were used to show that differences in phenotype between transgenic and wild genotypes depend on rearing conditions and, critically, that such genotype-by-environment interactions may influence subsequent ecological effects in nature (Sundstrom et al., 2007). Genetically wild and growth hormone transgenic coho salmon (*Oncorhynchus kisutch*) were reared from the fry stage under either standard hatchery conditions or under naturalized stream conditions. When reared under standard hatchery conditions, the transgenic fish grew almost three times longer than wild conspecifics and had (under simulated natural conditions) stronger predation effects on prey than wild genotypes (even after compensation for size differences). In contrast, when fish were reared under naturalized stream conditions, transgenic fish were only 20% longer than the wild fish, and the magnitude of difference in relative predation effects was quite reduced.

A major cause of concern regarding aquaculture is the escapement of farmed organisms into the wild and subsequent interaction with native populations, possibly leading to significant alterations in the properties of the natural ecosystem (Ramirez and Morrissey, 2003). The escapees can be particularly harmful to wild populations, especially when farming occurs in the native habitat of the escaped fish, when there is a proportionally high number of farmed fish compared to the wild stock, or when the wild population is exposed to pathogens occurring in farmed fish (Naylor et al., 2005).

When transgenic fish breed with wild populations, the resulting fish may acquire transgenes that could alter natural behaviour in areas such as reproduction, anti-predator response, and feeding (Galli, 2002). Fish containing artificial genes such as those coding for AFPs, increased growth, or increased disease resistance have the potential to outcompete native populations. Studies involving comparisons of GH-transgenic and non-transgenic coho salmon have shown contradictory results, with reports of GH-transgenic salmon showing increased competitive feeding abilities (Devlin et al., 1999), greater mortality in the fry (Sundstrom et al., 2004), and equal competitive feeding abilities without increased mortality rates (Tymchuk et al., 2005). Besides the possibility of increased survivability, disease resistant transgenic fish also have the potential of carrying certain bacteria, parasites, or viruses that may be harmful to natural populations. Although advocates of transgenesis argue that genetically modified organisms (GMOs) are not too different from species that have been genetically altered by breeding techniques, the general population and many environmental groups remain wary of the concept of artificial gene insertion (FAO, 2000).

Application of transgenic fish in environmental toxicology remains at an early stage. However, while theoretically feasible, the idea of using fish as canaries for detection of contaminants in water sounds intriguing. Progress in this field has been reviewed by Carvan et al. (2000) and Winn

(2001). Several transgenic lines of zebrafish carrying reporter genes driven by pollutant-inducible DNA response elements were tailored to be utilized as aquatic biomonitors for detection of hazardous substances in water (Carvan et al., 2001). Briefly, the fish is placed in water containing the environmental pollutant to be tested. Following uptake, distribution, and accumulation of the pollutant in fish tissues, then integrated with the genome response elements that respond to the selected substance are capable of activating a reporter gene. The reporter gene activity that is proportional to the concentration of the chemical can be easily assayed in the intact live zebrafish. Transgenic fish have been also developed to perform mutation assays to assess potential DNA damage after exposure to chemicals in aquatic environments (Amanuma et al., 2000). These fish are transgenic for specific genes, originally developed for rodents, spliced to prokaryotic vectors.

The effects of long-term consumption of GM foods are unknown. Also insertion of foreign genes into species might result in production of toxins or allergens that were not present previously (Galli, 2002). Another potential area of concern is that increased disease resistance of transgenic fish might make them better hosts for new pathogens, which could then be passed on to humans through consumption (FAO, 2000). Allergens or toxins may be produced as a result of gene transfer if the transgene codes for an allergenic protein or a protein that induces expression of a previously inactive toxin (Kelly, 2005). In addition to production of allergens and toxins, there is also some concern over the expression of bioactive proteins such as GH and cecropins, which may continue to possess bioactive properties following consumption. For example, the antimicrobial properties of cecropins have the potential to alter the normal intestinal flora in humans and/or selectively promote the development of human pathogens with increased resistance (NAS, 2002).

According to the information reported by the World Health Organization (WHO), the GM products that are currently on the international market have all passed risk assessments conducted by national authorities. These assessments have not indicated any risk to human health. It is quite amazing to note that the review articles published in international scientific journals during the current decade did not find, or the number was particularly small, references concerning human and animal toxicological/health risks studies on GM foods. Domingo (2007) referred to the potential toxicity of GM/transgenic plants. A summary of results concerning the most relevant studies is given in Table 1.2.

Food safety trials with tilapia GH and transgenic GH tilapia meat have been conducted in Cuba using both non-human primates and healthy human volunteers. Six non-human primates were intravenously

Table 1.2 Dietary administration of a number of GM plants to various animal species

Plant	Animal species	Main effects	Length of study	Reference
Soybean (glyphosate-tolerant)	Catfish, Rats, Dairy cows, Broiler chickens	No significant effects in the concentrations of nutrients and antinutrients	10 wk (catfish), 4 wk (rats), 4 wk (cows), 6 wk (chickens)	Hammond et al., 1996
Soybean (GM 40-3-2)	Rats	The hepatocyte membrane function and enzymatic activity were modified within physiological standards	5 mon	Tutel'ian et al., 1999
Soybean (glyphosate-tolerant)	Rats	No adverse effects of GM soybean meal were seen even at levels as high as 90% of the diet	13 wk	Zhu et al., 2004
Maize (GM CBH351)	Rats, Mice	No immunotoxicity was detected. No other specific toxicity tests were included	13 wk	Teshima et al., 2002
Maize (round-up ready)	Rats	No adverse effects were reported on overall health, body weight, food consumption, clinical pathology parameters, organ weights and microscopic appearance of tissues	13 wk	Hammond et al., 2004
Potato (GM)	Rats	Increase in the number of bacteria phagocytized by monocytes, percentage of neutrophils producing ROS and oxygen-dependent bactericidal activity of neutrophils	5 wk	Winnicka et al., 2001
Potato (GM delta-endotoxin treated)	Mice	Mild changes in the structural configuration of the ileum. Potential hyperplastic development of the ileum	2 wk	Fares and El-Sayed, 1998
Potato (GM)	Rats	Proliferation of the gastric mucosa. Effects on the small intestine and caecum	10 d	Ewen and Pusztai, 1999
Rice (transgenic, cowpea trypsin inhibitor)	Rats	Some alternations on haematological parameters	90 d	Zhuo et al., 2004
Rice (transgenic KMD1)	Rats	Although only minor changes were detected, additional test groups are required	90 d	Schroder et al., 2007

administered recombinant tilapia GH daily for 30 d. Blood samples were examined before and after the treatment period for clinical and biochemical parameters, including haemoglobin, total serum proteins, glucose, creatinine, leukocytes, and erythrocytes (Guillen et al., 1999). Even if transgenic fish become approved by federal regulators, consumers will be the ultimate determinants of the success of these products (Aerni, 2004). Therefore, in addition to the evaluation of environmental and human health effects of GMOs, surveys into consumer acceptance of transgenic products are also important. Interestingly, in a study concerning the opinions of 1365 Canadian consumers on sales of transgenic animal products, the majority showed a favourable attitude toward the concept of genetically modified (GM) salmon, while a lesser percentage revealed positive attitudes toward intent to purchase transgenic salmon products (Castle et al., 2005).

CONCLUSION

Transgenesis is a rapid way of producing fish with new genetic traits. Hence, it is a desirable method for broodstock development for aquaculture and marine resource management. However, before transgenesis can be applied to a wider range of organisms for aquaculture, the identification and isolation of species-specific promoter sequences is pivotal for designing gene constructs. This would minimize contamination of the gene pool of the wild population through the introduction of novel DNA sequences of another species (Sin, 1997).

Transgenic fish technology provides a key biotechnological opportunity to enhance global production and quality of aquatic foods, particularly at the local level in developing nations where aquaculture has an important role in producing animal protein for human consumption. The issues raised above regarding the estimation of natural fitness present significant obstacles but also raise challenges that require resolution if transgenic fish technology is to reach its potential in the coming decades. Although improved facilities are needed to enable the culture of transgenic strains with natural phenotypes for risk assessments, this approach will only partially solve the problems associated with fitness assessments. As such, it is clear that biotechnological solutions are urgently required for containment, to prevent interactions between transgenic fish and wild fish populations (Devlin et al., 2006). In response to the needs of a burgeoning human population, food production activities have destroyed many forest and prairie ecosystems, globally. The future application of safe biotechnological solutions to enhance the efficiency of food production, including transgenic fish, might stem the need for continued growth of conventional agricultural production systems to the benefit of natural ecosystems.

ABBREVIATIONS

AFP	Antifreeze protein
AFGP	Antifreeze glycoproteins
ALA	A-Linolenic acid
CMV	Cytomegalovirus
CNS	Central nervous system
DHA	Decosahexaenoic acid
EPA	Eicosapentaenoic acid
GAP	GnRH-associated peptide
GFP	Green fluorescence protein
GH	Growth hormone
GHR	Growth hormone receptor
GLO	L-gulono-γ-lactone oxidase
GnRH	Gonadotropin-releasing hormone
HEW	Hen egg white
hGLUI	Human glucose transporter
hLF	Human lactoferrin
IGF	Insulin-like growth factor
IRES	Internal ribosomal entry sites
ISAV	Infectious salmon anaemia virus
L-AAB	L-ascorbic acid biosynthesis
PUFA	Omega-3-polyunsaturated fatty acids
rHKII	Rat hexokinase

REFERENCES

Aerni, P. (2004). Risk, regulation and innovation: the case of aquaculture and transgenic fish. Aquat. Sci. 66: 327-341.

Alimuddin, G.Y., Kiron, V., Satoh, S. and Takeuchi, T. (2005). Enhancement of EPA and DHA biosynthesis by over-expression of masu salmon n-6-desaturase-like gene in zebrafish. Transgenic Res. 14(2):159-165.

Allen, S.K. (1998). Commercial applications of bivalve genetics: not a solo effort. World Aquaculture March: 38-43.

Amanuma, K., Takeda, H., Amanuma, H. and Aoki, Y. (2000). Transgenic zebrafish for detecting mutations caused by compounds in aquatic environments. Nat. Biotechnol. 18: 62-65.

Ashihara, M., Suzuki, M., Kubokawa, K., Yoshiura, Y., Kobayashi, M., Urano, A. and Aida, K. (1995). Two differing precursor genes for the salmon-type gonadotropin-releasing hormone exist in salmonids. J. Mol. Endocrinol. 15: 1-9.

Assem, S.S. and El-Zaeem, S.Y. (2005). Application of biotechnology in fish breeding. II: production of highly immune genetically modified redbelly tilapia, *Tilapia zillii*. Afr. J. Biotechnol. 4(5): 449-459.

Bao, D.Q., Mori, T.A., Burke, V., Puddey, I.B. and Beilin, L.J. (1998). Effects of dietary fish and weight reduction on ambulatory blood pressure in overweight hypertensives. Hypertension 32: 710-717.

Beaumont, A.R. and Fairbrother, J.E. (1991). Ploidy manipulation in molluscan shellfish: a review. J. Shellfish Res. 10: 1-18.

Bevin, C.L. and Zasloff, M. (1990). Peptides from frog skin. Annu. Rev. Biochem. 59: 395-414.

Blanc, J.M., Maunas, P. and Vallee, F. (2005). Effect of triploidy on paternal and maternal variance components in brown trout, *Salmo trutta* L. Aquaculture Res. 36(10): 1026-1033.

Brem, G., Brenig, B., Horstgen-Schwark, G. and Winnacker, E.L. (1988). Gene transfer in tilapia (*Oreochromis niloticus*). Aquaculture 68: 209-219.

Buchanan, J.T., Nickens, A.D., Cooper, R.K. and Tiersch, T.R. (2001). Transfection of eastern oyster (*Crassotrea virginica*) embryos. Mar. Biotechnol. 3(4): 322-335.

Caelers, A., Maclean, N., Hwang, G., Eppler, E. and Reinecke, M. (2005). Expression of endogenous and exogenous growth hormone (GH) messenger (m) RNA in a GH-transgenic tilapia (*Oreochromis niloticus*). Transgenic Res. 14(1): 95–104.

Carvan, M.J., Dalton, T.P., Stuart, G.W. and Nebert, D.W. (2000). Transgenic zebrafish as sentinels for aquatic pollution. Ann. NY Acad. Sci. 919: 133-147.

Carvan, M.J., Sonntag, D.M., Cmar, C.B., Cook, R.S., Curran, M.A. et al. (2001). Oxidative stress in zebrafish cells: potential utility of transgenic zebrafish as a deployable sentinel for site hazard ranking. Sci. Total Environ. 274: 183-196.

Castle, D., Finlay, K. and Clark, S. (2005). Proactive consumer consultation: the effect of information provision on response to transgenic animals. J. Publ. Aff. 5: 200-216.

Chatakondi, N., Lovell, R., Duncan, P., Hayat, M., Chen, T., Powers, D., Weete, T., Cummins, K. and Dunham, R.A. (1995). Body composition of transgenic common carp, *Cyprinus carpio*, containing rainbow trout growth hormone gene. Aquaculture 138: 99-109.

Chen, T.T. and Powers, D.A. (1990). Transgenic fish. Trends Biotechnol. 8: 209-214.

Chen, T.T., Lin, C.M., Dunham, R.A. and Powers, D.A. (1992). Integration, expression and inheritance of foreign fish growth hormone gene in transgenic fish. In: Transgenic Fish. C.L. Hew and G.L. Fletcher (eds). World Scientific Publishing Co., River Edge, NJ., USA, pp. 164-175.

Chen, T.T., Lu, J.K., Shamblott, M.J., Cheng, C.M., Lin, C.M., Burns, J.C., Reimschuessel, R., Chatakondi, N. and Dunham, R. (1996). Transgenic fish: ideal models for basic research and biotechnological applications. In: Molecular Zoology: Advances, Strategies, and Protocols. J.D. Ferraris and S.R. Palumbi (eds). Wiley-Liss Inc., New York, USA, pp. 401-433.

Chevassis, B. and Dorson, M. (1990). Genetics of resistance to disease in fishes. Aquaculture 85(1-4): 83-107.

Chourrout, D. (1986). Techniques of chromosome manipulation in rainbow trout: a new evaluation with karyology. Theoret. Appl. Genet. 72: 627-632.

Collares, T., Bongalhardo, D.C., Deschamps, J.C. and Moreira, H.L.M. (2005). Transgenic animals: the melding of molecular biology and animal reproduction. Anim. Reprod. 2(1): 11-27.

Cook, J.T., McNiven, M.A., Richardson, G.F. and Sutterlin, A.M. (2000). Growth rate, body composition and feed digestibility/conversion of growth-enhanced transgenic Atlantic salmon. Aquaculture 188: 15-32.

Davies, P.L., Fletcher, G.L. and Hew, C.L. (1989). Fish antifreeze protein genes and their use in transgenic studies. In: Oxford Surveys on Eukaryotic Genes, Volume 6. N. Maclean (ed). Oxford University Press, Oxford, UK, pp. 85-109.

Davies, P.L. and Hew, C.L. (1990). Biochemistry of fish antifreeze proteins. FASEB J. 4: 2460-2468.

de la Fuente, J., Guillen, I. and Estrada, M.P. (1998). The paradox of growth acceleration in fish. In: New Developments in Marine Biotechnology. Y. Le Gal and H.O. Halvorson (eds). Plenum Press, New York, USA, pp. 7-10.

Delgado, C., Rosegrant, M., Meijer, S., Wada, N. and Ahmed, M. (2002). Fish as food: projections to 2020. The Biennial Meeting of International Institute for Fisheries Economics and Trade (IIFET), Wellington, New Zealand.

Devlin, R.H. (1997). Transgenic salmonids. In: Transgenic Animals: Generation and Use. L.M. Houdebine (ed). Harwood Academic Publishers, Amsterdam, The Netherlands, pp. 105-117.

Devlin, R.H., Yesaki, T.Y., Donaldson, E.M., Du, S.J. and Hew, C.L. (1995). Production of germline transgenic Pacific salmonids with dramatically increased growth performance. Can. J. Fish. Aquat. Sci. 52: 1376-1384.

Devlin, R.H., Johnsson, J.I., Smailus, D.E., Biagi, C.A., Jonsson, E. and Bjornsson, B.T. (1999). Increased ability to compete for food by growth hormone-transgenic coho salmon *Oncorhynchus kisutch* (Walbaum). Aquaculture Res. 30: 479-482.

Devlin, R.H., Biagi, C.A., Yesaki, T.Y., Smailus, D.E. and Byatt, J.C. (2001). Growth of domesticated transgenic fish. Nature 409: 781-782.

Devlin, R.H., Sundstrom, L.F. and Muir, W.M. (2006). Interface of biotechnology and ecology for environmental risk assessment of transgenic fish. Trends Biotechnol. 24(2): 89-97.

Dhar, A.K., Moss, R.J. and Allnut, F.C.T. (2006). Internal ribosomal entry site (IRES)-based shrimp expression vector for heterologous protein production in shrimp. Meeting Abstract #487. AQUA, Florence, Italy, May 10-13.

Domingo, J.L. (2007). Toxicity studies of genetically modified plants: a review of the Published literature. Crit. Rev. Food Sci. Nutr. 47: 721-733.

Du, S.J., Gong, Z., Fletcher, G.L., Shears, M.A., King, M.J., Idler, D.R. and Hew, C.L. (1992). Growth enhancement in transgenic Atlantic salmon by the use of an "all fish" chimeric growth hormone gene constructs. Bio/Technology 10: 176-181.

Dunham, R.A. (2004). Aquaculture and Fisheries Biotechnology: Genetic Approaches. CABI Publishing, Wallingford, UK.

Dunham, R.A. (2005). Cecropin transgenic catfish and studies toward commercial application. Transgenic Animal Research Conference V, Tahoe City, California, USA, August 14-18.

Dunham, R.A. and Devlin, R. (1998). Comparison of traditional breeding and transgenesis in farmed fish with implications for growth enhancement and fitness. In: Transgenic Animals in Agriculture. J.D. Murray, G.B. Anderson, A.M. Oberbauer and M.N. McGloughlin (eds). CAB International, Wallingford, UK, pp. 209-229.

Dunham, R.A. and Liu, Z. (2002). Gene mapping, isolation and genetic improvement in catfish. In: Aquatic Genomics: Steps Toward a Great Future. N. Shimizu, T. Aoki, I. Hirono and F. Takashima (eds). Springer-Verlag, New York, USA, pp. 45-60.

Dunham, R.A., Eash, J., Askins, J. and Townes, T.M. (1987). Transfer of the metallothionein human growth hormone fusion gene into channel catfish. Trans. Am. Fish. Soc. 116: 87-91.

Dunham, R.A., Ramboux, A.C., Duncan, P.L., Hayat, M., Chen, T.T., Lin, C.M., Kight, K., Gonzalez-Villasenor, I. and Powers, D.A. (1992). Transfer, expression, and inheritance of salmonid growth hormone genes in channel catfish, *Ictalurus punctatus*, and effects on performance traits. Mol. Mar. Biol. Biotechnol. 1(4-5): 380-389.

Dunham, R.A., Warr, G.W., Nichols, A., Duncan, P.L., Argue, B., Middleton, D. and Kucuktas, H. (2002). Enhanced bacterial disease resistance of transgenic channel catfish *Ictalurus punctatus* possessing cecropin genes. Mar. Biotechnol. 4(3): 338-344.

El-Zaeem, S.Y. and Assem, S.S. (2004). Application of biotechnology in fish breeding I: production of highly immune genetically modified Nile tilapia, *Oreochromis niloticus*, with accelerated growth by direct injection of shark DNA into skeletal muscles. Egypt. J. Aquat. Biol. Fish. 8(3): 67-92.

Enikolopov, G.N., Benyumov, A.O., Barmintsev, A., Zelenina, L.A., Sleptsova, L.A., Doronin, Y.K., Golichenkov, V.A., Grashchuk, M.A., Georgiev, G.P., Rubtsov, P.M., Skryabin, K.G. and Baev, A.A. (1989). Advanced growth of transgenic fish containing human somatotropin gene. Doklady Akademii Nauk SSSR 301: 724-727.

Estrada, M.P., Herrera, F., Cabezas, L., Martinez, R., Arenal, A., Tapanes, L., Vazquez, J. and de la Fuente, J. (1999). Culture of transgenic tilapia with accelerated growth under different 18 intensive culture conditions. In: Special Adaptations of Tropical Fish. J. Nelson and D. MacKinley (eds). pp. 93-100.

Ewen, S.W. and Pusztai, A. (1999). Effects of diets containing genetically modified potatoes expressing *Galanthus nivalis* lectin on rat small intestine. Lancet 354: 1353-1354.

Exadactylos, A. and Arvanitoyannis, I.S. (2006). Aquaculture biotechnology for enhanced fish production for human consumption. In: Microbial Biotechnology in Agriculture and Aquaculture, Volume 2. R.C. Ray (ed). Science Publishers, Inc., Enfield, NH, USA, pp. 453-500.

FAO (2000). The state of the world fisheries and aquaculture (SOFIA). FAO, Rome. Available from: http://www.fao.org/sof/sofia/index en.htm.

Fares, N.H. and El-Sayed, A.K. (1998). Fine structural changes in the ileum of mice fed on delta-endotoxin-treated potatoes and transgenic potatoes. Nat. Toxins 6: 219-233.

FDA (Food and Drug Administration). Questions and answers about transgenic fish. Available from: http://www.fda.gov/cvm/transgen.htm (6 October 2007).

Figueiredo, M.A., Lanes, C.F.C., Almeida, D.V., Proietti, M.C. and Marins, L.F. (2007). The effect of GH overexpression on GHR and IGF-I gene regulation in different genotypes of GH-transgenic zebrafish. Comp. Biochem. Physiol. Part D 2: 228-233.

Fletcher, G.L. and Davies, P.L. (1991). Transgenic fish for aquaculture. Genetic Eng. 13: 331-369.

Fletcher, G.L., Shears, M.A., King, M.J., Davies, P.L. and Hew, C.L. (1988). Evidence for antifreeze protein gene transfer in Atlantic salmon (*Salmo salar*). Can. J. Fish. Aquat. Sci. 45: 352-357.

Fletcher, G.L., Shears, M.A., King, M.J., Goddard, S.V., Kao, M.H., Du, S.J., Davies, P.L. and Hew, C.L. (1992). Biotechnology for aquaculture: transgenic salmon with enhanced growth and freeze resistance. Bull. Aquacult. Assoc. Can. 92: 31-33.

Fletcher, G.L., Shears, M.A., Yaskowiak, E.S., King, M.J. and Goddard, S.V. (2004). Gene transfer: potential to enhance the genome of Atlantic salmon for aquaculture. Aust. J. Exp. Agri. 44: 1095-1100.

Foresti, F. (2000). Biotechnology and fish culture. Hydrobiologia 420(1): 45-47.

Fu, C., Cui, Y., Hung, S.S.O. and Zhu, Z. (1998). Growth and feed utilization by F4 human growth hormone transgenic carp fed diets with different protein levels. J. Fish. Biol. 53: 115-129.

Galli, L. (2002). Genetic modification in aquaculture—a review of potential benefits and risks. Bureau of Rural Sciences, Canada. Available from: http://www.affa.gov.au/corporate docs/publications/pdf/rural science/landuse/GM in Aquaculture.pdf

Gong, Z., Wan, H., Ju, B., He, J., Wang, X. and Yan, T. (2002). Generation of living color transgenic zebrafish. In: Aquatic Genomics: Steps toward a Great Future. N. Shimizu, T. Aoki, I. Hirono and F. Takashima (eds). Springer-Verlag, New York, USA, pp. 329-339.

Gordon, M.S., Andur, B.N. and Scholander, P.F. (1962). Freezing resistance in some northern fishes. Biol. Bull. 122: 52-62.

Grimm, H., Mayer, K., Mayser, P. and Eigenbrodt, E. (2002). Regulatory potential of n-3 fatty acids in immunological and inflammatory processes. Brit. J. Nutr. 87(1): S59-S67.

Gross, M.L., Schneider, J.F., Moav, N., Moav, B., Alvarez, C., Myster, S.H., Liu, Z., Hallerman, E.M., Hackett, P.B., Guise, K.S., Faras, A.J. and Kapuscinski, A.R. (1992). Molecular analysis and growth evaluation of northern pike (*Esox lucius*) microinjected with growth hormone genes. Aquaculture 103(3-4): 253-273.

Guillen, I., Berlanga, J., Valenzuela, C.M., Morales, A., Toledo, J., Estrada, M.P., Puentes, P., Hayes, O. and de la Fuente, J. (1999). Safety evaluation of transgenic tilapia with accelerated growth. Mar. Biotechnol. 1: 2-14.

Hallerman, E.M., McLean, E. and Fleming, I.A. (2007). Effects of growth hormone transgenes on the behaviour and welfare of aquacultured fishes: a review identifying research needs. Appl. Animal Behav. Sci. 104: 265-294.

Hammond, B.G., Vicini, J.L., Hartnell, G.F., Naylor, M.W., Knight, C.D., Robinson, E.H., Fuchs, R.L. and Padgette, S.R. (1996). The feeding value of soybeans fed to rats, chickens, catfish and dairy cattle is not altered by genetic incorporation of glyphosate tolerace. J. Nutr. 26: 717-727.

Hammond, B., Dudek, R., Lemen, J. and Nemeth, M. (2004). Results of a 13-week safety assurance study with rats fed grain from glyphosate tolerant corn. Food Chem. Toxicol. 42: 1003-1014.

Hew, C.L. and Fletcher, G.L. (1992). Transgenic Fish. World Scientific Pub. Co., River Edge, NJ., USA.

Hew, C.L., Fletcher, G.L. and Davies, P.L. (1995). Transgenic salmon: tailoring the genome for food production. J. Fish. Biol. 46(suppl A): 1-19.

Hew, C.L., Poon, R., Xiong, F., Gauthier, S., Shears, M., King, M., Davies, P. and Fletcher, G. (1999). Liver-specific and seasonal expression of transgenic Atlantic salmon harbouring the winter flounder antifreeze protein gene. Transgenic Res. 8: 405-414.

Horrocks, L.A. and Yeo, Y.K. (1999). Health benefits of docosahexaenoic acid (DHA). Pharmacol. Res. 40(3): 211-225.

Hostetler, H.A., Collodi, P.R., Devlin, R.H. and Muir, W.M. (2003). Ecological risks and benefits of fish transgenic for the phytase gene. Transgenic Animal Research Conference IV, Tahoe City, California, USA, August 10-14.

Houdebine, L.M. and Chourrout, D. (1991). Transgenesis in fish. Experientia 47: 891-897.

Hsih, M.H., Kuo, J.C. and Tsai, H.J. (1997). Optimization of the solubilization and renaturation of fish growth hormone produced by *Escherichia coli*. Appl. Microbiol. Biotechnol. 48: 66-72.

Hu, W., Wang, Y. and Zhu, Z. (2006). A perspective on fish gonad manipulation for biotechnical applications. Chin. Sci. Bull. 51(1): 1-7.

Hu, W., Li, F., Tang, B., Wang, Y., Lin, H., Liu, X., Zou, J. and Zhu, Z. (2007). Antisense of gonadotropin-releasing hormone reduces gonadotropin synthesis and gonadal development in transgenic common carp (*Cyprinus carpio*). Aquaculture 271: 498-506.

Huang, J.M., Nandi, S., Wu, L.Y., Yalda, D., Bartley, G., Rodriguez, R., Lonnerdal, B. and Huang, N. (2002). Expression of natural antimicrobial human lysozyme in rice grains. Mol. Breed. 10: 83-94.

Jensen, I., Albuquerque, A., Sommer, A.I. and Robertsen, B. (2002). Effect of poly I:C on the expression of Mx proteins and resistance against infection by infectious salmon anaemia virus in Atlantic salmon. Fish Shellfish Immunol. 13(4): 311-326.

Jhingan, E., Devlin, R.H. and Iwama, G.K. (2003). Disease resistance, stress response and effects of triploidy in growth hormone transgenic coho salmon. J. Fish Biol. 63: 806-823.

Kapuscinski, A.R. (2005). Current scientific understanding of the environmental biosafety of transgenic fish and shellfish. Rev. Sci. Tech. 24(1): 309-322.

Kelly, L. (2005). The safety assessment of foods from transgenic and cloned animals using the comparative approach. Rev. Sci. Tech. 24(1): 61-74.

Krasnov, A., Agren, J.J., Pitkanen, T.I. and Molsa, H. (1999). Transfer of growth hormone (GH) transgenes into Arctic charr (*Salvelinus alpinus* L.) II. Nutrient partitioning in rapidly growing fish. Genet. Anal. 15(3-5): 99-105.

Leaf, A., Kang, J.X., Xiao, Y.F. and Billman, G.E. (2003). Clinical prevention of sudden cardiac death by n-3 polyunsaturated fatty acids and mechanism of prevention of arrhythmias by n-3 fish oils. Circulation 107: 2646-2652.

Leggatt, R.A. and Iwama, G.K. (2003). Occurrence of polyploidy in the fishes. Rev. Fish Biol. 13(3): 237-246.

Leggatt, R.A., Brauner, C.J., Iwama, G.K. and Devlin, R.H. (2007). The glutathione antioxidant system is enhanced in growth hormone transgenic coho salmon (*Oncorhynchus kisutch*). J. Comp. Physiol. B 177: 413-422.

Leong, J.C., Bootland, L., Anderson, E., Chiou, P.W., Drolet, B., Kim, C., Lorz, H., Mourich, D., Ormonde, P., Perez, L. and Trobridge, G. (1995). Viral vaccines for aquaculture. J. Mar. Biotech. 3: 16-23.

Levy, J.A., Marins, L.F. and Sanchez, A. (2000). Gene transfer technology in aquaculture. Hydrobiologia 420: 91-94.

Liu, S., Zang, X., Liu, B., Zhang, X., Arunakurama, K.K.I.U., Zhang, X. and Liang, B. (2007). Effect of growth hormone transgenic *Synechocystis* on growth, feed efficiency, muscle composition, haematology and histology of turbot (*Scophthalmus maximus* L.). Aquacult. Res. 38: 1283-1292.

Low, W.K., Miao, M., Ewart, K.V., Yang, D.S., Fletcher, G.L. et al. (1998). Skin-type antifreeze protein from the shorthorn sculpin, *Myoxocephalus scorpius*. Expression and characterization of A M_r 9,700 recombinant protein. J. Biol. Chem. 273: 23098-23103.

Lu, J.K., Fu, B.H., Wu, J.L. and Chen, T.T. (2002). Production of transgenic silver sea bream (*Sparus sarba*) by different gene transfer methods. Mar. Biotechnol. 4(3): 328-337.

Lu, Y. and Sun, P.S. (2005). Viral resistance in shrimp that express an antisense Taura syndrome virus coat protein gene. Antiviral Res. 67(3): 141-146.

Lutz, C.G. (2003). Practical Genetics for Aquaculture. Blackwell Publishing Co., Oxford, UK.

Maclean, N. and Laight, R.J. (2000). Transgenic fish: an evaluation of benefits and risks. Fish and Fisheries 1: 146-172.

Maclean, N., Penman, D. and Talwar, S. (1987). Introduction of novel genes into fish. Biotechnology 5: 257-261.

Maisse, G., Labbe, C., de Baulny, B.O., Calvi, S.L. and Haffray, P. (1998). Fish sperm and embryos cryopreservation. Production Animals 11(1): 57-65.

Mao, W., Wang, Y., Wang, W., Wu, B., Feng, J. and Zhu, Z. (2004). Enhanced resistance to *Aeromonas hydrophila* infection and enhanced phagocytic activities in human lactoferrin-transgenic grass carp. Aquaculture 242: 93-103.

Martinez, R., Estrada, M.P., Berlanga, J., Guillen, I., Hernandez, O., Cabrera, E., Pimentel, R., Morales, R., Herrera, F., Morales, A., Pina, J.C., Abad, Z., Sanchez, V., Melamed, P., Lleonart, R. and de la Fuente, J. (1996). Growth enhancement in transgenic tilapia by ectopic expression of tilapia growth hormone. Mol. Mar. Biol. Biotechnol. 5(1): 62-70.

Martinez, R., Juncal, J., Zaldivar, C., Arenal, A., Guillen, I., Morera, V., Carrillo, O., Estrada, M., Morales, A. and Estrada, M.P. (2000). Growth efficiency in transgenic tilapia (*Oreochromis* sp.) carrying a single copy of a homologous cDNA growth hormone. Biochem. Biophys. Res. Commun. 267: 466-472.

Mason, A.J., Hayflick, J.S., Zoeller, R.T., Young, W.S., Phillips, H.S. et al. (1986). A deletion truncating the gonadotropin-releasing hormone gene is responsible for hypogonadism in the HPG mouse. Science 234: 1366-1371.

Mitra, A., Foster-Frey, J., Rexroad, C.E., Wells, K.D. and Wall, R.J. (2003). Molecular characterization of lysozyme type II gene in rainbow trout (*Oncorhynchus mykiss*): evidence of gene duplication. Anim. Biotechnol. 14(1): 7-12.

Morvan, A., Bachere, E., Da Silva, P.P., Pimenta, P. and Mialhe, E. (1994). *In vitro* activity of the antimicrobial peptide magainin 1 against *Bonamia ostreae*, the intrahemocytic parasite of the flat oyster, *Ostrea edulis*. Mol. Mar. Biol. Biotech. 3: 327-333.

Nam, Y.K., Cho, Y.S., Cho, J.C. and Kim, D.S. (2002). Accelerated growth performance and stable germline transmission in androgenetically derived homozygous transgenic mud loach, *Misgurnus mizolepis*. Aquaculture 209: 257-270.

NAS (National Academy of Science). (2002). Animal biotechnology: science-based concerns. The National Academy Press, Washington, D.C., USA.

Naylor, R.L., Goldburg, R.J., Primavera, J.H., Kautsky, N., Beveridge, M.C., Clay, J., Folke, C., Lubchenco, J., Mooney, H. and Troell, M. (2000). Effect of aquaculture on world fish supplies. Nature 405(6790): 1017-1024.

Naylor, R., Hindar, K., Fleming, I.A., Goldburg, R., Williams, S., Volpe, J., Whoriskey, F., Eagle, J., Kelso, D. and Mangel, M. (2005). Fugitive salmon: assessing the risks of escaped fish from net-pen aquaculture. BioScience 55(5): 427-437.

Ozato, K., Kondoh, H., Inohara, H., Iwamatsu, T., Wakamatsu, Y. and Okada, T.S. (1986). Production of transgenic fish: introduction and expression of chicken delta-crystallin gene in medaka embryos. Cell Differ. Dev. 19: 237-244.

Pandian, T.J. and Venugopal, T. (2005). Contribution to transgenesis in Indian major carp *Labeo rohita*. In: Fish Genetics and Aquaculture Biotechnology. T.J. Pandian, C.A. Strussmann and M.P. Marian (eds). Science Publishers Inc., Enfield, NH, USA, pp. 1-20.

Pitkanen, T.I., Krasnov, A., Reinisalo, M. and Molsa, H. (1999a). Transfer and expression of glucose transporter and hexokinase genes in salmonid fish. Aquaculture 173(1): 319-332.

Pitkanen, T.I., Krasnov, A., Teerijoki, H. and Molsa, H. (1999b). Transfer of growth hormone (GH) transgenes into Arctic charr (*Salvelinus alpinus* L.) I. Growth response to various GH constructs. Genet. Anal. 15(3-5): 91-98.

Powers, D.A., Cole, T., Creech, K., Chen, T.T., Lin, C.M., Kight, K. and Dunham, R. (1992). Electroporation: a method for transferring genes into the gametes of zebrafish, *Brachydanio rerio*, channel catfish, *Ictalurus punctatus*, and common carp, *Cyprinus carpio*. Mol. Mar. Biol. Biotechnol. 1: 301-309.

Rahman, A. and Maclean, N. (1998). Production of lines of growth enhanced transgenic tilapia (*Oreochromis niloticus*) expressing a novel piscine growth hormone gene. In: New Developments in Marine Biotechnology. Y. Le Gal and H.O. Halvorson (eds). Plenum Press, New York, USA, pp. 19-28.

Rahman, M.A. and Maclean, N. (1999). Growth performance of transgenic tilapia containing an exogenous piscine growth hormone gene. Aquaculture 173: 333-346.

Rahman, M.A., Mak, R., Ayad, H., Smith, A. and Maclean, N. (1998). Expression of a novel piscine growth hormone gene results in growth enhancement in transgenic tilapia (*Oreochromis niloticus*). Transgenic Res. 7: 357-369.

Rahman, M.A., Ronyai, A., Engidaw, B.Z., Jauncey, K., Hwang, G.L., Smith, A., Roderick, E., Penman, D., Varadi, L. and Maclean, N. (2001). Growth and nutritional trials on transgenic Nile tilapia containing an exogenous fish growth hormone gene. J. Fish. Biol. 59: 62-78.

Ramirez, J.C. and Morrissey, M.T. (2003). Marine biotechnology. First Joint Trans-Atlantic Fisheries Technology Conference (TAFT), Reykjavik, Iceland.

Rasmussen, R.S. and Morrissey, M.T. (2007). Biotechnology in aquaculture: transgenics and polyploidy. Compr. Rev. Food Sci. Food Safety 6: 2-16.

Raymond, J.A. (1991). Inhibition of ice crystal growth by fish antifreezes. ACS Symposium series — American Chemical Society 444: 249-255.

Razak, S.A., Hwang, G.L., Rahman, M.A. and Maclean, N. (1999). Growth performance and gonadal development of growth enhanced transgenic tilapia *Oreochromis niloticus* (L.) following heat-shock-induced triploidy. Mar. Biotechnol. 1: 533-544.

Robert, S.S. (2006). Production of eicosapentaenoic and docosahexaenoic acid-containing oils in transgenic land plants for human and aquaculture nutrition. Mar. Biotech. 8(2): 103-109.

Sarmasik, A., Jang, I.K., Chun, C.Z., Lu, J.K. and Chen, T.T. (2001). Transgenic live-bearing fish and crustaceans produced by transforming immature gonads with replication-defective pantropic retroviral vectors. Mar. Biotechnol. 3(5): 470-477.

Sarmasik, A., Warr, G. and Chen, T.T. (2002). Production of transgenic medaka with increased resistance to bacterial pathogens. Mar. Biotechnol. 4(3): 310-322.

Scholander, P.F., Van Dam, L., Kanwisher, J.W., Hammel, H.T. and Gordon, M.S. (1957). Supercooling and osmoregulation in Arctic fish. J. Cell. Comp. Physiol. 49: 5-24.

Schroder, M., Poulsen, M., Wilcks, A., Kroghsbo, S., Miller, A., Frenzel, T., Danier, J., Rychlik, M., Emami, K., Gatehouse, A., Shu, Q., Engel, K.H., Altosaar, I. and Knudsen, I. (2007). A 90-day safety study of genetically modified rice expressing Cry1Ab protein (*Bacillus thuringiensis* toxin) in Wistar rats. Food Chem. Toxicol. 45: 339-349.

Sellars, M., Degnan, B.M. and Preston, N.P. (2006). Recent advances in *Marsupenaeus japonicus* polyploid induction: mitotic tetraploidy and polar body I triploidy. Meeting Abstract # 55. AQUA, Florence, Italy, May 10-13.

Shatkin, G., Shumway, S.E. and Hawes, R. (1997). Considerations regarding the possible introduction of the Pacific oyster (*Crassostrea gigas*) to the gulf of Maine: A review of global experience. J. Shellfish Res. 16: 463-477.

Shears, M.A., Fletcher, G.L., Hew, C.L., Gauthier, S. and Davies, P.L. (1991). Transfer, expression, and stable inheritance of antifreeze protein genes in Atlantic salmon (*Salmo salar*). Mol. Mar. Biol. Biotechnol. 1: 58-63.

Sin, F.Y.T. (1997). Transgenic fish. Rev. Fish Biol. Fisheries 7: 417-441.

Smith, K. and Spadafora, C. (2005). Sperm-mediated gene transfer: applications and implications. BioEssays 27(5): 551-562.

Stevens, E.D. and Devlin, R.H. (2000). Intestinal morphology in growth hormone transgenic Pacific coho salmon, *Oncorhynchus kisutch* Walbaum. J. Fish Biol. 56: 191-195.

Sundstrom, L.F., Lohmus, M., Johnsson, J.I. and Devlin, R.H. (2004). Growth hormone transgenic salmon pay for growth potential with increased predation mortality. Proc. Royal Soc. Lond. B. Biol. Sci. 271 (Biology Letters suppl. 5): S350-S352.

Sundstrom, L.F., Lohmus, M., Tymchul, W.E. and Devlin, R.H. (2007). Gene-environment interactions influence ecological consequences of transgenic animals. Biol. Sci./Ecol. 104(10): 3889-3894.

Teshima, R., Watanabe, T., Okunuki, H., Isuzugawa, H., Akiyama, H., Onodera, H., Imai, T., Toyoda, M. and Sawada, J. (2002). Effect of subchronic feeding of genetically modified corn (CBH351) on immune system in BN rats and B10A mice. Shokuhin Eiseigaku Zasshi 43: 273-279.

Thresher, R.E., Hinds, L., Grewe, P., Patil, J., McGoldrick, D., Nesbitt, K., Lumb, C., Whyard, S. and Hardy, C. (2001). Repressible sterility in aquaculture species: A genetic system for preventing the escape of genetically improved stocks. Aquaculture 2001: Book of Abstracts. World Aquaculture Society, Baton Rouge, Louisiana, USA.

Tutel'ian, V.A., Kravchenko, L.V., Lashneva, N.V., Avren'eva, L.I., Guseva, G.V., Sorokina, E.I. and Chernysheva, O.N. (1999). Medical and biological evaluation of safety of protein concentrates from genetically modified soybeans. Biochemical Studies. Vopr. Pitan. 68: 9-12.

Tymchuk, W.E.V., Abrahams, M.V. and Devlin, R.H. (2005). Competitive ability and mortality of growth enhanced transgenic coho salmon fry and parr when foraging for food. Trans. Am. Fish. Soc. 134: 381-389.

Udvadia, A.J. and Linney, E. (2003). Windows into development: historic, current, and future perspectives on transgenic zebrafish. Dev. Biol. 256: 1-17.

Uzbekova, S., Chyb, J., Ferriere, F., Bailhache, T., Prunet, P., Alestrom, P. and Breton, B. (2000). Transgenic rainbow trout expressed sGnRH-antisense RNA under the control of sGnRH promoter of Atlantic salmon. J. Mol. Endocrinol. 25: 337-350.

von Schalburg, K.R., Harrower, W.L. and Sherwood, N.M. (1999). Regulation and expression of gonadotropin-releasing hormone in salmon embryo and gonad. Mol. Cell. Endocrinol. 157: 41-54.

Wang, R., Zhang, P., Gong, Z. and Hew, C.L. (1995). Expression of the antifreeze protein gene in transgenic goldfish (*Carassius auratus*) and its implication in cold adaptation. Mol. Mar. Biol. Biotech. 4(1): 20-26.

Wang, Y., Hu, W., Wu, G., Sun, Y., Chen, S., Zhang, F., Zhu, Z., Feng, J. and Zhang, X. (2001). Genetic analysis of "all-fish" growth hormone gene transferred carp (*Cyprinus carpio* L.) and its F1 generation. Chin. Sci. Bull. 46(14): 1174-1177.

Wang, Z., Li, Y., Yu, R., Gao, Q., Tian, C., Zheng, X. and Wang, R. (2003). Growth comparison between triploid and diploid Pacific oyster during the reproductive season. Am. Fish. Soc. Symp. 38: 285-289.

Wargelius, A., Fjelldal, P.G., Benedet, S., Hansen, T., Bjornsson, B.T. and Nordgarden, U. (2005). A peak in GH-receptor expression is associated with growth activation in Atlantic salmon vertebrae, while upregulation of IGF-I receptor expression is related to increased bone density. Gen. Comp. Endocrinol. 142(1-2): 163-168.

Winn, R.N. (2001). Trangenic fish as models in environmental toxicology. Ilar. J. 42: 322-329.

Winnicka, A., Sawosz, E., Klucinski, W., Kosieradzka, I., Szopa, J., Malepszy, S. and Pastuszewska, A. (2001). A note on the effect of feeding genetically modified potatoes on selected indices of non-specific resistance in rats. J. Anim. Feed Sci. 10(2): 13-18.

Wu, G., Sun, Y. and Zhu, Z. (2003). Growth hormone gene transfer in common carp. Aquat. Living Resour. 16: 416-420.

Yazawa, R., Hirono, I. and Aoki, T. (2005). Characterization of promoter activities of four different Japanese flounder promoters in transgenic zebrafish. Mar. Biotechnol. 7(6): 625-633.

Yazawa, R., Hirono, I. and Aoki, T. (2006). Transgenic zebrafish expressing chicken lysozyme show resistance against bacterial diseases. Transgenic Res. 15: 385-391.

Yee, J.K., Miyanohara, A., LaPorte, P., Bouic, K., Burns, J.C. and Friedmann, T. (1994). A general method for the generation of high-titer, pantropic retroviral vectors: highly efficient infection of primary hepatocytes. Proc. Nat. Acad. Sci., USA 91: 9564-9568.

Zbikowska, H.M. (2003). Fish can be first—advances in fish trangenesis for commercial applications. Transgenic Res. 12: 379-389.

Zhang, F., Wang, Y., Hu, W., Cui, Z. and Zhu, Z. (2000). Physiological and pathological analysis of the mice fed with "all-fish" gene transferred yellow river carp. High Technol. Lett. 7: 17-19.

Zhang, P., Joyce, C., Gozalez-Villasenor, L.I., Lin, C.M., Dunham, R.A., Chen, T.T. and Power, D.A. (1990). Gene transfer, expression, and inheritance of pRSV-rainbow trout-GH cDNA in common carp (Linnaeus). Mol. Reprod. Dev. 25: 3-13.

Zhang, P., Xu, Y., Liu, Z., Xiang, Y., Du, S. and Hew, C.L. (1998). Gene transfer in red sea bream (*Pagrosomus major*). In: New Developments in Marine Biotechnology. Y. Le Gal and H.O. Halvorson (eds). Plenum Press, New York, USA, pp. 15-18.

Zhong, J., Wang, Y. and Zhu, Z. (2002). Introduction of the human lactoferrin gene into grass carp (*Ctenopharyngodon idellus*) to increase resistance against GCH virus. Aquaculture 214: 93-101.

Zhu, Y., Li, D., Wang, F., Yin, J. and Jin, H. (2004). Nutritional assessment and fate of DNA of soybean meal from Roundup Ready or conventional soybeans using rats. Arch. Anim. Nutr. 58: 295-310.

Zhu, Z. (1992). Generation of fast growing transgenic fish: methods and mechanisms In: Transgenic Fish. C.L. Hew and G.L. Fletcher (eds). World Scientific Publishing Co., River Edge, NJ., USA, pp. 92-119.

Zhu, Z., Li, G., He, L. and Chen, S. (1985). Novel gene transfer into the fertilized eggs of gold fish (*Carassius auratus* L. 1758). J. Appl. Ichthyol. 1: 31-34.

Zhu, Z., Xu, K., Li, G., Xie, Y. and He, L. (1986). Biological effects of human growth hormone gene microinjected into the fertilized eggs of loach *Misgurnus anguillicaudatus* (Cantor). Kexue Tongbao 31(14): 988-990.

Zhuo, Q., Chen, X., Piao, J. and Gu, L. (2004). Study on food safety of genetically modified rice which expressed cowpea trypsin inhibitor by 90-day feeding test rats. Wei Sheng Yan Jiu 33: 176-179.

2

Application of PCR-DGGE Method in Determining Origin of Fish: Case Studies of Pangasius Fish from Vietnam, Tilapia from Thailand and Sea Bass from France

Didier Montet[1*], Le Nguyen Doan Duy[1,2], Ratanaporn Leesing[1,3], Thierry Goli[1] and Gérard Loiseau[1]

INTRODUCTION

The issues surrounding food safety and security continue to be hot topics of concern throughout the supply chain. BSE (Bovine Spongiform Encephalopathy), *Salmonella* and avian influenza remain embedded in the memories of European consumers. Regulations across Europe continue to be tightened in order to provide a greater degree of assurance in quality and safety. Meanwhile, the traceability and labelling of imported products in European countries remains a compulsory issue (UE regulation 178/2002). With similar scares occurring globally, the need for vigilance and strict monitoring is necessary.

[1]*Centre de Coopération Internationale en Recherche Agronomique pour le Développement, CIRAD, UMR Qualisud, TA 95B/16, 34398 Montpellier Cedex 5, France*

[2]*Cantho University, College of Agriculture and Applied Biology Rue 3-2, Campus 2, Ninh Kieu District, Cantho Province, Vietnam*

[3]*Department of Microbiology, Faculty of Science, Khon Kaen University, Khon Kaen 40002, Thailand*

*Corresponding author: Tel.: 33 467615728; Fax: 33 467614444; E-mail: didier.montet@cirad.fr

One of the great concerns for customers is the traceability of the food products. Traceability is the capacity to find the history, use or origin of a food (activity, process and product) by registered methods (ISO 8402, 1994). For a long time the food industry has had simple traceability systems, but with the increasing implementation of current Good Manufacturing Practice and ISO 9000 quality management in food manufacture, traceability systems have become more important in the production chain. In view of the difficulties of installing these documentary systems in developing countries such as in South-East Asia, and to follow the product during processing, the identification and validation of some pertinent biological markers which come from the environment of fish is proposed to assure the traceability of aquaculture products during processing and international trade. Currently, there are no existing scientific methods which determine or follow the geographical origin of food precisely. The determination of geographical origin is one such demand for the traceability of import-export products.

One hypothesis of tracing the source of a product is by analyzing in a global way the bacterial communities on the food samples. The predominant bacterial flora would permit the determination of the capture area, production process or sanitary or hygienic conditions during post harvest operations (Montet et al., 2004). Numerous studies of the microbiota in fish captured from various geographical locations have been done (Grisez et al., 1997; Spanggaard et al., 2000; Al-Harbi and Uddin, 2004; Leesing, 2005). Microorganisms are found on all the outer surfaces (skin and gills) and in the intestines of live and newly caught fish. The water composition, temperature and weather conditions can influence the bacterial communities of fish. The research of de Sousa and Silva-Sousa (2001) on the bacterial communities of the Congonhas River in Brazil showed that there was a direct relation between the bacterial community of the river and the fish commensal bacteria. Wong et al. (1999) showed that some species of *Vibrio parahaemolyticus* isolated from the shrimps imported from Asia were specific for the geographical capture zone.

In this chapter, principles, applications and advantages of PCR-DGGE (Polymerase chain reaction-Denaturing gradient gel electrophoresis) method to food and agricultural systems with particular emphasis on tracing the origin of fish have been studied.

PRINCIPLE OF PCR-DGGE

The principle of the DGGE is based on the partial separation of a fragment of a double stranded DNA under the action of heat or a denaturing agent (Fischer and Lerman, 1983). A molecule made up partly with the traditional double helix and partly of two simple strands is strongly slowed down compared to a molecule double-stranded or completely denatured (Fig. 2.1).

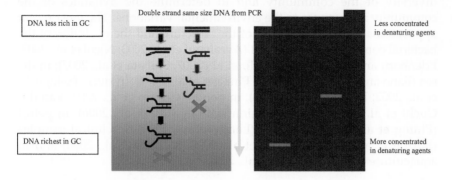

Fig. 2.1   Principle of DGGE migration (Leesing, 2005)

The DGGE allows to separate DNA fragments of identical size but with different sequences of which their percentage of Guanine (G) and Cytosine (C) permits their separation. Separation is based on the electrophoretic mobility of the DNA double strands in a polyacrylamide gel containing a linear gradient of denaturing agents (urea and formamide). The electrophoresis is carried out at a constant temperature between 55 and 65°C, preferably at 60°C.

However, if the gradient increases in a linear way, the denaturation of the DNA is not made in a progressive way throughout the molecule, but from area to area, which opens more or less quickly according to their composition in bases, Adenine (A)–Thyamine (T) and Guanine (G)–Cytosine (C). The space conformation of the DNA fragment changes in parallel with the importance of the denatured areas which limits its electrophoretic mobility (Muyzer et al., 1993).

An optimal resolution is obtained when the fragments are not completely denatured. To avoid the total denaturation of the double strand, a structure at very high melting point called GC clamp (very rich in GC) is associated during amplification. The PCR is then carried out with the primer forward bearing into 5′ a sequence of 40 GC: the GC clamp (Sheffield et al., 1989).

APPLICATIONS OF PCR-DGGE TO FOOD, AGRICULTURAL AND ENVIRONMENTAL SYSTEMS

Study of the Microbial Biodiversity in the Environment

Since its introduction by Muyzer et al. (1993) in microbial ecology, the PCR-DGGE is employed successfully to study the microbial communities in the environment. This technique is usually employed to evaluate the diversity of the community and to determine the dynamics of the community in response to the environmental variations. The most recent applications relate to the study of the structure and the evolution of the bacterial communities in the soil (Avrahami et al., 2003; Nicol et al., 2003; Edenborn and Sexstone, 2007; Hu et al., 2007; Shibata et al., 2007); in the sea (Bano and Hollibaugh, 2002; Pinhassi et al., 2004), in rivers (Sekiguchi et al., 2002; Lyautey et al., 2005), in ponds (Crump et al., 2003; Van der Gucht et al., 2005), in purification stations (Liu et al., 2006), in petrol (Phung et al., 2004), in wood (Landy et al., 2007), in clinical samples (Burton et al., 2003; Donskey et al., 2003; McBain et al., 2003) and in aquaculture fish (Le Nguyen et al., 2008).

Follow-up and Identification of the Microbial Community during Fermentation

The great potential shown by the PCR-DGGE to analyze the environmental samples stimulates the microbiologists to study the possibility of its application in the follow-up of microbial fermentations. Among the first, Ampe et al. (1999) studied the spatial distribution of the microorganisms in Pozol balls, a Mexican food containing fermented corn. Lactic bacteria were identified by the sequencing of DNA fragments purified from DGGE gel. The DGGE permitted to identify the various lactic acid bacteria and to specify their role in the process of fermentation. The authors concluded that the culture-independent methods seemed to be better than the traditional methods to study fermented food.

For the same product Pozol, Ben Omar and Ampe (2000) studied the microbial flora responsible for the fermentation by examining the dynamics of the microbial community during the pozol production. They identified the dominant microbial species at different steps of fermentation, and showed the responsibility of the important species in the fermentation and the production of the essential compounds. Thus, *Streptococcus* spp. dominated the total process while the heterofermentative lactic bacteria such as *Lactobacillus fermentum* were present from the very beginning of the fermentation but were then replaced by the heterofermentative lactic bacteria (*Lactobacillus plantarum*, *Lactobacillus casei* and *Lactobacillus delbrueckii*).

The genetic fingerprint thus gives an indication of the bacterial community at a given moment. It is the succession of these images which makes it possible to establish the dynamics of fermentation. This application is interesting since it allows one to follow the bacterial fluctuations during a technological treatment. This aspect could in the long term enable in controlling the manufacture of fermented foodstuffs (Ercolini, 2004). In another study concerning the fermentation of cassava paste, Miambi et al. (2003) identified 10 bacterial species by using the same culture-dependent and culture-independent techniques. The authors showed that *Lactobacillus manihotivorans*, *Lactobacillus fermentum* and *Lactobacillus crispatus*, which were identified by sequencing of DGGE bands, could not be found by traditional techniques. The dairy products were studied among the foodstuffs and the approach of PCR-DGGE was recently exploited to supervise their fermentation. Ercolini et al. (2001, 2003) worked successively on the microflora of two cheeses: Mozzarella of Italian origin and Stilton of English origin. By sequencing the DGGE fragments, they could identify, in Italian cheese, some thermophilous lactic bacteria like *Streptococcus thermophilus*, mesophilic lactic bacteria such as *Lactococcus lactis*, and some bacteria like *Alishewanella fetalis* usually linked to contaminations. The British cheese seems to contain a much more complex flora, whose distribution is not homogeneous between the heart and the crust. This method was also applied to characterize dominant microbial populations in Spanish Cabrales cheese (Flórez and Mayo, 2006) and Domiati cheese, a traditional Egyptian cheese (El-Baradei et al., 2007). PCR-DGGE was also used for the study of fermented drinks. The bacterial communities present during the fermentation of malt for preparing whisky was studied (Van Beek and Priest, 2002). The authors showed that the whole spectrum of *Lactobacillus* flora plays an important part in the first steps of the fermentation, whereas at the end of the fermentation, only two strains of homofermentative *Lactobacillus* dominate.

The microbial community intervening during the maturation of vanilla in Indonesia was also studied by this approach (Röhling et al., 2001). The authors could find a significant presence of *Bacillus* and other microorganisms theoretically non-cultivable during the high temperature treatment of the pod.

Quality Control of the Foodstuffs

PCR-DGGE can be applied to determine the total colony counts in finished products, and to control their biological quality. Silvestri et al. (2007) used PCR-DGGE by combining it with a method of traditional culture for the monitoring of the hygienic quality of Ciauscolo, a traditional Italian salami. This method allowed one to prove the quality of this product by showing the absence of pathogenic bacteria in the product at the artisanal level as well as at the industrial level. In the same way, PCR-DGGE enabled the supervision of the hygienic quality of salad (Handschur et al., 2005). On this product, the authors found bacteria of the Microbacteriaceae family such as *Microbacterium testaceumm*, *Acietobacter junii*, *Rathayibacter tritici* and of Enterobacteriaceae like *Klebsiella pneumoniae* and *Enterobacter aerogenes*. Moreover, analytical laboratories could use this technique for the detection of the pathogens possibly present in the foodstuffs (Ji et al., 2004). The comparison between the food microflora and a mixture of reference bacteria should permit in giving the sanitary quality of the product, more quickly than with traditional techniques. Indeed, the duration of a total analysis of the bacteria by DGGE, for about 30 samples, requires less than 24 h which is the minimal time necessary for the majority of the traditional analyses.

The PCR-DGGE was also applied to control the hygienic quality of mineral water (Dewettinck et al., 2001). The authors showed that it was possible to obtain genetic fingerprints from underground water and commercial mineral water. The presence of non-cultivable microorganisms was also proven by this method. With this technique, Fasoli et al. (2003) identified the microorganisms present in probiotic yoghourts and freeze-dried probiotic preparations. References of migrations were created by using the bacteria generally met in the probiotic products. The comparison of commercial products to these models showed that certain not-declared microorganisms are present in the products.

In the seafood industry, there are a few applications of the PCR-DGGE method for the analysis of the bacterial community of fish. Hovda et al. (2007a, 2007b) used this technique to characterize the dominant bacteria in cod and halibut fillets preserved under controlled atmosphere. They showed that the method is applicable to identify the microorganisms responsible for the fish deterioration. The principal bacteria found in packing under controlled and normal atmosphere were *Photobacterium phosphoreum*, *Bronchothrix thermosphacta* and *Pseudomonas* spp. These authors also used this technique to measure the influence of an ozone treatment on the microflora of the cod fillets.

Authentification and Identification of Geographical Origin

This global analysis could be useful for the control of appellations. Coppola et al. (2001) compared various cheese processes: a method known as "artisanal" and a more industrial method. Their studies showed that the cheese microflora depends partly on the mode of production. So the technique of PCR-DGGE could be considered as a tool for authentification of products marked as "artisanal".

An original and effective application of the PCR-DGGE to group foodstuffs was described by Mauriello et al. (2003). This work on the initiating flora of the production of Mozzarella, a traditional Italian cheese, showed that the composition of the starters strongly depended of their geographical origin. Moreover, the organoleptic characters of various cheeses are also dependent of the geographical origin of the initiating flora. Genetic fingerprints could be a good means to protect the European labels PDO (Protected Designation of Origin) or PGI (Protected Geographical Indication) whose principle of attribution is requirements centred primarily on the zone of production and human know-how. Leesing (2005) optimized the conditions of PCR-DGGE for the characterization and the comparison of the bacterial communities of fish coming from various areas of France and Thailand. The author showed that it is possible to differentiate fish of various origins by the DGGE profiles obtained. Le Nguyen et al. (2008) worked with PCR-DGGE for the determination of the geographical origin of catfish (Pangasius) from Vietnam. The fishes were taken in different fish farms and during two different seasons, the rainy season and the dry season. DGGE profiles of the bacterial populations of the various farms were different and specific for each site. The concept of genetic code bars thus appeared because DGGE gel could be used as a bar code to certify the geographical origin of the fish (Montet, 2003).

ADVANTAGES OF PCR-DGGE IN STUDYING BACTERIAL COMMUNITY IN FISH

The fish microbiota in aquaculture has previously been studied for many purposes. Detection of bacterial communities in fish samples has been mostly done by cultivation-dependent plating techniques and their identification by phenotypical and biochemical characteristics (Grisez et al., 1997; Syvokienë and Mickënienë, 1999; Al Harbi and Uddin, 2003, 2004; El-Shafai et al., 2004). This is time-consuming, and it is tempting to use a limited number of key characteristics. In order to provide a better insight into the diversity of bacterial community in fish, there is a need for cultivation-independent techniques enabling the direct information of these populations. Separation of PCR products in DGGE is based on the decrease of the electrophoretic mobility of partially melted doubled-stranded DNA molecules in polyacrylamide gels containing a linear gradient of DNA denaturants like formamide and urea at 60°C. Molecules with different sequences will have a different melting behaviour and will stop migrating at a different position in the gel (Muyzer et al., 1993; Leesing, 2005). PCR-DGGE has been already used to investigate several patterns of distribution of marine bacterial assemblages (Murray et al., 1996; Øvreas et al., 1997; Moeseneder et al., 1999; Riemann et al., 1999) but this technique had not been previously applied to the study of bacteria community on freshwater fish for traceability.

A specific advantage of this technique is that it permits the analysis of cultivable and non-cultivable, anaerobic and aerobic bacteria and provides a rapid method to observe the changes in community structure in response to different environmental factors (Yang et al., 2001).

APPLICATION OF PCR-DGGE IN TRACING ORIGIN OF FISH

The PCR-DGGE method is applied for analyzing the bacteria in aquaculture fish in order to create a technique to link bacterial communities to the geographical origin and avoid the individual analysis of each bacterial strain. The acquired band patterns for the bacterial species of different fish are compared and analyzed statistically to determine the fish origin.

The succession of the electrophoretic bands in these images is described as "biological bar code". In products of large consumption, this digitized code contains information related to the traceability of the geographical origin of the studied fish. One of the principal originality of this technique is concentrated on the study of the fish commensal flora but without any culture of microorganism, all the steps of extraction are realized directly starting from fish.

A case study comprising the following fish species is discussed below.

1. Pangasius from Vietnam,
2. Tilapia from Thailand, and
3. Sea bass from France.

Fish Sampling

Tilapias were collected from ponds of three freshwater fish farms in Thailand (Fig. 2.2, Photo 2.1): Khon Kaen, 450 km North-East of Bangkok; Nakorn Pratom, 50 km West of Bangkok; Ladkrabang (Bangkok). Sea bass (*Dicentrarchus labrax*) were collected from two fish farms in France: Aquanord (Gravelines) and Viviers du Gois (Beauvoir sur Mer), North-West of France (Fig. 2.3). The fish were placed immediately in ice and then stored at –20°C. The Pangasius fish samples (*Pangasius hypophthalmus*) were collected from five aquaculture farms of five different districts from South Vietnam, namely Chau Phu, An Phu, Phu Tan, Chau Doc and Tan Chau of An Giang Province (Fig. 2.4, Photo 2.2). This province supplies about two-third (about 80,000 MT in 2005) of Pangasius fish for export (Ministry of Aquaculture, Vietnam, 2005). The samples were collected during two seasons in Vietnam: the rainy season (October 2005) and the dry season (February 2006). Samples were taken from one pond in each farm and aseptically transferred to storage bags. The samples were maintained on ice and transported to the laboratory. Then the skin, gills and intestines were aseptically removed from each fish specimen and put

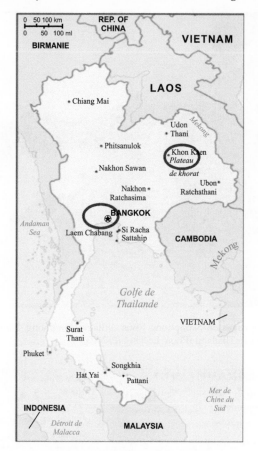

Fig. 2.2   Place of sample collection in Thailand.

in separate sealed plastic bags, and kept frozen at –20°C until analysis at the Cirad laboratory, Montpellier, France.

Operational Mechanism of PCR-DGGE

The operational mechanism of PCR-DGGE is described as follows. Samples containing approximately equal amounts of PCR amplicons are loaded into 8% (wt/v) polyacrylamide gels (Table 2.1) (acrylamide/NN'-methylene bisacrylamide) in 1X TAE buffer (40 mM Tris-HCl pH 7.4, 20 mM sodium acetate, 1.0 mM Na_2-EDTA). All electrophoresis experiments are performed at 60°C using a denaturing gradient ranging from 30 to 60% (100% corresponded to 7 M urea and 40% [v/v] formamide). The gels are electrophoresed at 20 V for 10 min and then at 80 V for 12 h. After

Photo 2.1 Flotting cages on Namphong River, village Namphong, Namphong district, Khon Kaen province in Thailand (Photo Leesing R., 2003)

Fig. 2.3 Place of sample collection in France.

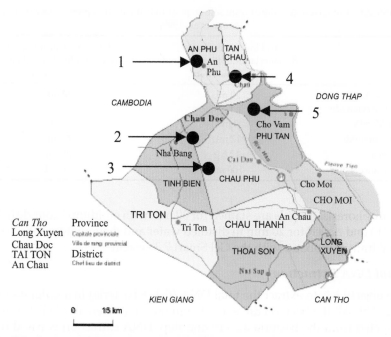

Fig. 2.4 The five different sampling locations in An Giang province, Vietnam : (1) An Phu; (2) Chau Doc; (3) Chau Phu; (4) Tan Chau; (5) Phu Tan.

Photo 2.2 Flotting cages on Mekong River, An Giang province in Vietnam (Photo Montet D., 2007)

Table 2.1 Composition of denaturant solutions 30 and 60% used to prepare DGGE gels (Ampe et al., 1999)

	Gel low concentration 8% Acrylamide/Bisacrylamide 30% Denaturant	Gel high concentration 8% Acrylamide/Bisacrylamide 60% Denaturant
40% Acrylamide/ bisacrylamide	20 mL	20 mL
TAE 50X	2 mL	2 mL
Formamide	12 mL	24 mL
Urea	12.6 g	25.2 g
Balance water	100 mL	100 mL

electrophoresis, the gels are stained for 30 min with ethidium bromide (0.5 mg/L) and rinsed for 20 min in distilled water and then photographed on a UV transilluminator with the Gel Smart 7.3 system.

Total DNA Extraction

The objective is to extract the total DNA (fish + bacteria) in a unique step (Fig. 2.5). All the bacteria are extracted in one step, then all the DNA are extracted from the bacteria also in one step. DNA extraction is based on various methods (Ampe et al., 1999; Leesing, 2005; Le Nguyen et al., 2007). The method followed by Le Nguyen et al. (2007) is given here. Around 2 g each of gills, skin and intestine are homogenized for 3 min by vortexing after addition of 6 mL sterile peptone water (pH 7.0). Four 1.5 mL tubes containing the resulting suspension are then centrifuged at 10,000 g for 10 min. 100 μL of lysis buffer TE (10 mM Tris-HCl; 1 mM EDTA; pH 8.0) and 100 μL of lysozyme solution (25 μg·μL^{-1}) and 50 μL of proteinase K solution (10 μg·μL^{-1}) are added to each pellet. Samples are vortexed for 1 min and incubated at 42°C for 30 min. Then 50 μL of 20% SDS (sodium dodecyl sulphate) is added to each tube, and the tubes are incubated at 42°C for 10 min. MATAB (300 μL) (mixed alkyltrimethyl ammonium bromide) is added to each tube, and the tubes are incubated at 65°C for 10 min. The lysates are then purified by repeated extractions with 700 μL of phenol-chloroform-isoamyl alcohol (25:24:1), and the residual phenol is removed by extraction with an equal volume of chloroform-isoamyl alcohol (24:1). The DNA is precipitated with isopropanol, washed with 70% ethanol and is air-dried at room temperature, 25°C. Finally, the DNA is resuspended in 100 μL of ultra pure water and stored at –20°C until analysis.

Polymerase Chain Reaction (PCR)

The development of the PCR-DGGE technique depends primarily on the choice of the primers. The primers used for amplification determine the

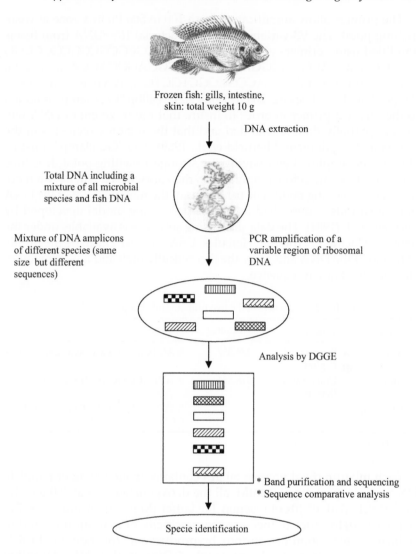

Fig. 2.5 Protocol for isolating bacterial flora from frozen fish and different steps of PCR-DGGE analysis.

smoothness and the specificity of the method. The most commonly employed target for PCR amplification prior to DGE is the ribosomal DNA. As this is a very conserved region of the genome that also includes variable regions. Therefore, primers can be designed by hybridizing to conserved regions but spanning variable regions in order to obtain PCR amplicons with species-specific difference in base pair composition that can be separated by DGGE.

The primers allow amplification from 100 to 500 Pb in a zone at weak melting point. The V3 variable region of bacterial 16S rDNA from fish is amplified using primers gc338f (5′CGCCCGCCGCGCGCGGCGGGCGG GGCGGGGGCACGGGGGGACTCCTACGGGAGGCAGCAG, Sigma, France) and 518r (5′-ATTACCGCGGCTGCTGG) (Øvreas et al., 1997; Ampe et al., 1999; Leesing, 2005) (Table 2.2). A 40bp "GC-clamp" is added to the forward primer in order to insure that the fragment of DNA will remain partially double-stranded and that the region screened is in the lowest melting domain (Sheffield et al., 1989). The "GC clamp" must be close to the amplified end having the strongest melting point. It is thus necessary to know before amplification the zones which will have a more or less high melting point. The behaviour of the melting point of the DNA double strands is described by a data-processing model developed by Lerman et al. (1984). The data-processing software is available, under the name of MacMelt^{TM} (Biorad, Hercules, USA) and can calculate the profile of fusion of the DNA and show the theoretically high and low domains of stability of a known sequence.

Table 2.2 Sequences of the two primers and clamp used in PCR

Primers	Targets	Positions*	Sequences
338f	Bacteria, V3 Region	338-357	5′ AC TCC TAC GGG AGG CAG CAG 3′
518r	Universal, V3 Region	518-534	5′ ATT ACC GCG GCT GCT GG 3′
GC clamp			5′ CGC CCG CCG CGC GCG GCG GGC GGG GCG GGG GCA CGG GGG G 3′

*Number established for *Escherichia coli*

Each mixture (final volume 50 μL) contains about 100 ng of template DNA, all the primers at 0.2 μM, all the deoxynucleosides at 200 μM, 1.5 mM MgCl$_2$, 5 μL of 10x of reaction *Taq* buffer (MgCl$_2$ free) and 5U of Taq polymerase. In order to increase the specificity of amplification and to reduce the formation of spurious by-products, a "touchdown" PCR is performed according to the protocol of Díez et al. (2001). An initial denaturation at 94°C for 1 min and 10 touchdown cycles of denaturation at 94°C for 1 min, then annealing at 65°C (with the temperature decreasing 1°C per cycle) for 1 min, and extension at 72°C for 3 min, followed 20 cycles of 94°C for 1 min, 55°C for 1 min and 72°C for 3 min, are followed sequentially. During the last cycle, the extension step is increased to 10 min. Aliquots (5 μL) of PCR products are analyzed first by conventional electrophoresis in 2% (w/v) agarose gel with TAE 1X buffer (40 mM Tris-

HCl pH 7.4, 20 mM sodium acetate, 1.0 mM Na$_2$-EDTA), stained with ethidium bromide, 0.5 µg/mL in TAE 1X and quantified by using a standard (DNA mass ladder 100 bp).

Denaturing Gradient Gel Electrophoresis (DGGE) Analysis

The PCR products are analyzed by DGGE and the procedure was first described by Muyzer et al. (1993) and improved by Leesing (2005).

Image and Statistical Analysis

Individual lanes of the gel images are straightened and aligned using ImageQuant TL software (Amersham Biosciences, Buckinghamshire, UK). Banding patterns are standardized with the two reference patterns included in all gels which are the patterns of Gram-*Escherichia coli* DNA and Gram + *Lactobacillus plantarum* DNA. This *Lactobacillus* was given by IRD France (Fig. 2.6). This software permitted in identifying the bands and their relative position compared with standard patterns. In DGGE analysis, the generated banding pattern is considered as an "image" of all the major bacterial species in the population. An individual discrete band refers to a unique "sequence type" or phylotype (Muyzer et al., 1995; van Hannen et al., 1999), which is treated as a discrete bacterial population. It is expected that PCR fragments generated from a single population will display an identical electrophoretic mobility in the analysis. This was confirmed by Kowalchuk et al. (1997) who showed that co-migrating bands generally corresponded to identical sequence.

The DGGE fingerprints are manually scored by the presence and absence of co-migrating bands, independent of intensity. Pairwise community similarities are quantified using the Dice similarity coefficient (S_D) (Heyndrickx et al., 1996).

$$S_D = 2 N_c / N_a + N_b$$

where N_a represents the number of bands detected in the sample A, N_b represents the number of bands detected in the sample B, and N_c represents the numbers of bands common to both samples. Similarity indexes are expressed within a range of 0 (completely dissimilar) to 1 (perfect similarity). Dendograms are constructed using the Statistics version 6 software (StatSoft, France). Similarities in community structure are determined using the cluster analysis by the single linkage method with the Euclidean distance measure. Significant differences of bacterial communities of fish between seasons are determined by factorial correspondence analysis using the first two factors which describes most of the variance in the data set.

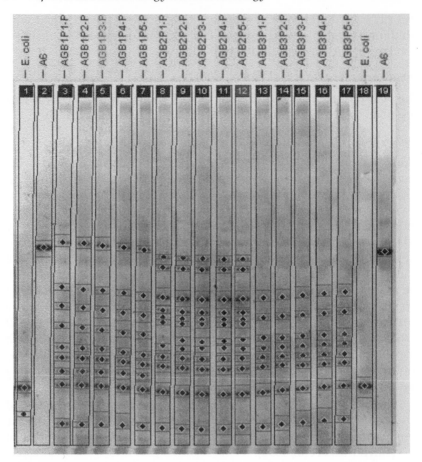

Fig. 2.6 Individual lanes of DGGE gel images were straightened and aligned using ImageQuant software (Amersham Biosciences, USA).

Lines 2 and 19: *Lactobacillus plantarum* A6

Lines 1 and 18: *Escherichia coli*

Use of PCR-DGGE to Differentiate Locations within the Same Sampling Period

Analysis of bacterial community composition by PCR-DGGE was first performed on tilapia and sea bass under electrophoresis condition at 80 V for 12 h, a combination that was considered to be the best compromise to differentiate the DNA bands directly extracted from fish. DGGE band patterns for the 16S rDNA of dominant bacteria from tilapia and sea bass collected from different geographical locations in France and Thailand are shown in Fig. 2.7. DGGE band patterns from all fish samples showed some

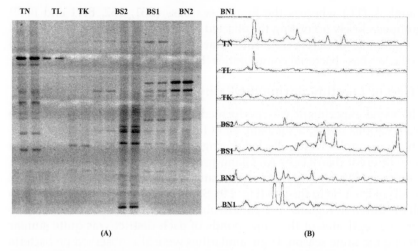

(A)   (B)

Fig. 2.7 DGGE gel analysis of 16S rDNA fragments of the representative bacteria from fish samples collected from Thailand and France; BN1-BN2: sea bass provided by Viviers du Gois, France; BS1-BS2: sea bass provided by Aquanord, France; TK: tilapia from Khon Kaen, Thailand; TN: tilapia from Nakorn Pratom, Thailand; TL: tilapia from Ladkrabang, Bangkok, Thailand.

(A) DGGE profiles (negatively converted). (B) DGGE chromatogram obtained from (A) by using ImageJ software.

differences. The number of bands per lane varied from 1 to 9. Some differences were noted in band position, intensity, and number of bands present in the fish bacterial DGGE patterns throughout the collected locations. Each fish had its own unique profile, indicating variations within fish species but also depending on the sampling location as well. The Dice similarity coefficient (S_D) was used to quantify the similarity of these community fingerprints between each fish samples. There was only a moderate similarity ($S_D = 0.4$) between sea bass BS1 and BS2 from France. The fingerprints from sea bass (BN1 and BN2) were highly similar but not identical, community fingerprints ($S_D = 0.82$). There was a complete dissimilarity between tilapia from Thailand (TN, TL, TK) and BN1, BN2, BS1 and BS2 ($S_D = 0$). This pair-wise comparison of community structures suggested a larger community shift in response to different geographical locations where the fishes have been collected than within the fish species.

The number of bands in an individual lane ranged from 1 to 9, corresponding to a total of 24 unique band positions. Cluster analysis was performed with the UPGMA algorithm to study general patterns of community similarity among the different geographic locations where the fish samples were collected. The bacterial communities of the Thai tilapia from Nakorn Pratom (TN) were closely related to the tilapia from

Bangkok (TL) whereas the French sea bass (BS1, BS2, BN1 and BN2) grouped closely according to collected locations. Otherwise, the bacterial communities of tilapia from Khon Kaen (TK) were only distantly related to the sea bass bacterial communities (Fig. 2.8).

Effect of Location during the Rainy Season on DGGE Patterns

The same experiment was performed to differentiate fish (Pangasius) from various farms in Vietnam, all of them located on the Mekong delta. The fish samples were collected during the rainy season (October 2005) in five different districts of An Giang province, Vietnam. The PCR-DGGE patterns of five replicates for each location revealed the presence of 8 to 12 bands of bacteria in the fish (Fig. 2.9). Some of the bands were common to all the different regions. The pattern obtained for the bacterial community for five replicates of the same ponds of each district was quite similar among the same season. High similarities were also observed on bacteria patterns for the samples from the same districts, as well as the neighbouring district where the water was supplied by the same branch of the Mekong River.

The statistical analysis of the DGGE gel patterns for the five replicates of fish samples from five different districts of An Giang province harvested in the rainy season (25 samples), showed the community similarity among the different geographical locations where the fish samples were collected (Fig. 2.10). At 70% similarities level, two main

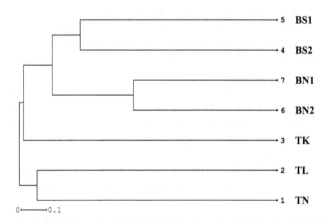

Fig. 2.8 Dendrogram revealing the relatedness of DGGE fingerprints from fish samples collected from different geographical locations.

BN1-BN2: sea bass provided by Viviers du Gois, France; BS1-BS2: sea bass provided by Aquanord, France; TK: tilapia from Khon Kaen, Thailand; TN: tilapia from Nakorn Pratom, Thailand; TL: tilapia from Ladkrabang, Bangkok, Thailand.

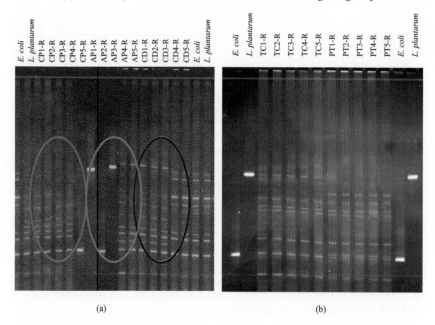

(a)   (b)

Fig. 2.9   Gel of PCR-DGGE 16S rDNA banding profiles of fish bacteria from 5 districts of An Giang province, Vietnam in the rainy season 2006 (a): CP (red circle): Chau Phu district; AP (green circle): An Phu district; CD (blue circle): Chau Doc district; (b): TC: Tan Chau district, PT: Phu Tan district.

1, 2, 3, 4, 5: replicate of fish.

R = rainy season.

clusters were observed (Fig. 2.11): (1) the first cluster included the samples from Chau Doc and Chau Phu districts; and (2) the second cluster comprised the samples from An Phu, Tan Chau and Phu Tan districts. The bacterial communities of Chau Doc and Chau Phu districts were then closely related, as well as the bacterial communities from An Phu, Tan Chau and Phu Tan districts.

Effect of the Seasons on DGGE Pattern of Fish within the Same Location

To know if the identified markers could be stable for two different seasons and in particular with the influence of the dilution due to rain in the ponds. For this purpose, samples were taken from the same ponds as previous experiments in the dry (February 2006) and rainy seasons (October 2005). Due to the closing down of the Phu Tan pond, only four districts were studied for the dry season. Differences on the DGGE band patterns of the same ponds at two different seasons could be clearly noted (Fig. 2.11a, b, c, d). It was observed that some dominant DNA bands

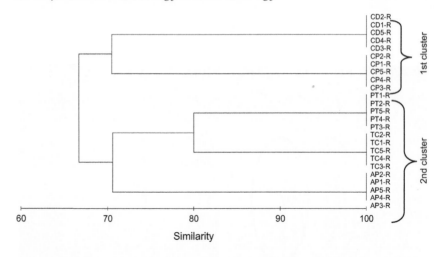

Fig. 2.10 Cluster analysis of 16S rDNA banding profiles for fish bacterial communities from 5 districts of An Giang province, Vietnam in the rainy season 2006: CP: Chau Phu district; AP: An Phu district; CD: Chau Doc district; TC: Tan Chau district; PT: Phu Tan district.

1, 2, 3, 4, 5: replicate of fish.   R = rainy season.

remained intact during the two seasons. Moreover, cluster analysis by Statistical software showed that the bacterial communities of the same ponds were quite similar for the two seasons. At 75% similarity level, two main clusters were also observed (Fig. 2.12): (1) The first cluster included the samples from An Phu and Tan Chau districts; and (2) The second cluster comprised all the samples from Chau Doc and Chau Phu districts. It was also shown that the number of common bands throughout the two seasons might vary from one location to another one. For example, for the farm in An Phu district, 10 bands were observed in the dry season and 9 bands in the rainy season but only 7 bands were common and the calculation by Statistics showed 75% similarity between the two seasons. The similarities were higher for the other farms (88% for Than Chau, 84% for Chau Phu and 86% for Chau Doc).

In addition, Factorial Correspondence Analysis (FCA) was used to compare the similarities of the bacterial communities from the four different geographical locations for two seasons. The first two factors described 76% of the variance in the data set and differences between the bacterial communities are shown in Fig. 2.12. Four different groups of four different districts were observed regardless of the seasons (Fig. 2.13).

Identification of Bacteria by DGGE Bands Sequencing

The bands were cut directly from the DGGE gel using a sterile scalpel. The DNA fragment of each band was diluted in 20 µL sterile water and stored

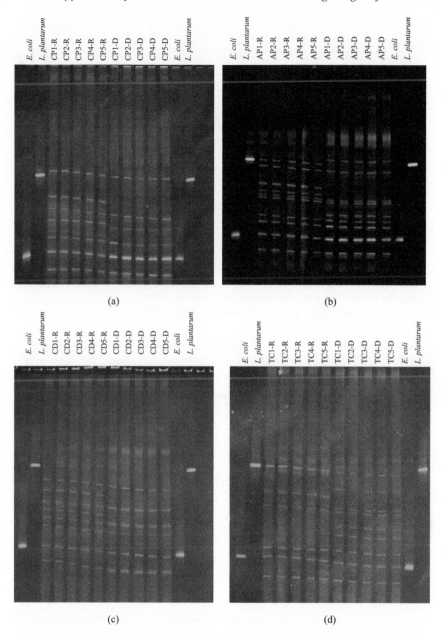

Fig. 2.11 Gel of PCR-DGGE 16S rDNA banding profiles of fish bacteria from 4 districts of An Giang province, Vietnam in the dry season and rainy season 2006: (a) CP: Chau Phu district, (b) AP: An Phu district, (c) CD: Chau Doc district; (d): TC: Tan Chau district.
1, 2, 3, 4, 5: replicate of fish.   R = rainy season; D = dry season

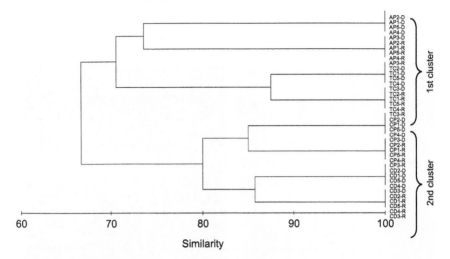

Fig. 2.12 Cluster analysis of 16S rDNA banding profiles of fish bacteria from 4 districts of An Giang province, Vietnam in the dry season and rainy season 2006: CP: Chau Phu district; AP: An Phu district; CD: Chau Doc district; TC: Tan Chau district.

1, 2, 3, 4, 5: replicate of fish.   R = rainy season; D = dry season.

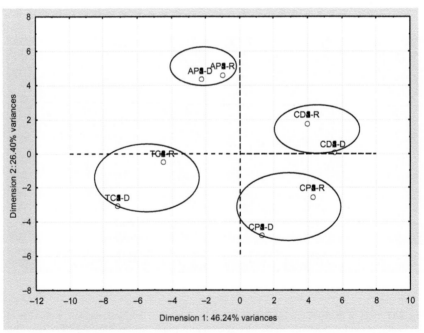

Fig. 2.13 Factorial variance analysis of 16S rDNA banding profiles of fish bacteria from 4 districts of An Giang province, Vietnam in the dry season and rainy season 2006: CP: Chau Phu district; AP: An Phu district; CD: Chau Doc district; TC: Tan Chau district.

R = rainy season; D = dry season.

during the night at 4°C. During this storage, the DNA fragment was solubilized in sterile water. This solution containing PCR amplicons and some impurities (BET, gel, loading dye) was purified with the Kit Wizard PCR Preps DNA Purification system (Promega, France).

The effectiveness of this procedure was checked by an electrophoresis of PCR amplicons by using the usual primers with GC clamp on DGGE gel. DNA solution (1 µL) was then reamplified by PCR under the same conditions but by using the primers without GC-clamp. These amplicons were then sent for sequencing at GATC Biotech (Germany). The sequences of 16S rDNA obtained were then compared using the programme BLAST with the available data bank on the NCBI internet site (National Center for Biotechnology Information Databases, USA) in order to find the known closest sequences and thus to be able to obtain the specie name of the bacteria with a percentage of similarity.

The results of sequencing showed that bacterial flora found on aquaculture fish in Vietnam were split in nine different species (Table 2.3):

Table 2.3 Bacteria identified on Vietnamese fish by sequencing of DGGE bands

Bands	Species, 16S rRNA gene, partial sequence	% similarity	Accession number
1	*Escherichia coli*	98%	EU014689.1
2	*Lactobacillus plantarum*	97%	EU074834.1
3	*Klebsiella pneumoniae*	97%	CP000641.1
4	*Acinetobacter* sp.	97%	EU029245.1
5	*Enterococcus faecalis*	96%	AE016830.1
6	*Aeromonas hydrophila*	96%	CP000462.1
7	*Streptophomonas* sp.	97%	AB306290.1
8	*Pseudomonas aeruginosa*	96%	DQ768706.1
9	*Pseudomonas* sp.	96%	X86625.1
10	*Burkholderia vietnamiensis*	98%	EU074833.1
11	*Lactococcus garvieae*	96%	EU081016.1

- *Lactococcus garviae* is a pathogenic zoonic bacterium that was isolated from many fish species as rainbow trout species (Carson et al., 1993). This bacterium is one of the most important Gram positive pathogen in fish (Eldar et al., 1996).
- Enterobacteriaceae are Gram negative bacteria that constitute one of the most important families of bacteria (more than 40 genera). These bacteria are ubiquitous, that means that they can be found in numerous ecosystems such as, in soil, water, vegetal, food and also in animals and human. This family includes in particular the species *Escherichia coli* and *Klebsiella pneumoniae*.

1. *E. coli* is the dominant specie of the thermotolerant coliform group. This microorganism is present in the digestive tract of warm blood animals and is abundant in the faeces. This is why it constitutes a determinant indicator of water contamination in faecal matters. Even if most of the strains are not pathogen, some serotypes like *E. coli* o157 H7 could cause severe diarrhoeic infections in humans.

2. *Klebsiella* are Gram negative Enterobacteriaceae, non-mobile and incapsulated. They are ubiquitous bacteria presented in the digestive tube and in the respiratory system of human and animals as commensal bacteria. They are abundant in soils and water. *Klebsiella* spp. are known to be able to produce histamines and other biogenic amines specially in fish.

The specie *Klebsiella pneumoniae* could cause bacterial pneumonia, in general due to its aspiration during alcohol ingestion or by infection. It is also an opportunist microorganism for patients with a Chronic Obstructive Pulmonary Disease, and enteric pathogenicity, and nasal atrophy or rhinoscleroma.

- *Aeromonas hydrophila* is a facultative anaerobic bacillus, in the shape of a stick, non-sporulated and gram-negative which belongs to the family of Vibrionaceae. This specie is omnipresent in the environment. It can be in freshwater, mud and sediments. In general, the watery environment constitutes the principal habitat of this bacterium. Various food of animal origin such as fish, meat of beef, pig and poultry can be contaminated by this strain. Regarded for a long time as simple opportunist bacteria for man, *A. hydrophila* is more and more suspected of causing gastroenteritis and even septicaemias.

- The strains of the genus *Acinetobacter* consist of Gram negative bacteria, non-sporulated, sometimes capsulated, non-mobile (but being able to have mobility by jerks resulting from the presence of polar fimbriae), and strict aerobes. They are regarded as ubiquitous bacteria having for principal habitat grounds, water, plants and skin of men and animals. Strains of *Acinetobacter* sp. are frequently isolated from waste water and the activated sludge of the purification stations. Gonzalez et al. (2000) showed the presence of *Acinetobacter* on freshwater fish (*Salmo trutta*, *Esox lucius* and *Oncorhynchus mykiss*).

- *Enterococcus faecalis* forms part of Gram positive enterococcus presented in the form of chains. It is a commensal flora of fish and is found in particular in the digestive and genital tracts.

- The genus *Stenotrophomonas* is part of the Xanthomonadaceae family. It was reclassified on July 10, 1993. It was known under the names of *Pseudomonas maltophila* then of *Xanthomonas maltophila* (Palleroni and Bradbury, 1993). Later on, four other species were included in the *Stenotrophomonas* genus. This genus is ubiquitous; present in ground, plants, water, etc. The specie *Strenotrophomonas maltophilia* was underlined in fish like the tuna (*Thunnus alaluga*) by Ben-Gigirey et al. (2002). This specie is responsible for nosocomial infections mainly on immuno-depressed individuals (cancers, virus HIV infections, neutropenias).

- The genus *Burkholderia* is part of the Burkholderiaceae family. They are Gram negative bacilli, mobile thanks to one or several polar whips. This genus was part of the *Pseudomonas* genus until Yabuuchi et al. (1992), while being based on the sequence of 16S rRNA, on homologies DNA-DNA, on the lipid and fatty acid composition and on the phenotypical characters, transferred the seven species from the group of homology II of the *Pseudomonas* genus to the new *Burkholderia* genus. *Burkholderia vietnamiensis* is a N_2-fixing proteobacteria of the rice rhizosphere isolated from the sulphated acid grounds (Tran Van et al., 1996). The ground of the Mekong delta is naturally sulphated and is thus a favourable habitat for this bacterium.

- The bacteria of the *Pseudomonas* genus are Gram negative bacilli, strict aerobes and generally mobiles thanks to one polar whip. They are widespread in the environment and live as saprophytes in ground and water. They are found in food and are sometimes responsible for the deterioration of food. In humans, the specie *Pseudomonas aeruginosa* frequently intervenes as a pathogenic opportunist. It is found as transit flora on the skin and the mucous membrane and is the cause of a lot of infection of wounds or burns.

These results agree rather well with those obtained by Leesing (2005) who observed that the genus *Escherichia*, *Klebsiella*, *Enterobacter*, *Acinetobacter* and *Pseudomonas* are dominant in Vietnamese *Pangasius hypophthalamus*.

CONCLUSION AND PERSPECTIVES

Analysis of bacterial communities in fish samples has been often investigated using culture dependent methods and culture-independent methods by random amplified polymorphic DNA (RAPD) (Spanggaard et al., 2000). There are only a few published works that analyzed the bacterial communities in fish samples by PCR-DGGE methods (Spanggaard et al., 2000; Huber et al., 2004).

From this study, it was found that the band pattern of the bacterial communities obtained by DGGE was strongly linked to the microbial environment of the fish. The skin is in direct contact with water and the gills that filter the air from water which goes to the lung of fish and is a good accumulator of the environmental bacteria. The intestine contains also a high amount of bacteria which is affected by the feeding habits of the fish.

The analysis of fish samples from different locations within the same period (rainy season) showed some significant differences in the migration patterns on the DGGE gel. However, the five replicates for each sampling location had statistically similar DGGE patterns throughout the study. The differences in the band profiles could be attributed to the differences of the feeding methods in between farms and the type of aquaculture system applied. The variations might also be due to the water supply which could be affected by the pollution from urban life. Furthermore, the antibiotics needed to cure diseases and stress factors could also affect the microbial communities of the fish. However, some common bands obtained by DGGE have been found in all the profiles within the same sampling periods and origin.

In fact, when seeing the different locations on the map of An Giang province in South Vietnam, there is a separation of the Mekong River in two main branches which are then divided into many small canals. Chau Phu and Chau Doc are on the same branch in the west of the Mekong River and An Phu, Tan Chau and Phu Tan are on another branch in the east of the river. It could be concluded that there were enough differences in the water quality and the environment of the fish to obtain a major effect on the bacterial ecology.

The study of the fish samples from the same locations at two different seasons showed that there were some significant statistical variations in the DGGE bands profiles. This fact could result from the important variations due to the heavy rains in the rainy season which could greatly affect the salinity and the pH of the pond water. These factors would greatly affect the microbial communities of the fish which are dependent on the outside environment. These results suggested that the DGGE band profile of the bacterial community of the fish was unique and representative of a particular farm and a season. However, there was a relatively higher similarity between the samples of the same location across different sampling seasons than the samples from different locations. The DGGE profiles showed some dominant bands of the same locations which are present throughout the whole sampling time.

In conclusion, the PCR-DGGE analysis of fish bacterial community suggests that this technique could be applied to differentiate geographical locations. It was shown that the biological markers for the specific locations stayed stable among the different seasons and that they show sufficient statistical specificity per farm. This global technique is quicker (less than 24 h) than all of the classical microbial techniques and avoids the precise analysis of bacteria by biochemistry or molecular biology (sequencing). This method can be proposed as a rapid analytical traceability tool for fish products and could be considered as a provider of a unique biological bar code.

ABBREVIATIONS

DGGE   Denaturing Gradient Gel Electrophoresis
dNTP   Deoxynucleotide triphosphate
PCR   Polymerase chain reaction
PDO   Protected designation of origin
PGI   Protected geographical indication
TAE   Buffer of Tris-acetate-EDTA
TE   Buffer Tris/EDTA
TEMED   N,N,N',N'-Tetramethylethylenediamine
UPGMA   Unweighted Pair-wise Grouping with Mathematic Averages

REFERENCES

Al Harbi, A.H. and Uddin, M.N. (2003). Quantitative and qualitative studies on bacterial flora of hybrid tilapia (*Oreochromis niloticus* × *O. aureus*) cultured in earthen ponds in Saudi Arabia. Aquaculture 34: 43-48.

Al-Harbi, A.H. and Uddin, M.N. (2004). Seasonal variation in the intestinal bacterial flora of hybrid tilapia (*Oreochromis niloticus* × *O. aureus*) cultured in earthen ponds in Saudi Arabia. Aquaculture 229: 37-44.

Ampe, F., Omar, N.B., Moizan, C., Wacher, C. and Guyot, J.P. (1999). Polyphasic study of the spatial distribution of microorganisms in Mexican pozol, a fermented maize dough, demonstrates the need for cultivation-independent methods to investigate traditional fermentations. Appl. Environ. Microbiol. 65: 5464-5473.

Avrahami, S., Liesack, W. and Conrad, R. (2003). Effect of temperature and fertilizer on activity and community structure of soil ammonia oxidizers. Environ. Microbiol. 5: 691-705.

Bano, N. and Hollibaugh, J.T. (2002). Phylogenetic composition of bacterioplankton assemblages from the Arctic Ocean. Appl. Environ. Microbiol. 68: 505-518.

Ben-Gigirey, B., Vieites, J.M., Kim, S.H., An, H., Villa, T.G. and Barros-Velazquez, J. (2002). Specific detection of *Stenotrophomonas maltophilia* strains in albacore tuna (*Thunnus alalunga*) by reverse dot-blot hybridization. Food Control 13: 293-299.

Ben Omar, N. and Ampe, F. (2000). Microbial community dynamics during production of Mexican fermented maize product pozol. Appl. Environ. Microbiol. 66: 3664-3673.

Burton, J.P., Cadieux, P.A. and Reid, G. (2003). Improved understanding of the bacterial vaginal microbiota of women before and after probiotic instillation. Appl. Environ. Microbiol. 69: 97-101.

Carson, J., Gudkows, N. and Austin, B. (1993). Characteristic of *Enterococcus*-like bacterium from Australia and South African, pathogenic for rainbow trout (*Oncorhynchus mykisss* Walbaum). J. Fish Dis. 16: 381-388.

Coppola, S., Blaiotta, G., Ercolini, D. and Moschetti, G. (2001). Molecular evaluation of microbial diversity occurring in different types of Mozzarella cheese. J. Appl. Microbiol. 90: 414-420.

Crump, B.C., Kling, G.W., Bahar, M. and Hobbie, J.E. (2003). Bacterioplankton community shits in an arctic lake correlate with seasonal changes in organic matters source. Appl. Environ. Microbiol. 69: 2253-2268.

de Sousa, J.A. and Silva-Sousa, A.T. (2001). Bacterial community associated with fish and water from Congohas river, Sertaneja, Parana, Brazil. Braz. Arch. Biol. Technol. 44: 373-381.

Dewettinck, T., Hulsbosch, W., Van Hege, K., Top, E.M. and Verstraete, W. (2001). Molecular fingerprinting of bacterial populations in ground water and bottled mineral water. Appl. Microbiol. Biotechnol. 57: 412-418.

Díez, B., Pedrós–Alió, C., Marsh, T.L. and Massana, R. (2001). Application of denaturing gradient gel electrophoresis (DGGE) to study the diversity of marine picoeukaryotic assemblage and comparison of DGGE with other molecular techniques. Appl. Environ. Microbiol. 67: 2942-2951.

Donskey, C.J., Hujer, A.M., Das, S.M., Pultz, N.J., Bonomo, R.A. and Rice, L.B. (2003). Use of denaturing gradient gel electrophoresis for analysis of the stool microbiota of hospitalized patients. J. Microbiol. Methods 54: 249-256.

Edenborn, S.L. and Sexstone, A.J. (2007). DGGE fingerprinting of culturable soil bacterial communities complements culture-independent analyses. Soil Biol. Biochem. 39: 1570-1579.

El-Baradei, G., Delacroix-Buchet, A. and Ogier, J.C. (2007). Biodiversity of bacterial ecosystems in traditional Egyptian Domiati cheese. Appl. Environ. Microbiol. 73: 1248-1255.

Eldar, A., Ghittino, C., Asanta, L., Bozetta, E., Goria, M., Prearo, M. and Bercovier, H. (1996). *Enterococcus seriolicida* is a junior synonym of *Lactococcus garviae*, a causative agent of septicemia and meningoecephalitis in fish. Curr. Microbiol. 32: 85-88.

El-Shafai, S.A., Gijzen, H.J., Nasr, F.A. and El-Gohary, F.A. (2004). Microbial quality of tilapia reared in fecal-contaminated ponds. Environ. Res. 95: 231-238.

Ercolini, D. (2004). PCR-DGGE fingerprinting: novel strategies for detection of microbes in food. J. Microbiol. Methods 56: 297-314.

Ercolini, D., Moschetti, G., Blaiotta, G. and Coppola, D. (2001). The potential of a polyphasic PCR-DGGE approach in evaluating microbial diversity of Natural Whey Clusters for water-buffalo Mozzarella cheese production: bias of "culture dependent" and "culture independent" approaches. Syst. Appl. Microbiol. 24: 610-617.

Ercolini, D., Hill, P.J. and Dodd, C.E.R. (2003). Bacterial community structure and location in Stilton cheese. Appl. Environ. Microbiol. 69: 3540-3548.

Fasoli, D., Marzotto, M., Rizzotti, L., Rossi, F., Dellaglio, F. and Torriani, S. (2003). Bacterial composition of commercial probiotic products as evaluated by PCR-DGGE analysis. Int. J. Food. Microbiol. 82: 59-70.

Fisher, S.G. and Lerman, L.S. (1983). DNA fragments differing by single base pair substitutions are separated in denaturing gradient gels: correspondence with melting theory. Proc. Natl. Acad. Sci. USA 80: 1579-1583.

Flórez, A.B. and Mayo, B. (2006). PCR-DGGE as a tool for characterizing dominant microbiol populations in the Spanish blue-veined Cabrales cheese. Int. Dairy J. 16: 1205-1210.

Grisez, L., Reyniers, J., Verdonck, L., Swings, J. and Ollevier, F. (1997). Dominant intestinal microbiota of sea bream and sea bass larvae, from two hatcheries, during larval development. Aquaculture 155: 387-399.

Gonzalez, C.J., Santos, J.A., Garcia-Lopez, M.L. and Otero, A. (2001). Mesophilic aeromonads in wild and aquacultured fresh water fish. J. Food Protec. 64: 687-691.

Handschur, M., Pinar, G., Gallist, B., Lubitz, W. and Haslberger, A.G. (2005). Culture free DGGE and cloning based monitoring of changes in bacterial communities of salad due to processing. Food Chemical Toxicol. 43: 1595-1605.

Heyndrickx, M., Vauterin, L., Vandamme, P., Kersters, K. and De Vos, P. (1996). Applicability of combined amplified ribosomal DNA restriction analysis (ARDRA) patterns in bacterial phylogeny and taxonomy. J. Microbiol. Methods 26: 247-259.

Hovda, M.B., Sivertsvik, M., Lunestad, B.T., Lorentzen, G. and Rosnes, J.T. (2007a). Characterisation of the dominant bacterial population in modified atmosphere packaged farmed halibut (*Hippoglossus hippoglossus*) based on 16S rDNA-DGGE. Food Microbiol. 24: 362-371.

Hovda, M.B., Lunestad, B.T., Sivertsvik, M. and Rosnes, J.T. (2007b). Characterization of the bacterial flora of modified atmosphere packaged farmed Atlantic cod (*Gadus morhua*) by PCR-DGGE of conserved 16S rRNA gene regions. Int. J. Food Microbiol. 117: 68-75.

Hu, Q., Qi, H-y., Zeng, J-h. and Zhang, H-x. (2007). Bacterial diversity in soil around a lead and zinc mine. J. Environ. Sci. 19: 74-79.

Huber, I., Spanggaard, B., Appel, K.F., Rossen, L., Nielsen, T. and Gram, L. (2004). Phylogenetic analysis and *in situ* identification of the intestinal microbial community of rainbow trout (*Oncorhynchus mykiss*, Walbaum). J. Appl. Microbiol. 96: 117-132.

ISO 8402 (1994). Quality management and quality assurance.

Ji, N., Peng, B., Wang, G., Wang, S. and Peng, X. (2004). Universal primer PCR with DGGE for rapid detection of bacterial pathogens. J. Microbiol. Methods 57: 409-413.

Kowalchuk, G.A., Stephen, J.R., de Boer, W., Prosser, J.I., Embley, T.M. and Woldendorp, J.W. (1997). Analysis of ammonia-oxydizing bacteria of the beta sub-division of the class *Proteobacteria* in coastal sand dunes by denaturing gradient gel electrophoresis and sequencing of PCR amplified 16S ribosomal DNA fragments. Appl. Environ. Microbiol. 63: 1489-1497.

Landy, E.T., Mitchell, J.I., Hotchkiss, S. and Eaton, R.A. (2007). Bacterial diversity associated with archaeological waterlogged wood: Ribosomal RNA clone libraries and denaturing gradient gel electrophoresis (DGGE). Int. Biodet. Biodegrad. 61: 106-116.

Leesing, R. (2005). Identification et validation de marqueurs spécifiques pour la traçabilité de poisons d'aquaculture lors de leur import/export. Thèse de doctorat. Université Montpellier 2, France.

Le Nguyen, D.D., Ha, N.H., Dijoux, D., Loiseau, G. and Montet, D. (2008). Determination of fish origin by using 16S rDNA fingerprinting of bacterial communities by PCR-DGGE: An application on Pangasius fish from Vietnam. Food Control 19: 454-460.

Lerman, L.S., Fisher, S.G., Hurley, I., Silverstain, K. and Lumelsky, N. (1984). Sequence-determined DNA separations. Annu. Rev. Biophys. Bioeng. 13: 399-423.

Liu, X.C., Zhang, Y., Yang, M., Wang, Z.Y. and Lu, W.Z. (2006). Analysis of bacterial community structures in two sewage treatment plants with different sludge properties and treatment performance by nested PCR-DGGE method. J. Environ. Sci. 19: 60-66.

Lyautey, E., Lacoste, B., Ten-Hage, L., Rols, J.L. and Garabetian, F. (2005). Analysis of bacterial diversity in river biofilms using 16S rDNA PCR-DGGE: methodological settings and fingerprints interpretation. Water Res. 39: 380-388.

McBain, A.J., Bartolo, R.G., Catrenich, C.E., Charbonneau, D., Ledder, R.G. and Gilbert, P. (2003). Growth and molecular characterization of dental plaque microcosms. J. Appl. Microbiol. 94: 655-664.

Mauriello, G., Moio, L., Genovese, A. and Ercolini, D. (2003). Relationship between flavouring capabilities, bacterial composition and geographical origin of Natural Whey Cultures (NWCs) used for water-buffalo Mozzarella cheese manufacture. J. Dairy Sci. 86: 486-497.

Miambi, E., Guyot, J.P. and Ampe, F. (2003). Identification, isolation and quantification of representative bacteria from fermented cassava dough using an integrated approach of culture-dependent and culture-independent methods. Int. J. Food Microbiol. 82 : 111-120.

Moeseneder, M.M., Arrieta, J.M., Muyzer, G., Winter, C. and Herndl, G.J. (1999). Optimization of terminal-restriction length polymorphism analysis for complex marine bacterioplankton communities and comparison with denaturing gradient gel electrophoresis. Appl. Environ. Microbiol. 65: 3518-3525.

Montet, D. (2003). Un 'code barre' biologique pour les poissons d'Asie du Sud-Est. Tracenews net (www.tracenews.net/dossiers/codebarre2.php)

Montet, D., Leesing, R., Gemrot, F. and Loiseau, G. (2004). Development of an efficient method for bacterial diversity analysis: Denaturing Gradient Gel Electrophoresis (DGGE). In: Seminar on Food Safety and International Trade, Bangkok, Thailand.

Murray, A.E., Hollibaugh, J.T. and Orrego, C. (1996). Phylogenetic composition of bacterioplankton from two California estuaries compared by denaturing gradient gel electrophoresis of 16S rDNA fragments. Appl. Environ. Microbiol. 6: 2676-2680.

Muyzer, G., De Waal, E.C. and Uitterlinden, A.G. (1993). Profiling of complex microbial populations by denaturing gradient gel electrophoresis analysis of polymerase chain reaction-amplified genes coding for 16S rRNA. Appl. Environ. Microbiol. 59: 695-700.

Muyzer, G., Teske, A., Wirsen, C.O. and Jannasch, H.W. (1995). Phylogenetic relationships of *Thiomicrospira* species and their identification in deep-sea hydrothermal vent sample by denaturing gradient gel electrophoresis of 16S rDNA fragment. Arch. Microbiol. 164: 165-172.

Nicol, G.W., Glover, L.A. and Prosser, J.I. (2003). The impact grassland management on archaeal community structure in upland pasture rhizosphere soil. Environ. Microbiol. 5: 152-162.

Øvreas, L., Forney, L., Dae, F.L. and Torsvik, V. (1997). Distribution of bacterioplankton in Meromictic Lake Sælenvannet, as determined by Denaturing Gradient Gel Electrophoresis of PCR-amplified gene fragment coding for 16S rRNA. Appl. Environ. Microbiol. 63: 3367-3373.

Palleroni, N.J. and Bradbury, J.F. (1993). *Stenophomonas*, a new bacterial genus for *Xanthomonas maltophila* . Int. J. Syst. Bacteriol. 43: 606-609.

Phung, N.T., Lee, J., Kang, K.H., Chang, I.S., Gadd, G.M. and Kim, B.H. (2004). Analysis of microbial diversity in oligotrophic microbial fuel cells using 16S rDNA sequences. FEMS Microbiol. Lett. 233: 77-82.

Pinhassi, J., Sala, M.M., Havskum, H., Peters, F., Guadayol, O., Malits, A. and Marrasé, C. (2004). Changes in bacterioplankton composition under different phytoplankton regimens. Appl. Environ. Microbiol. 70: 6753-6766.

Riemann, L., Steward, G.F., Fandino, L.B., Campbell, L., Landry, M.R. and Azam, F. (1999). Bacterial community composition during two consecutive North- East Monsoon periods in the Arabian Sea studied by denaturing gradient gel electrophoresis (DGGE) of rRNA genes. Deep Sea Res. 46: 1791-1811.

Röhling, W.F.M., Kerler, J., Braster, M., Apriyantono, A., Stam, H. and van Verseveld, H.W. (2001). Microorganisms with a taste for Vanilla: microbial ecology of traditional Indonesian Vanilla curing. Appl. Environ. Microbiol. 67: 1995-2003.

Sekiguchi, H., Watanabe, M., Nakahara, T., Xu, B. and Uchiyama, H. (2002). Succession of bacterial community structure along the Changjiang River determined by denaturing gradient gel electrophoresis and clone library analysis. Appl. Environ. Microbiol. 68: 5142-5150.

Shibata, A., Toyota, K., Miyake, K. and Katayama, A. (2007). Anaerobic biodegradation of 4-alkylphenols in a paddy soil microcosm supplemented with nitrate. Chemosphere 68: 2096-2103.

Sheffield, V.C., Beck, J.S., Stone, E.M. and Myers, R.M. (1989). Attachment of a 40 bp G+C rich sequence (GC-clamp) to genomic DNA fragments by polymerase chain reaction results in improved detection of single-base changes. Proc. Natl. Acad. Sci., USA 86: 232-236.

Silvestri, G., Santarelli, S., Aquilanti, L., Beccaceci, A., Osimani, A., Tonucci, F. and Clementi, F. (2007). Investigation of the microbial ecology of Ciauscolo, a traditional Italian salami, by culture-dependent techniques and PCR-DGGE. Meat Science 77: 413-423.

Spanggaard, B., Huber, I., Nielsen, T.J., Nielsen, T., Appel, K. and Gram, L. (2000). The microbiota of rainbow trout intestine: a comparison of traditional and molecular identification. Aquaculture 182: 1-15.

Syvokienë, J. and Mickënienë, L. (1999). Micro-organisms in the digestive tract of fish as indicators of feeding condition and pollution. ICES J. Marine Sci. 56: 147-149.

Tran Van, V., Berge, O., Balandreau, J., Ngo Ke, S. and Heulin, T. (1996). Isolement et activité nitrogénasique de *Burkholderia vietnamiensis*, bactérie fixatrice d'azote associée au riz (*Oryza sativa* L). cultivé sur un sol sulfaté du Viêt-nam. Agronomie 16: 479-491.

Van Beek, S. and Priest, F.G. (2002). Evolution of the lactic acid bacterial community during malt whisky fermentation: a polyphasic study. Appl. Environ. Microbiol. 68: 297-305.

Van der Gucht, K., Vandekerckhove, T., Vloemans, N., Cousin, S., Muylaert, K., Sabbe, K., Gillis, M., Declerk, S., De Meester, L. and Vyverman, W. (2005). Characterization of bacterial communities in four freshwater lakes differing in nutrient load and food web structure. FEMS Microbiol. Ecol. 53: 205-220.

van Hannen, E.J., Zwart, G., van Agterveld, M.P., Gons, H.J., Ebert, J. and Laanbroek, H.J. (1999). Changes in bacterial and eukaryotic community structure after mass lysis of filamentous cyanobacteria associated with viruses. Appl. Environ. Microbiol. 65: 795-801.

Wong, H.C., Chen, M.C., Liu, S.H. and Liu, D.P. (1999). Incidence of highly genetically diversified *Vibrio parachaemolyticus* in seafood imported from Asian countries. Int. J. Food Microbiol. 52: 181-188.

Yang, C.H., Crowley, D.E. and Menge, J.A. (2001). 16S rDNA fingerprinting of rhizosphere bacterial communities associated with healthy and Phytophthora infected avocado roots. FEMS Microbiol. Ecol. 35: 129-136.

Yabuuchi, E., Kosako, Y., Oyaizu, H., Yano, I., Hotta, H., Hashimoto, Y., Ezaki, T. and Arakawa, M. (1992). Proposal of *Burkholderia* gen. nov. and transfer of seven species of the genus *Pseudomonas* homology group II to the new genus, with the type species *Burkholderia cepacia* (Palleroni and Holmes, 1981) comb. nov. Microbiol. Immunol. 36: 1251-1275.

3

Bacterial Fish Diseases and Molecular Tools for Bacterial Fish Pathogens Detection

Ana Roque[1]*, Sonia A. Soto-Rodríguez[2] and Bruno Gomez-Gil[2]

INTRODUCTION

The interest in fish bacteria can be seen by the available literature, where their importance as pathogens and also as spoilage bacteria for human consumption is reflected. Bacteria are ubiquitous in the aquatic environment and almost all bacterial fish pathogens can live outside the fish. Reasons for a fish disease to emerge are not completely understood but they include: changes in the environment favouring the potential pathogen, lowering of resistance of the host to infection, introduction of new species sensitive to a local microbe, introduction of a new species carrying a microbe virulent to local species, improvement of diagnostic tools giving oportunity to describe new pathogens (Colorni, 2004).

One of the focuses of this chapter is the available molecular tools for the detection of pathogenic fish bacteria. These techniques are mainly based on the detection of a more or less specific DNA fragment and include all the tests and methods which on a molecular level are capable of defining the presence/absence of a pathogen or test the predisposition of an individual towards a particular disease. The benefits of using molecular tools are their high sensitivity and specificity. The disadvantages are that they detect nucleic acid in general and not necessarily a viable pathogen.

[1]*IRTA-Sant Carles de la Rápita, Carr al Poblenou SN km 5.5 Sant Carles de la Rapita. Spain*
[2]*CIAD, A.C. Mazatlan Unit for Aquaculture and Environmental Management, A.P. 711 Mazatlan, Sinaloa 82010, Mexico*
*Corresponding author: E-mail: Ana.Roque@irta.es

For this chapter the authors selected the two bacterial diseases which are listed by the World Organization for Animal Health (OIE):

- Bacterial kidney disease (BKD),
- Enteric septicaemia of catfish (ESC) and other important bacterial diseases,

since less literature is available on these diseases. In some cases, where a bacterial pathogen was initially associated with disease in freshwater species and later was detected as a causative agent of disease in marine species, were also reviewed (e.g. streptococcosis in seawater cultured fish).

BACTERIAL KIDNEY DISEASE (BKD)

BKD is a systemic, chronic bacterial infection of salmonids in both fresh and seawater stages of the life cycle, which was initially described in the Atlantic salmon (Smith, 1964). The causative agent of BKD is *Renibacterium salmoninarum*, a small, non-motile, non-spore forming, non-acid-fast, Gram positive diplobacillus (Fryer and Sanders, 1981), which is capable of multiplying within the host macrophages (Gutenberger et al., 1997).

The presence of this disease has been recorded in most countries where salmonid culture occurs or where there are wild salmonids, including North and South America, Japan, Europe (Fryer and Lannan, 1993). Most salmonids are susceptible to BKD, though their susceptibility varies with host species, and Pacific salmon species of the genus *Oncorhynchus* are generally considered to be the most susceptible (Starliper et al., 1997). *R. salmoninarum* has been isolated from several non-salmonid fish species (Traxler and Bell, 1988). *R. salmoninarum* can be transmitted both horizontally via the faecal-oral route (Balfry et al., 1996) or through wound sites (Elliott and Pascho, 2001), and vertically in association with eggs from infected parents (Pascho et al., 1991).

BKD typically causes a slowly progressing systemic infection, with the overt disease rarely evident until fish are 6-12 mon old (Evelyn, 1993). It can cause serious mortality in juvenile salmonids in both freshwater and seawater, and also in prespawning adults. Clinical signs of the disease may vary and include lethargy, skin darkening, abdominal distension due to ascites, pale gills associated with anaemia, exophthalmia, haemorrhages around the vent, and cystic cavities in the skeletal muscle. Internally there are focal to multifocal greyish-white nodular lesions present in the kidney, and sometimes in the spleen and liver. There may be turbid fluid in the abdominal cavity, haemorrhages on the abdominal wall and in the viscera, and a diffuse white membranous layer (pseudomembrane) on one or more of the internal organs. In tissue sections of BKD lesions, *R. salmoninarum* is frequently observed within phagocytic cells, particularly macrophages (Gutenberger et al., 1997).

Diagnostic Methods Including Molecular Tools for the Detection and Identification of Causative Pathogen

Bacteriological culture (Austin et al., 1983) remains the benchmark method for determining the viability of *R. salmoninarum* in a sample, and may also be used to quantify the number of bacteria in samples. Immunodiagnostic procedures (Hoffman et al., 1989; Bruno and Brown, 1999), however, have become the most widely used for screening large numbers of fish in aquaculture facilities or elsewhere in the field. The two principal immunodiagnostic methods are the enzyme-linked immunosorbent assay (ELISA; Pascho et al., 1987, 1991) and the fluorescent antibody test (FAT) and quantitative FAT (Rhodes et al., 2006). Nucleic-acid-based diagnostic tests, such as the polymerase chain reaction (PCR) (Konigsson et al., 2005), are now acceptable for confirmatory identification of *R. salmoninarum* in bacteriological cultures and fish tissue or body fluid samples (OIE, 2006). ELISA followed by PCR on positive samples are routinely used for screening and confirmation, respectively (Bruno et al., 2007).

For the molecular tools of detection of BKD, the target gene is almost always the gene coding for p57 protein which is an antigen produced by *R. salmoninarum* (Cook and Lynch, 1999), a highly conserved gene. Among the molecular techniques to detect BKD, the OIE manual (2006) describes a nested polymerase chain reaction for testing tissue samples, whole blood, and coelomic fluid which is based on protocols described by Chase and Pascho (1998) and Pascho et al. (1998). Other methods reported include PCR, RT-PCR (reverse transcriptase PCR), a probe (Brown et al., 1994; Hariharan et al., 1995; McIntosh et al., 1996; Rhodes et al., 1998; Cook and Lynch, 1999) and quantitative PCR (Bruno et al., 2007). For the nested polymerase chain reaction when testing tissue samples, whole blood, and coelomic fluid, two pairs of oligonucleotide primers are used in the protocol. The primers were designed from the published sequence of the p57 protein of *R. salmoninarum* (Chien et al., 1992). Deoxyribonucleic acid (DNA) can be extracted from kidney tissue, ovarian fluid, whole blood or *R. salmoninarum* cells. The amplicon must produce a band of 320 bp. PCR method also targets p57 protein gene, with the size of the amplicon between 149 and 376 bp depending on the pair of primers selected (McIntosh et al., 1996; Miriam et al., 1997; Pascho et al., 1998). There is also a nested RT-PCR (Cook and Lynch, 1999), where the target gene is the p57 mRNA (messenger ribonucleic acid) gene, producing a 349 bp fragment. This assay is completed in 72 h. A probe method was developed targeting a 149 bp fragment of the p57 protein coding gene, using CSPD (disodium 3-(4-methoxyspirol (1,2-dioxetane-3,2′-(5′-chloro)-tricyclo(3.3.1.1) decan)-4-yl) phenyl phosphate) as the chemiluminescent alkaline phosphatase substrate (Miriam et al., 1997).

Existing Q PCR (quantitative PCR, or realtime PCR) tests use as target gene the *msa* gene mRNA, coding for the p57 protein of *R. salmoninarum* (Powell et al., 2005; Bruno et al., 2007). The probes were labelled with the fluorescent tag 6-FAM and the MGB quencher. Probes were TaqMan and Ct value was calculated using an internal PCR control (Powell et al., 2005; Bruno et al., 2007). Suzuki and Sakai (2007) developed a protocol for detection of BKD which included an RT followed by a Q PCR protocol for *msa* gene mRNA and a Q PCR for *msa* gene DNA. Both assays used a TaqMan probe. Signal amplification was plotted against PCR cycle number to generate Ct value.

ENTERIC SEPTICAEMIA OF CATFISH (ESC): EDWARDSIELLIOSIS

The genus *Edwardsiella* includes two species of bacteria that cause major diseases in fish: *Edwardsiella tarda*, which infects fish and other animals including man and *E. ictaluri* which infects only fish (Plumb, 1999). There is a third species in the genus, *E. hoshinae* which infects birds and reptiles (Plumb, 1999). *Edwardsiella ictaluri* and *E. tarda* produce very different diseases, thus they will be discussed separately.

(a) E. ictaluri

Edwardsiella ictaluri is the causative agent of enteric septicaemia of catfish (ESC, Hawke et al., 1981) and it is one of the most important diseases of aquaculture where channel catfish is cultured. The main species affected is the channel catfish (*Ictalurus punctatus*), though *E. ictaluri* has been isolated from blue catfish (*Ictalurus furcatus*), white catfish (*Ameiurus catus*), and brown bullhead (*A. nebulosus*) (Hawke et al., 1981). ESC has also been reported from *Clarias batrachus* in Thailand (Kasornchandra et al., 1987), *Pangasius hypophthalmus* in Vietnam (Crumlish et al., 2002) and from several ornamental species (Waltman et al., 1985). *E. ictaluri* has been confirmed in the USA, Thailand and Australia. Acute outbreaks of ESC occur within a limited temperature range, from 18 to 28°C. *E. ictaluri* is a short pleomorphic Gram negative rod, weakly motile at 25-30°C, but not at higher temperatures. It is catalase positive and oxidase negative, it is also glucose fermentative and reduces nitrate to nitrite (Shotts and Teska, 1989). Growth on media is slow, 36 to 48 h to form colonies on brain-heart infusion (BHI) agar at 28-30°C. Several studies have shown that *E. ictaluri* is a very biochemically, antigenically and genetically homogeneous species (Waltman et al., 1986; Bertolini et al., 1990; Panangala et al., 2005, 2006).

Two clinical forms of ESC occur in channel catfish, a chronic encephalitis and an acute septicaemia (Shotts et al., 1986; Newton et al.,

1989). In the chronic form the bacterium infects the olfactory sacs, and migrates to the brain, forming a granulomatous inflammation, which causes abnormal swimming behaviour. In later stages of this disease, swelling develops on the dorsum of the head which ulcerates exposing the brain. In the acute form of ESC the bacterium is thought to infect through the intestinal mucosa (Baldwin and Newton, 1993), and then to establish a bacteraemia. Fish show petechial haemorrhages around the mouth, on the throat, the abdomen and at the base of the fins. They present multifocal distinct 2-mm diameter raised haemorrhagic cutaneous lesions that progress to depigmented ulcers. Common signs are anaemia, moderate gill inflammation and exophthalmia. Internally, haemorrhages and necrotic foci are spread in the liver and other internal organs. Mortalities can vary from 10 to 50% of the population (Plumb, 1999).

Diagnostic Methods Including Molecular Tools for the Detection and Identification of Causative Pathogen

The identification of *E. ictaluri* is based on the isolation of the causative agent and characterization by biochemical tests (Waltman et al., 1985, 1986). *E. ictaluri* can easily be differentiated from *E. tarda* by its inability to produce indole and hydrogen sulphide (*E. tarda* produces both). Additionally, the two species do not cross-react serologically (OIE, 2006). The samples should be streaked for isolation on to blood agar plates, BHI (brain heart infusion) agar or nutrient agar plates. Following incubation for 36-48 h, *E. ictaluri* appears as smooth, circular (1-2 mm diameter), slightly convex non-pigmented colonies with entire edges. This bacterium grows slowly or not at all at 37°C. This species is negative for acid production from D-mannitol, sucrose, treahalose, L-arabinose, malonate utilization, indole and hydrogen sulphide production, motility and citrate (OIE, 2006). Slide agglutination with specific antisera against *E. ictaluri*, fluorescent antibody techniques (FATs), enzyme-linked immunostaining and enzyme-linked immunosorbent assays (ELISAs) have been used to provide confirmatory diagnosis. Specific monoclonal antibodies (MAbs) are generally used in these assays (Ainsworth et al., 1986), but polyclonal antisera may also be used (Rogers, 1981). Assays based on PCR amplification from bacterial colonies and direct sequencing are among the nucleic acid-based techniques to detect *E. ictaluri*.

PCR amplification of structural RNA sequences from bacterial colonies and direct sequencing of the products can be used. Species confirmation can be done by amplifying and sequencing the 16S portion of the ribosomal RNA operon and comparing the sequence with GenBank accession AF310622. A Multiplex protocol targeting *Flavobacterium columnaris, F. psychrophilum* (Wiklund et al., 2000) *E. ictaluri* and *Aeromonas hydrophila* has been published (Panangala et al., 2007), where a fragment of 407 bp is amplified in the 16S rRNA gene of *E. ictaluri*.

A loop-mediated isothermal amplification (LAMP) method for rapid detection has also been proposed for the detection of *E. ictaluri* (Yeh et al., 2005). This method is based on autocycling strand displacement DNA synthesis in the presence of exonuclease negative *Bst* DNA polymerase under isothermal conditions within 1 h (Notomi et al., 2000). An initial pair of primers (FIP—foward inner primers) anneal and hybridize with the target DNA and the first strand is synthesized by Bst polymerase, the outer primers (at a lower concentration) hybridize and displace the first strand linked with the FIP-linked complementary strand, which forms a looped structure in the end. This strand initiates a process in which another primer (BIP—backward inner primer) hybridizes the complementary strand and initiates strand synthesis and later displacement by the backward outer primers (Savan et al., 2004). Because there are four specific primers that recognize six sequences on the target-DNA that are used, this technique amplifies DNA with high specificity. The primers designed for this assay target the *eip* 18 gene, producing a ladder-like specific pattern from 234 bp up to the loading wells (Yeh et al., 2005).

A rapid (4.5 h) and realtime assay based on polymerase chain reaction (PCR) was developed to facilitate the early detection of *E. ictaluri* in channel catfish *Ictalurus punctatus*, amplifying a 129-base-pair fragment of a sequence specific to *E. ictaluri* (Bilodeau et al., 2003; Wise et al., 1998).

(b) E. tarda

Edwardsiella tarda is the causative agent of edwardselliosis, a disease of cultured (Kusuda and Kawai, 1998) and wild fish species (Baya et al., 1997). This species has been reported in tropical and subtropical areas of Africa, America, Asia, Australia and in Europe (Alcaide et al., 2006). *Edwardsiella tarda* has been isolated from very different groups of fish including eels, salmonids, tilapia, turbot flounder, channel catfish amongst others (Plumb, 1999). It survives best at 30-37°C and with an optimum sodium chloride (NaCl) of 0.5-1%. The bacterium is a small straight Gram negative rod, usually motile with a peritrichous flagella and facultative anaerobe. It is catalase positive, oxidase negative, lactose negative, indole positive, produces hydrogen sulphide (Table 3.1). Two biogroups have been defined in the Bergey's Manual (1984) by their biochemical and biophysical characteristics, the wild type and the biogroup 1 (Table 3.1). A third biogroup has been proposed conforming strains which have only been isolated from humans (Walton et al., 1993).

A variety of clinical signs occur in the different fish species affected by *E. tarda* (Plumb, 1999). For example, eels with acute infection of *E. tarda* develop hyperaemia with bloody congestion of the fins, haemorrhages on the body surfaces, gas-filled pockets in the skin and necrosis of the muscle.

Table 3.1 Differentiation of the species and biogroups of the genus *Edwardsiella* (OIE, 2006)

Characteristics	E. tarda		E. hoshinae	E. ictaluri
Acid production from:	Wildtype	Biogroup 1		
D-Manitol	−	+	+	−
Sucrose	−	+	+	−
Tetrahalose	−	−	+	−
L-Arabinose	−	+	(−)	−
Malonate utilization	−	−	+	−
Indole production	+	+	−	−
Hydrogen sulphide production in triple sugar iron (agar)	+	−	−	−
Motility	+	+	+	−
Citrate (Christensen's)	+	+	(+)	−

Internally there is general hyperaemia of the peritoneum, liver is oedematous and abscessed (Plumb, 1999). In the turbot, affected individuals present tumefaction around the eyes and at the base of the dorsal and anal fins. These tumefacted areas are full of a purulent fluid. Some fish also present haemorrhages in the muscles, specially in the head region. The abdomen is distended due to the presence of ascitic fluid. Internally, there are haemorrhages in the liver and congestion of the intestine spleen and kidneys. Kidneys present abscess-like lesions (Padros et al., 2006).

Diagnostic Methods Including Molecular Tools for the Detection and Identification of Causative Pathogen

As clinical signs vary with the species of fish affected, they are of little use. It is therefore necessary to isolate and identify the bacteria. Isolation can be performed in triptone soy agar (TSA) +1% NaCl or onto BHI agar followed by characterization with API 20E incubated at 25°C, expected profile 4344000. Other techniques such as FAT (fluorescent antibody technique) or ELISA can be used to perform the diagnostic. A protocol for detection of *E. tarda* directly in fish using capillary electrophoresis with blue light-emitting diode-induced fluorescence is available (Yu et al., 2004). Some nucleic acid detection assays are also published.

Few molecular protocols have been published to detect *E. tarda*. A PCR protocol exists targeting a 1109 bp fragment from the haemolysin gene between the beginning of the region ORF (Open Reading Frame) II and the end of the region ORF III (Chen and Lai, 1998), since this gene has low homology to the same gene from other bacteria (Savan et al., 2004). This same fragment was used to produce a probe for direct hybridization.

RAPD (randomly amplified polymorphic DNA) analyses were also used to characterize *E. tarda* strains isolated from turbot (Castro et al., 2006). Primers came from a commercial kit and generated reproducible patterns for an accurate analysis.

Finally, a LAMP protocol is available where primers were designed to target the haemolysin gene from *E. tarda* (accessation number BAA21097), producing a ladder-like specific pattern from 100 bp up to the loading wells (Savan et al., 2004). This protocol can be applied directly on fish tissues such as kidney and spleen and on seawater.

VIBRIOSIS

Vibriosis has been recognized as a disease for over 20 years. It affects most aquatic organisms under culture, including fish. The species of Vibrionaceae which are most often associated with losses are *Listonella anguillarum* (previously *Vibrio anguillarum*), *Vibrio ordalii*, *V. salmonicida*, *V. harveyi*, *V. alginolyticus*, *Moritella viscosa* (previously, *V. viscosus*) and *V. vulnificus* biotyope II (Toranzo et al., 2005). Vibriosis has been recognized worldwide. Vibrios are short Gram negative rods, which are facultative anaerobics, mobile, oxidase positive and in most cases are sensitive to the vibriostactic agent O/129. Most fish species cultured nowadays are susceptible to one or another *Vibrio* species (Table 3.2, Toranzo et al., 2005). Fish affected with a classical vibriosis show signs of generalized septicaemia with haemorrhage on the base of the fins, exophthalmia and corneal opacity and they are normally anorexic (Toranzo et al., 2005). *Moritella viscosa* is the causative agent for winter ulcer in salmon, which is characterized by skin ulcers confined to scale on covered parts of the body surface and often diffuse or petechial haemorrhages in internal organs (Bruno et al., 1998). This disease has only been recorded in Norway, Iceland and Scotland. It does not cause high mortalities but it is of concern due to the loss of quality of the salmon.

Table 3.2 Example of *Vibrio* species and susceptible fish species

Vibrio species	Fish affected
Listonella anguillarum	Eels, Atlantic and Pacific salmon, rainbow trout, turbot, seabass, striped bass, seabream, cod and ayu
V. vulnificus biotype II	Eel
V. harveyi/alginolyticus	Seabream and sea bass
V. ordalii	Salmonids and cod
V. salmonicida	Salmon and cod
Moritella viscosa	Atlantic salmon

Diagnostic Methods Including Molecular Tools for the Detection and Identification of Causative Pathogen

Depending on the *Vibrio* species, it can be easily recovered on plates of tryptone soya agar or other general purpose media supplemented with NaCl (0.5-3.5% w/v) or seawater and Thiosulphate Citrate Bile Sucrose (TCBS) agar (Austin and Austin, 1993) with incubation at 15-30°C for up to 7 d. Further characterization can be done using either phenotypic or genotypic tests. Not all species can be identified with certainty using only biochemical tests but profiles can be recorded in order to compare with future outbreaks of disease independently from the identification to species level of the presumptive pathogen.

A very recent and complete review of molecular techniques available to detect vibrios was made by Nishibushi (2006). It is clear that for this family (Vibrionaceae), biochemical identification is insufficient. There are numerous techniques used for typing the bacteria. For identification targeting specific genes include the 16S rRNA, 23S rRNA, the ITS (internal transcribed spacer) region, a capsular polysaccharide, the urease gene (*ureC*), *lux* A (luciferase subunit), the *Tox* R, several haemolysins. The majority of these target genes have the species as their target group, though some are targeting virulent strains of particular species, normally to humans and not fish. The techniques vary from probes to different PCR protocols (multiplex, one step, real time). Several protocols exist for some fish pathogens such as *Vibrio alginolyticus* (Di Pinto et al.,. 2005), *Listonella anguillarum* (Gonzalez et al., 2003), *L. anguillarum* to be distinguished from *V. ordalii* (Ito et al., 1995). A PCR protocol targeting the ITS region can identify eight different species of aquatic pathogens: *Salinivibrio costicola, V. diazotrophicus, V. fluvialis, V. nigripulchritudo, V. proteolyticus, V. salmonicida, V. splendidus* and *V. tubiashii* (Lee et al., 2002). There are some protocols published for real time PCR but mainly they are to detect and quatify vibrios with importance in public health (Lyon, 2001; Campbell and Wright, 2003; Kauffman et al., 2004; Takahashi et al., 2005).

PHOTOBACTERIOSIS (FORMERLY PASTEURELOSIS)

Since 1969, photobacteriosis has been recognized as an important disease in Japan affecting yellowtail (Egusa, 1983). Since then it has been recorded in several Mediterranean fish species, sea bream, sea bass, common sole, Senegalese sole and hybrid striped bass (Hawke et al., 1987; Toranzo et al., 1991, 2005; Salati et al., 1997; Zorrilla et al., 1999). This disease is caused by *Photobacterium damselae* subsp. *piscicida*. This is a pronounced bipolar staining Gram negative, non-motile, fermentative rod, which can be pleomorphic. Photobacteriosis has been reported in Japan, several Mediterranean countries and the USA. Clinical signs involve mainly an acute septicaemia with granulomatous-like deposits on the kidney and spleen. Purulent material may accumulate in the abdominal cavity (Austin and Austin, 1993). High mortalities occur when temperatures are above 20°C; below this temperature fish can harbour the bacterium for long periods of time (Magariños et al., 2001). Despite the origin and source of isolation, strains of *Ph. damselae* subsp. *piscicida* are very homogenous phenotypically (Bakopoulos et al., 1997), though molecular techniques have shown the existence of two biotypes, one grouping the European strains and the other including the Japanese and American strains (Magariños et al., 2000).

Diagnostic Methods Including Molecular Tools for the Detection and Identification of Causative Pathogen

The presumptive identification of *Photobacterium damselae* subsp. *piscicida* is relatively easy since all strains exhibit the same morphological features (non-motile Gram negative rods) and API 20E profile (2005004) (Magariños et al., 1992). All of them are catalase positive, fermentative, sensitive to O/129 (Magariños et al., 1992). Isolation can be performed on marine agar or other general purpose media at 25-28°C for 48-72 h (Austin and Austin, 1993). There are other commercial kits developed for the identification of this pathogen (Romalde et al., 1999). There are several DNA-based protocols which will be mentioned below in more detail. These include a multiplex PCR targeting the 16S rRNA and the *ureC* genes in order to distinguish between *Ph. damselae* subsp. *piscicida* and subsp. *damselae* (Osorio et al., 2000).

Photobacterium damselae subsp. *piscicida* has very similar nucleotide sequences of the rRNA genes and their intergenic sequences to *Ph. damselae* subsp. *damselae* (Osorio et al., 2004, 2005). A multiplex protocol detecting the 16S rRNA sequence and the *ureC* gene can differentiate between the two subspecies (Osorio et al., 2000). With this protocol, *Ph. damselae* subsp. *damselae* yields two amplification products, 1 of 267 bp corresponding to an internal fragment of the 16S rRNA gene and the other of 448 bp, corresponding to an internal fragment of the *ureC* gene. On the other hand, with this protocol *Ph. damselae* subsp. *pisicicida* only presents the amplification of a band of 267 bp (16S rRNA fragment) indicating the absence of the urease gene in its genome. Other PCR protocols have been developed but they are not capable of distinguishing between the two subspecies of *Ph. damselae* (Dalla Valle et al., 2002; Rajan et al., 2003). The protocol proposed by Rajan et al. (2003) overcomes this problem by amplifying the expected fragment of 410 bp, indicating *Ph. damselae* and then advises plating of the bacterial isolate onto TCBS agar. If there is growth (green colonies) it is a *Ph. damselae* subsp. *damselae* and if there is no growth, then the isolate is a *Ph. damselae* subsp. *piscicida*.

There is also a protocol available to identify *Ph. damselae* subsp. *piscicida* by PCR-RFLP (restriction fragment length polymorphism) analysis, where the PCR amplifies non-conserved sites of two genomic regions (GenBank AY191120 and AY191121). Expected fragments are of 713 and 201 bp, respectively. RFLP is then carried out on these fragments. For the 713 bp fragment, expected fragments are of 459, 158, 63, and 33 bp whereas for the fragment of 210 bp, expected fragments are of 125 and 76 bp (Zapulli et al., 2005).

Finally, another protocol uses AFLP (amplified fragment length polymorphism) to detect *Ph. damselae* subsp. *piscicida* (Kvitt et al., 2002). This protocol is based on a PCR targeting the 16S rRNA gene combining low annealing temperature that detects low quantities of *Ph. damselae* (and related species) and high annealing temperature for the specific identification of *Ph. damselae*. The isolate is then further identified to subsp. level by AFLP.

Random Amplification of Polymorphic DNA (RAPD) analysis has been used to distinguish between strains of *Ph. damselae* subsp. *piscicida* isolated from Japan and from Europe (Magariños et al., 2000). In this study six pairs of primers were tested though only p4, p5 and p6 yielded results which produced an adequate pattern of bands, and only p4 and p6 actually separated the isolates into two subgroups that could be related to the geographical origin of the isolates, regardless of their host species or year of isolation. With p4, the European subgroup produced a pattern with 9 major bands whereas the Japanese subgroup produced a pattern with 8 bands. Similarity within each group was 100%.

FURUNCULOSIS

Furunculosis has been recognized for over 100 years. The first report of an outbreak was on a German brown trout hatchery. Furunculosis is caused by *Aeromonas salmonicida*, and most fish (salmonids, labrids, ciprinids, turbot and cod) are susceptible to this bacterium in different degrees, as it is the case with trouts, where rainbow trout is a lot less susceptible than are brook or brown trout (Traxler and Bell, 1988; Hiney and Olivier, 1999). The typical furunculosis is caused by a subspecies named *salmonicida*, whereas the forms of disease caused by atypical *A. salmonicida* are due to other subspecies namely *achromogenes*, *masoucida* or *smithia*. The geographic distribution of this pathogen is almost worldwide, with the possible exception of New Zealand. *Aeromonas salmonicida* is a non-motile fermentative Gram negative rod which produces a brown water soluble pigment on tryptone containg agar; it does not grow at 37°C and produces catalase and oxidase (Austin and Austin, 1993). *Aeromonas salmonicida* subsp. *salmonicida* is formed by a very homogenous group of isolates, whereas the atypical *A. salmonicida* form a very heterogenous group with respect to biochemical features, growth conditions and production of extracellular proteases (Wiklund and Dalsgaard, 1998).

The subacute and chronic form of furunculosis, more common in older fish, is characterized by the presence of furuncles in the muscles, lethargy, slight exophthalmia, bloodshot fins, multiple haemorrrhages in the muscle, internally haemorrhages in the liver, swelling of the spleen and necrosis of the kidney may occur (Austin and Austin, 1993). The acute form, more common in growing fish, presents a general septicaemia accompanied by darkening in colour, lack of appetite, lethargy and small haemorrhages at fin base. Internally there are haemorrhages over the abdominal walls, viscera and heart (Austin and Austin, 1993). In this form, fish usually die in 2 or 3 d. There are also atypical forms of furunculosis such as the carp erythrodermatitis, cutaneous ulcerative disease of goldfish, head ulcer disease of Japanese eel and ulcerative diseases of marine flat fish and salmonids (Toranzo et al., 1991, 2005).

Diagnostic Methods Including Molecular Tools for the Detection and Identification of Causative Pathogen

Aeromonas salmonicida can be recovered from the kidney but sampling from more organs increases the probabilities of recovery. Isolation can be made onto TSA (tryptone soya agar) or BHIA (brain heart infusion agar). The brown water soluble pigment produced on TSA appears after 2-4 d incubation at 20-25°C and can be used as presumptive identification. Pre-enrichment of the samples of kidney samples in TSB for 48 h greatly increases the detection rate. The use of CBB agar (TSA supplemented with 0.01% w/v of coomassie brilliant blue) as a differential medium can be useful since *A. salmonicida* will stand out from other bacteria by forming deep blue colonies (Cipriano et al., 1992). Nevertheless, the use of CBB agar as the isolation medium reduces the numbers of bacteria growing on the plate. Identification can be done with biochemical tests, Gram negative non-motile rods, producing catalase and oxidase, fermentative, not capable of growing at 37°C, producing gelatinase, degrading starch, not producing urease, producing arginine dyhydrolase, and not producing ornithine decarboxylase, also not oxidizing gluconate (McCarthy, 1976), ELISA (Adams and Thompson, 1990), immunodot-blot (Hiney and Olivier, 1999) and several PCR protocols.

There are several PCR and probe detection protocols to detect *A. salmonicida* subsp. *salmonicida* both from bacterial cultures and from fish tissues. Most of them target plasmid sequences, A-layer (paracrystalline surface protein layer) and the 16S rRNA genes. Apparently the sensitivity is higher using a protocol to amplify the A-layer gene but it can produce false positives by giving a cross reaction with *A. hydrophila*. The A-layer is essential to virulence therefore the presence of gene related to it gives information on the potential virulence of the identified strains (Gustafson et al., 1992). The protocols targeting the 16S rRNA use different pairs of primers: (a) after Dorsch and Stakebrandt (1992) with an expected amplified fragment of 1465 bp; (b) after O'Brien et al. (1994) with an expected amplified fragment of 423 bp; (c) after Gustafson et al. (1992) with an expected amplified fragment of 421 bp and (d) after Myata et al. (1996) with an expected amplified fragment of 512 bp. Most protocols based on the 16S rRNA detect *A. salmonicida* to species level. In order to make the identification to subsp. level the plasmid profile is useful. Hoie et al. (1997) propose two separate PCRs, one targeting the 16S rRNA gene amplifying a fragment of 271 bp and another targeting the plasmids amplifying approximately 710 bp. Plasmid profile can also be useful during outbreaks of atypical *A. salmonicida* infections in fish (Sorum et al. 2000).

The protocol developed by O'Brien et al. (1994) was used to develop a multiplex PCR to detect simultaneously the presence of *A. salmonicida, Flavobacterium psychrophilum* and *Yersinia ruckeri* (Cerro et al., 2002). A one tube RT PCR-enzyme hybridization protocol can also detect simultaneously *Aeromonas salmonicida, Tenicibaculum maritimum, Lactococcus garviae* and *Yersinia ruckeri*, targeting the 16S rRNA of each of them. Expected fragment for *A. salmonicida* is of 261 bp. For each RT- PCR, the reverse primer is used as the solid-phase primer. Hybridization is carried at 50°C and readings are taken on a plate reader at 405 nm, positive reading is defined as at least 1.2 times the reading of the control (Wilson and Carson, 2003).

TENACIBACILOSIS

Tenacibacilosis affects marine fish and is also known as eroded mouth syndrome, gliding bacterial disease of sea fish, black patch necrosis, salt water columnaris and bacterial stomatitis (Avendaño-Herrera et al., 2006). The disease is caused by *Tenibaculum maritimum* (formerly *Cytophaga marina* and *Flexibacter maritimus*). The disease is distributed worldwide in cultured and wild fish, having been detected in Europe, Japan, North America and Australia and is suspected in Chile (Avendaño-Herrera et al., 2004c). It has been reported to affect turbot, common sole, Senegalese sole, gilthead sea bream, sea bass, redseabream, balck sea bream, flounder, mullet and salmonids, affecting both juvenile and adults with prevalence and severity being higher at water temperatures above 15°C (Toranzo et al., 2005). In addition to temperature, the disease is also influenced by skin condition and stress in fish (Magariños et al., 1995).

Tenibaculum maritimum is a Gram negative filamentous bacterium with gliding motility, an absolute requirement for marine water; it does not grow in media simply supplemented with NaCl, with a growth temperature range of 15-34°C, with the optimum at 30°C. Typical colonies on specific growth medium, both *Flexibacterium maritimus* medium (FMM) and Anacker and Ordal agar (AOA) are flat, pale yellow with uneven edges and strongly adherent to the medium, whereas on marine agar the colonies are round and show a yellow pigment (Avendaño-Herrera et al., 2006). This is a very homogenous species, on solid media it absorbes Congo red but it does not contain the cell wall associated pigment, produces enzymes that degrade casein, tyrosine and tributyrin but it does not hydrolyse agar, carboximethyl cellulose, cellulose, starch, esculin or chitin. It is catalase and oxidase positive and it is not able to degrade most carbohydrates (Avendaño-Herrera et al., 2004a).

Clinical signs include characteristic lesions on the body surface, ulcers, necrosis, eroded mouth, frayed fins, and tail rot (McVicar and White, 1979). The disease can become systemic and the skin erosion can be become the portal of entry for other pathogens.

Diagnostic Methods Including Molecular Tools for the Detection and Identification of Causative Pathogen

Clinical signs together with the microscopical observation of acummulation of long rods in wet mounts or Gram stained preparations from gills or lesions set the basis for a presumptive diagnosis (Toranzo et al., 2005). Definitive diagnostic must be supported by the isolation of colonies of *T. maritimum* on appropriate specific media, followed by the determination of phenotypic or genotypic features. API zym profile is constant throughout most strains with positive results in the first enzymatic reactions. There are serological discrepancies and therefore these methodologies are not recommended. Culture is difficult since this is a microorganism with low nutrient requirements, which is easily overgrown by other bacterial species. Specific media have been designed and FMM is proposed as the most adequate for the successful isolation from fish tissues (Pazos et al., 1996).

Due to the difficulty of culturing this species and the serological discrepancies found, many efforts have been directed towards DNA-based protocols to identify this bacterium. Two PCR protocols targeting the 16S rRNA gene have been published, one amplifying a 1088 bp fragment (Toyama et al., 1996) and the other a 400 bp (Bader and Shottes, 1998). In a comparative study Avendaño-Herrera et al. (2004b) claim that the first protocol is more adequate for an accurate detection of *T. maritimum* from fish. This protocol can also distinguish *T. maritimum* from close species, *Flavobacterium branchiophilum* and *F. columnare*.

A nested PCR protocol has been developed using in the first PCR universal primers to amplify a fragment of 1500 bp of the 16S rRNA gene and from the fragment a second PCR using Toyama et al. (1996) primers amplified the 1088 bp fragment. This protocol increases the levels of detection by 2-3 times than the simple PCR (Avendaño-Herrera et al., 2004b).

Other protocols combine PCR with other methodologies, PCR-ELISA (Wilson et al., 2002), RT-PCR-EAH, referred in *A. salmonicida* section (Wilson and Carson, 2003) and a DNA microarray probe (Warsen et al., 2004). This microarray was designed to detect 15 different pathogens, including *A. salmonicida, E. ictaluri, T. maritimum, M. fortuitum, M. maritimum, M. chelonae, R. salmoninarum* and *S. iniae.*

The use of RAPD–PCR has shown the existence of two subgroups related to O-serotype and host species (Avendaño-Herrera et al., 2004c). Six different random 10-mer pairs of primers were used in this study, though only three, p2, p4 and p6 generated an adequate band pattern. Primer p2 generated two patterns which clearly seemed related to the host origin of the isolates, group I had four major bands and included gilthead sea bream and sole, group II incorporated isolates from turbot, yellowtail and Atlantic salmon. Primer p6 generated three subgroups which could also be related to host origin, group I with the isolates from sole and gilthead sea bream, group II with the rest of the isolates except two reference strains that constituted group III. With primer 4, practically every isolate generated its own pattern.

STREPTOCOCCOSIS (CAUSED BY *STREPTOCOCCUS INIAE*)

Streptococcosis of fish has increased in number during the last decade. From a clinical point of view streptococcosis refers to infection by at least six different Gram positive bacteria, *Streptococcus parauberis, S. iniae, S. difficilis, Lactococcus garvieae, Listonella piscium, Vagococcus salmoninarum,* and *Carnobacterium piscicola.* Temperature of the water is a predisposing factor for the onset of the disease, though for some, causative of cold water sptreptococcosis, the onset temperature is 15°C (Mata et al., 2004), whereas for others, warm water streptococcosis, the onset temperature is around 18°C (Akhlaghi and Mahjor, 2004). *Streptococcus iniae* causes septicaemia and meningoencephalitis in several cultured fish species, hybrid tilapia (*Oreochromis nilotica* × *O. aurea*), hybrid striped bass (*Morone saxatilis* × *M. chrysops*), rainbow trout and yellowtail. This bacterium has also emerged as a zoonotic agent causing localized cellulitis (Berridge et al., 1998), transmitted mainly by tilapia (Weinstein et al., 1997). Fish streptococcosis has been detected in most geographical areas where water temperature is adequate for an infection to start including North America,

Japan, South Africa and the Mediterranean countries. *Streptococcus iniae* is a Gram positive coccus, facultative anaerobe occurring mostly in large chains, it is Voges-Praskauer negative, does not produce acid from sorbitol, produces acid from sucrose and hydrolyses starch (Kusuda and Salati, 1999), catalase negative (Colorni et al., 2002). It is β-haemolytic on bood agar. This bacterium does not react with any of the Lancefields group antisera; it reacts only to serum raised to it specifically (Kitao, 1993).

The clinical signs of streptococcosis are similar to those induced by enterococcal agents, exophthalmia with haemorrhages in the eyes, ocular opacity, skin blackening and petechiae, inside the opercula, and congestion of the pectoral and caudal fins and mouth (Kusuda and Salati, 1999; Akhlaghi and Mahjor, 2004). Internally, congestion and haemorrhagic enteritis can be observed (Akhlaghi and Mahjor, 2004). Histology of rainbowtrout with streptococcosis shows abcesses in the abdominal muscles with coagulation necrosis in the centre, liver with focal necrosis and lymphocyte infiltration, eye tissue with hiperaemia and subretinal hemorrhages (Akhlaghi and Mahjor, 2004).

Diagnostic Methods Including Molecular Tools for the Detection and Identification of Causative Pathogen

Presumptive diagnostic is easy to suspect based on typical signs, however to confirm the species it is necessary to biochemically characterize the bacterium. API 20 strep expected profile is 4563117. Isolation can be performed from the brain, kidney or other internal organs onto TSA prepared with 25% aged seawater with or without outdated human bank blood. Incubation should be at 24°C. BHI agar or horse blood agar can also be used. Incubation lasts 48 h, after which period yellowish translucent slightly raised colonies can be observed. Definitive identification can later be done through molecular techniques such as PCR, RAPD or AFLP analysis. A fluorescent antibody technique is also available to detect *Streptococcus* sp. (Kawahara et al., 1986).

Several molecular tools have been developed to detect and characterize *Streptococcus iniae*. A PCR assay targets the 16S to amplify a 300 bp fragment (Zlotkin et al., 1998). RFLPs have been used by the same authors and all isolates of *S. iniae* clustered in a single ribotype, with *ECO* III digestion revealing five restriction fragments of 5 to 14 kb, and digestion with *Hind* III showing seven restriction fragments (Zlotkin et al., 1998). RAPD analysis (using primer p14, Bachrach et al., 2001) has shown five DNA fragments, 2500, 2000, 1400, 1100 and 850 bp long, whereas AFLP bands per primer set ranged from 70 to 134, which in total produced 461 bands. These results are consistent with the ATCC reference strain (Colorni et al., 2002). Another five primers have been used with RAPD analysis and the patterns formed using with each one are always constant

(Dodson et al., 1999), rep PCR using *box* A primer also gives a constant pattern to all biotypes of *S. iniae* (Dodson et al., 1999). A nested PCR protocol targeting the ITS rDNA has been developed, with a first set of primers amplifying a 550 bp fragment in this region, which is common with other published sequences for *Streptococcus* sp, and the second pair of primers amplifying a 373 bp fragment of the same region, corresponding to gen bank accessation number AF048773 (Berridge et al., 1998) and specific for *S iniae*. Finally, there is a multiplex PCR protocol to detect bacterial pathogens associated with warm water streptococcosis. This protocol can detect *S. difficilis, S. parauberis, S. iniae* and *L. garviae*. For the specific case of *S. iniae* the target gene is the *lct* O and the amplified fragment is of 870 bp (Mata et al., 2004).

PISCINE MYCOBACTERIOSIS

Piscine mycobateriosis causative agents were cultured from infections in fish 14 y before Robert Koch isolated the tuberculosis bacillus (Belas et al., 1995). Piscine mycobateriosis is assumed to affect all species of fish and it can be caused mainly by *Mycobacterium marinum, M. chelonae* and *M. fortuitum*, though other *Mycobacterium* species have been associated with this disease, mainly in ornamental fish. The three species mentioned are thought to be distributed worldwide (Jenkins, 1991). *Mycobacterium* sp. can be found in both freshwater and marine water environments. Incubation period is considered long, six or more wk. The disease is transmitted from fish to fish, during infection period due to the fish shedding large numbers of bacteria into the surrounding water, which can infect healthy fish. There is also the possibility of transmission from fish to man (Belas et al., 1995). Piscine mycobacteriosis is caused by a Gram positive, nonspore forming non-motile and acid fast bacteria (Frerichs, 1993). They are fastidious, growing slowly (2 d to 8 wk) and on special media. Optimum growth temperature varies with the specie, *M. marinum*, 25-35°C, the other two species, 30-37°C. *Mycobacterium marinum* is photochromogenic, if incubated in the dark its colonies are non-pigmented, but if grown in the light or exposed to light, colonies become yellow. *Mycobacterium chelonae* is pleomorphic, cultures less than 5 d old are strongly acid fast but afterwards lose this feature (Belas et al., 1995).

General symptoms and clinical signs of piscine mycobacteriosis include depression, fading of colour, lack of appetite, folded fins, lack of movement coordination, abdominal swelling, body deformation, anorexia, emaciation, nervous disorders, exophthalmia, dispigmentation and skin ulcers. Internally, there can be melanotic foci on the spleen, granulomatous caseous or necrotic lesions in the internal organs, mainly the liver and kidney (Gomez et al., 1993).

Diagnostic Methods Including Molecular Tools for the Detection and Identification of Causative Pathogen

In general, the diagnosis of piscine mycobacteriosis depends on clinical and histological signs and the identification of the bacterial pathogen (Chinabut, 1999). The disease is normally localized on the skin and internal organs. In *post mortem* exam, the presence of dirty grey or yellowish tuberculous nodules on the main organs (liver, spleen, gonads) is a good indicator of fish piscine mycobacteriosis. Smears from the surface of internal organs can be made and stained with Ziehl-Nielsen, under light microscopy there will appear acid-fast bacilli. Presumptive diagnostic is often made based on the histological findings since culture is difficult due to the fastidious nature of these bacteria. Nevertheless, specific diagnosis requires isolation and identification of the *Mycobacterium* from skin lesions, spleen or kidney. There are several media where *Mycobacterium* can be isolated, such as MacConkey agar or BLM (bacto-lowenstein medium). Purified isolates can be characterized using the following biochemical and growth tests (Kent and Kubrica, 1985): Production of arylsulphatase, nitrate reductase, niacin, pyrazinamidase, heat-stable catalase (68°C), Tween 80 hydrolysis, and urease.

As for molecular techniques of detection of *Mycobacterium* there exist several published PCR based protocols. A commercial test exists, Genotype *Mycobacterium* assay, which a DNA strip design to identify 13 species of *Mycobacterium*, the tests consists of three steps: (1) DNA extraction from culture, (2) amplification and (3) reverse hybridization (Pate et al., 2006). A RFLP method has been published (Pate et al., 2006), where amplification of a fragment of 924 bp of the 16S RNA gene is made and after cleaning digestion is made with restriction enzymes Ban *I* and Apa *I* to descriminate between *M. marinum*, *M. fortuitum* and *M. chelonae*. After digestion with Ban *I*, the fragment of *M. fortuitum* and *M. chelonae* yielded two DNA fragments of 562 and 362 bp, whereas the fragment from *M. marinum* remained intact. Digestion with Apa *I*, of the fragment of *M. chelonae* yielded two fragments 812 and 112 bp, while digestion of the fragments from *M. fortuitum* and *M. marinum* yielded three bands of 677, 132 and 115 bp.

A nested PCR protocol exists targeting the 16S RNA gene, amplifying during the first PCR a fragment of 924 to 940 bp followed by a second PCR where an internal fragment is amplified of 300 bp (Kaattari et al., 2005). This protocol detects *Mycobacterium* spp. Currently, there is a real time PCR assay; however it is designed to identify *Mycobacterium avium* (Rodriguez-Lazaro et al., 2005).

ABBREVIATIONS

AFLP	Amplified fragment length polymorphism
AOA	Anacker and Ordal agar
BHI	Brain heart infusion
BKD	Bacterial kidney disease
BLM	Bacto-Lowenstein medium
DNA	Deoxyribonucleic acid
EHA	Enzyme hybridization assay
ELISA	Enzyme-linked immunosorbent assay
ESC	Enteric septicaemia of catfish
FAT	Fluorescent antibody test
FMM	*Flexibacter maritimus* medium
ITS	Internal transcribed spacer
LAMP	Loop-mediated isothermal amplification
mRNA	Messenger ribonucleic acid
MAbs	Monoclonal antibodies
NaCl	Sodium chloride
OIE	World Organization for Animal Health
ORF	Open reading frame
Q PCR	Quantitative PCR
PCR	Polymerase chain reaction
RAPD	Random Amplification of Polymorphic DNA
RFLP	Restriction fragment length polymorphism
rRNA	Ribosomal RNA
RT-PCR	Reverse transcriptase polymerase chain reaction
TCBS	Thiosulphate Citrate Bile Sucrose
TSA	Triptone soy agar

REFERENCES

Adams, A. and Thompson, K. (1990). Development of an enzyme-linked immunosorbent assay (ELISA) for the detection of *Aeromonas salmonicida* in fish tissue. J. Aquat. Anim. Health 20: 281-288.

Ainsworth, A.J., Capley, G., Waterstreet, P. and Munson, D. (1986). Use of monoclonal antibodies in the indirect fluorescent antibody technique (IFA) for the diagnosis of *Edwardsiella ictaluri*. J. Fish Dis. 9: 439-444

Akhlaghi, M. and Majhor, A.A. (2004). Some histopathological aspects of streptococcosis in cultured rainbowtrout (*Oncorhynchus mykiss*). Bull. Eur. Assoc. Fish Pathologists 24: 132-143.

Alcaide, E., Herraiz, S. and Esteve, C. (2006). Occurrence of *Edwardsiell tarda* in wild European eels, *Anguilla anguilla* from Mediterranean Spain. Dis. Aquat. Organ. 73: 77-81.

Avendaño-Herrera, R., Magariños, B., Lopez-Romalde, S., Romalde, J.L. and Toranzo, A.E. (2004a). Phenotypic characterisation and description of two major O-serotypes in *Tenacibaculum maritimum* strains from marine fish. Dis. Aquat. Organ. 58: 1-8.

Avendaño-Herrera, R., Magariños, B., Toranzo, A.E., Beaz, R. and Romalde, J.L. (2004b). Species-specific polymerase chain reaction primer sets for the diagnosis *Tenacibaculum maritimum* infection. Dis. Aquat. Organ. 62: 75-83.

Avendaño-Herrera, R., Rodriguez, J., Magariños, B., Romalde, J.L. and Toranzo, A.E. (2004c). Intrasepcific diversity of the marine fish pathogen *Tenacibaculum maritimum* as determined by randomly amplified polymorphic DNA-PCR. J. Appl. Microbiol. 96: 871-877.

Avendaño-Herrera, R., Toranzo, A.E. and Magariños, B. (2006). Tenacibaculosis infection in marine fish caused by *Tenacibaculum maritimum*: a review. Dis. Aquat. Organ. 71: 255-266.

Austin, B. and Austin, D.A. (1993). Bacterial Fish Pathogens Disease in Farmed and Wild Fish, 2nd edition. Ellis Horwood Limited, Chichester, England.

Austin, B., Embley, T.M. and Goodfellow, M. (1983). Selective isolation of *Renibacterium salmoninarum*. FEMS Microbiol. Lett. 17: 111-114

Bachrach, G., Zlotkin, A., Hurvitz, A., Evans, D.L. and Eldar, A. (2001). Recovery of *Streptococcus iniae* from diseased fish previously vaccinated with a streptococcus vaccine. Appl. Environ. Microbiol. 67: 3756-3758.

Bader, J.A. and Shottes, E.B. (1998). Identification of *Flavobacterium* and *Flexibacter* species by species-specific polymerase chain reaction primers to the 16S ribosomal RNA gene. J. Aquat. Anim. Health 10: 311-319.

Bakopoulos, V., Volpatti, D., Papanagiotou, E., Richards, R., Galeotti, M. and Adams, A. (1997). Development of an ELISA to detect *Pasteurella piscicida* in cultured and spiked fish tissue. Aquaculture 156: 359-366.

Baldwin, T.J. and Newton, J.C. (1993). Pathogenesis of enteric septicemia of channel catfish, caused by *Edwardsiella ictaluri*: bacteriologic and light electron microscopy findings. J. Aquat. Anim. Health 5: 189-198.

Balfry, S.K., Albright, L.J. and Evelyn, T.P.T. (1996) Horizontal transfer of *Renibacterium salmoninarum* among farmed salmonids *via* the faecal-oral route. Dis. Aquat. Organ. 8: 7-11.

Baya, A.M., Romalde, J.L., Green, D.E., Navarro, R.B., Evans, J., May, E.B. and Toranzo, A.E. (1997). Edwardsielliosis in wild striped bass from Chesapeake bay. J. Wildlife Dis. 33: 517-525.

Belas, R., Faloon, P. and Hannaford, A. (1995). Potential implications of molecular biology to the study of fish mycobacteriosis. Annu. Rev. Fish Dis. 5: 133-173.

Berridge, B.R., Fuller, J.D., Azevedo, J., Low, D.E., Bercovier, H. and Frelier, P.F. (1998). Development of specific nested oligonucleotide PCR primers for *Streptococcus iniae* 16S-23S ribosomal DNA intergenic spacer. J. Clin. Microbiol. 36: 2778-2781.

Bertolini, J.M., Cipriano, R., Pyle, S.W. and McLaughlin, J.A. (1990). Serological investigation of the fish pathogen *Edwardsiella ictaluri*, cause of enteric septicemia of catfish. J. Wildlife Dis. 26: 246-252.

Bilodeau, A.L., Waldbieser, G.C., Terhune, J.S., Wise, D.J. and Wolters, W.R. (2003). A real-time polymerase chain reaction assay of the bacterium, *Edwardsiella ictaluri* in channel catfish. J. Aquat. Anim. Health 15: 80-86.

Brown, L.L., Iwama G.K., Evelyn, T.P.T., Nelson, W.S. and Levine, R.P. (1994). Use of the polymerase chain reaction (PCR) to detect DNA from *Renibacterium salmoninarum* within individual salmonid eggs. Dis. Aquat. Anim. 18: 165-171.

Bruno, D., Bertrand, C., Turnbull, A., Kilburnm R., Walkerm A., Pendrey, D., McIntosh, A., Urquhart, K. and Taylor, G. (2007). Evaluation and development of diagnostic methods for *Renibacterium salmoninarum* causing bacterial kidney disease (BKD) in the UK. Aquaculture 269: 114-122

Bruno, D.W. and Brown, L.L. (1999). The occurrence of *Renibacterium salmoninarum* within vaccine adhesion components from Atlantic salmon, *Salmo salar* L. and coho salmon, *Oncorhynchus kisutch* Walbaum. Aquaculture 170: 1-5.

Bruno, D.W., Griffiths, J., Petrie, J. and Hastings, T.S. (1998). *Vibrio viscosus* in farmed Atlantic salmon *Salmo salar* in Scotland: field and experimental observations. Dis. Aquat. Anim. 34: 161-166.

Campbell, M.S. and Wright, A.C. (2003). Real-time PCR analysis of *Vibrio vulnificus* from oysters. Appl. Environ. Microbiol. 69: 7137-7144.

Castro, N., Toranzo, A.E., Barja, J.L., Nuñez, S. and Magariños, B. (2006). Characterization of *Edwardsiella tarda* strains isolated from turbot, *Psetta maxima* (L.). J. Fish Dis. 29: 541–547.

Cerro, A., del Marquez, I. and Guijarro, J.A. (2002). Simultaneous detection of *Aeromonas salmonicida, Flavobacterium psychrophilum* and *Yersinia ruckeri*, three major fish pathogens by multiplex PCR. Appl. Environ. Microbiol. 68: 5177-5180.

Chase, D.M. and Pascho, R.J. (1998). Development of a nested polymerase chain reaction for amplification of a sequence of the p57 gene of *Renibacterium salmoninarum* that provides a highly sensitive method for detection of the bacterium in salmonid kidney. Dis. Aquat. Organ. 34: 223-229.

Chen, J.D. and Lai, S.Y. (1998). PCR for direct detection of *Edwardsiella tarda* from infected fish and environmental water by application of the hemolysin gene. Zool. Studies 37: 169-176.

Chien, M.S., Gilbert, T.L., Huang, C., Landolt, M.L., O'Hara ,P.J. and Winton, J. (1992). Molecular cloning and sequence analysis of the gene coding for the 57 kDa major soluble antigen of the salmonid fish pathogen *Renibacterium salmoninarum*. FEMS Microbiol. Lett. 96: 259-266.

Chinabut, S. (1999). Mycobacteriosis and nocardiosis. In: Fish Diseases and Disorders, Volume 3. Viral, Bacterial and Fungal Infections. P.T.K. Woo and D.W. Bruno (eds). CABI Publishing, New York, USA, pp. 319-340.

Cipriano, R.C., Ford, L.A., Teska, J.D. and Hale, L.E. (1992). Detection of *Aeromonas salmonicida* in the mucus of salmonid fish. J. Aquat. Anim. Health 4: 114-118.

Colorni, A. (2004). Diseases of Mediterranean fish species: problems, research and prospects. Bull. Eur. Assoc. Fish Pathologists 24: 22-32.

Colorni, A., Diamant, A., Eldar, A., Kvitt, H. and Zlotkin, A. (2002). *Streptococcus iniae* infections in red sea cage-cultured and wild fishes. Dis. Aquat. Organ. 49: 165-170.

Cook, M. and Lynch, W.H. (1999). A sensitive nested reverse transcriptase PCR assay to detect viable cells of the fish pathogen *Renibacterium salmoninarum* in Atlantic salmon (*Salmo salar* L.). Appl. Environ. Microbiol. 65: 3042-3047.

Crumlish, M., Dung, T.T., Turnbull, J.F., Ngoc, N.T.N. and Ferguson, H.W. (2002). Identification of *Edwardsiella ictaluri* from diseased freshwater catfish, *Pangasius hypophthalmus* (Sauvage), cultured in the Mekong Delta, Vietnam. J. Fish Dis. 25: 733-736.

Dalla Valle, L., Zanella, L., Belvedere, P. and Colombo, L. (2002). Use of random amplification to develop PCR detection method for the causative agent of fish pasteurellosis, *Photobacterium damselae* subsp. *piscicida* (Vibrionaceae). Aquaculture 207: 187-202.

Di Pinto, A., Ciccarese, G., Tantillo, G., Catalano, D. and Forte, V.T. (2005). A collagenase-targeted multiplex PCR assay for identification of *Vibrio alginolyticus, Vibrio cholerae* and *Vibrio parahaemolyticus*. J. Food Protec. 68: 150-153.

Dodson, D.V., Maurer, J.J. and Shotts, E.B. (1999). Biochemical and molecular typing of *Streptococcus iniae* isolated from fish and human cases. J. Fish Dis. 22: 331-336.

Dorsch, M. and Stackebrandt, E. (1992). Some modifications in the procedure for direct sequencing of PCR amplified 16S rDNA. J. Microbiol. Methods 16: 271-279.

Egusa, S. (1983). Disease problems in Japanese yellowtail, *Seriola quinqueradiata* culture: a review. In: Diseases of Commercially Important Marine Fish and Shellfish. J.E. Stewart (ed). Conseil International Pour l'Exploration de la Mer, Copenhagen, Denmark, pp. 10-18.

Elliott, D.G. and Pascho, R.J. (2001). Evidence that coded wire tagging procedures can enhance transmission of *Renibacterium salmoninarum* in Chinook salmon. J. Aquat. Anim. Health 13: 181-193.

Evelyn, P.T.P (1993). Bacterial kidney disease (BKD). In: Bacterial Diseases of Fish. V.J. Inglis, R.J. Roberts and N.R. Bromage (eds). Blackwell Scientific Publications, Oxford, England, pp. 177-195.

Frerichs, G.N. (1993). Mycobacteriosis: Nocardiosis. In: Bacterial Diseases of Fish. V.J. Inglis, R.J. Roberts and N.R. Bromage (eds). Blackwell Scientific Publications, Oxford, England, pp. 219-233.

Fryer, J.L. and Sanders, J.E. (1981). Bacterial kidney disease of salmonid fish. Annu. Rev. Microbiol. 35: 273-298.

Fryer, J.L. and Lannan, C.N. (1993). The history and current status of *Renibacterium salmoninarum*, the causative agent of bacterial kidney disease in Pacific salmon. Fish. Res. 17: 15-33.

Gomez, S., Bernabe, A., Gomez, M., Navarro, J. and Sanchez, J. (1993). Fish mycobacteriosis: morphopathological and immunocytochemical aspects. J. Fish Dis. 16: 137-141.

Gonzalez, S.F., Osorio, C.R. and Santos, Y. (2003). Development of a PCR-based method for detection of *Listonella anguillarum* in fish tissues and blood samples. Dis. Aquat. Organ. 55: 109-115.

Gustafson, C.E., Thomas, C.J. and Thrust T.J. (1992). Detection of *Aeromonas salmonicida* from fish by using polimerase chain reaction amplification of the virulence surface array protein gene. Appl. Environ. Microbiol. 58: 3816-3825.

Gutenberger, S.K., Duimstra, J.R., Rohovec, J.S. and Fryer J.L. (1997). Intracellular survival of *Renibacterium salmoninarum* in trout mononuclear phagocytes. Dis. Aquat. Organ. 28: 93-106.

Hariharan, H., Qian, B., Despres, B., Kibenge, F.S., Heaney, S.B. and Rainnie, D.J. (1995). Development of a specific biotinylated DNA probe for the detection of *Renibacterium salmoninarum*. Can. J. Vet. Res. 59: 306-310.

Hawke, J.P., McWhorter, A.C., Steigerwalt, A.G. and Brenner, D.J. (1981). *Edwardsiella ictaluri* sp. nov., the causative agent of enteric septicaemia of catfish. Int. J. Syst. Bacteriol. 31: 396-400.

Hawke, J.P., Plakas, S.M., Minton, R.V., McPhearson, R.M. and Snider, T.G. and Guarino, A.M. (1987). Fish pasteurellosis of cultured striped bass (*Morone saxatilis*) in coastal Alabama. Aquaculture 75: 193-204.

Hiney, M. and Olivier, G. (1999). Furunculosis (*Aeromonas salmonicida*). In: Fish Diseases and Disorders. Volume 3: Viral, Bacterial and Fungal Infections. P.T.K. Woo and D. W. Bruno (eds), Cabi Publishing, New York, USA, pp. 341-425.

Hoffman, R.W., Bell, G.R., Pfeil-Putzien, C. and Ogawa, M. (1989). Detection of *Renibacterium salmoninarum* in tissue sections by different methods—a comparative study with special regard to the indirect immunohistochemical peroxidase technique. Fish Pathol. 24: 101-104.

Hoie, S., Heum, M. and Thoresen, O.F. (1997). Evaluation of a polymerase chain reaction based assay for the detection of *Aeromonas salmonicida* subsp. *salmonicida* in Atlantic salmon *Salmo salar*. Dis. Aquat. Organ. 30: 27-35.

Ito, H., Uchida, I., Sekizaki, T. and Terakado, N. (1995). A specific oligonucleotide probe based on SS rRNA sequences for identification of *Vibrio anguillarum* and *Vibrio ordalii*. Vet. Microbiol. 43: 167-171.

Jenkins, P. (1991). Mycobacteria in the environment. Society of Applied Bacteriology Symposium Series 20, Blackwell Publishing Limited, Oxford, England, pp. 137S-141S.

Kaattari, I.M., Rhodes, M.W., Kator, H. and Kaattari, S.L. (2005). Comparative analysis of mycobacterial infections in wild striped bass *Morone saxatilis* from Chesapeake Bay. Dis. Aquat. Organ. 67: 125-132.

Kahawara, E., Nelson, J.S. and Kusuda, R. (1986). Fluorescent antibody technique compared to standard media culture for detection of pathogenic bacteria for yellowtail and amberjack. Fish Pathol. 21: 39-45.

Kasornchandra, J., Rogers, W.A. and Plumb, J.A. (1987). *Edwardsiella ictaluri* from walking catfish, *Clarias batrachus* L. in Thailand. J. Fish Dis. 10: 137-138.

Kauffman, G.E., Blackstone, G.M., Vickery, M.C.L., Bej, A.K., Bowers, J., Bowen, M.D., Meyer, R.F. and De Paola, A. (2004). Real-time PCR quantification of *Vibrio parahaemolyticus* in oysters using an alternative matrix. J. Food Protec. 67: 2424-2429.

Kent, P.T. and Kubrica, G.P. (1985). Public health mycobacteriology: a guide for the level III laboratory. Publication No 86, CDC, US Department of Health and Human Services, Atlanta, USA, p. 8230.

Kitao, T. (1993). Streptococcal infections. In: Bacterial Diseases of Fish. V.J. Inglis, R.J. Roberts and N.R. Bromage (eds). Blackwell Scientific Publications, Oxford, England, pp. 196-210.

Konigsson, M.H., Ballagi, A., Jansson, E. and Johansson, K.E. (2005) Detection of *Renibacterium salmoninarum* in tissue samples by sequence capture and fluorescent PCR based on the 16S rRNA gene. Vet. Microbiol. 105: 235–243.

Kusuda, R. and Kawai, K. (1998). Bacterial diseases of cultured marine fish in Japan. Fish Pathol. 33: 221-234.

Kusuda, R. and Salati, F. (1999). *Enterococcus seriolicida* and *Streptococcus iniae*. In: Fish Diseases and Disorders. Volume 3: Viral, Bacterial and Fungal Infections. P.T.K.Woo and D.W. Bruno (eds). Cabi Publishing, New York, USA, pp. 303-317.

Kvitt, H., Ucko, M., Colorni, A., Batargias, C., Zlotkin, A. and Knibb, W. (2002). *Photobacterium damselae* subsp. *piscicida*: detection by direct amplification of 16S rRNA gene sequences and genotypic variation as determined by amplified fragment length polymorphism (AFLP). Dis. Aquat. Organ. 48: 187-195.

Lee, S.K., Wang, H.Z., Law, S.H., Wu, R.S. and Kong, R.Y. (2002). Analysis of 16S-23S rDNA intergenic spacers (IGSs) of marine vibrios for species-specific signature DNA sequences. Mar. Pollut. Bull. 44: 412-420.

Lyon, W.J. (2001). TaqMan PCR for detection of *Vibrio cholerae* O1, O139, non-O1 and non-O139 in pure cultures, raw oysters and synthetic seawater. Appl. Environ. Microbiol. 67: 4685-4693.

Magariños, B., Romalde, J.L., Bandin, I., Fouz, B. and Toranzo, A.E. (1992). Phenotypic antigenic and molecular characterization of *Pasteurella piscicida* isolated from fish. Appl. Environ. Microbiol. 58: 3316-3322.

Magariños, B., Pazos, F., Santos, Y., Romalde, J.L. and Toranzo, A.E. (1995). Response of *Pasteurella piscicida* and *Flexibacter maritimus* to the skin mucus of marine fish. Dis. Aquat. Organ. 21:103-108.

Magariños, B., Toranzo, A.E., Barja, J.L. and Romalde, J.L. (2000). Existence of two geographically linked clonal lineages in the bacterial pathogen *Photobacterium damselae* subsp. *piscicida*. Epidemiol. Infect. 125: 213-219.

Magariños, B., Couso, N., Noya, M., Merino, P., Toranzo, A.E. and Lamas, J. (2001). Effect of temperature on the development of pasteurellosis in carrier gilthead seabream (*Sparus aurata*). Aquaculture 195: 17-21.

Mata, A.I., Gibello, A., Casamayor, A., Blanco, M.M., Dominguez, L. and Fernandez-Garayzabal, J.F. (2004). Multiplex PCR assay for detection of bacterial pathogens associated with warmwater streptococcosis in fish. Appl. Environ. Microbiol. 70: 3183-3187.

McCarthy, D.H. (1976). Laboratory techniques for the diagnosis of fish furunculosis and whirling disease. MAFF Technical Report, No. 23.

McIntosh, D., Meaden, P.G. and Austin, B. (1996). A simplified PCR-based method for the detection of *Renibacterium salmoninarum* utilizing preparations of rainbow trout (*Oncorhynchus mykiss* Walbaum) lymphocytes. Appl. Environ. Microbiol. 62: 3929-3932.

McVicar, A.H. and White, P.G. (1979). The prevention and cure of an infectious disease in cultivated juvenile Dover sole *Solea solea* (L.). J. Fish Dis. 2: 557-562.

Miriam, A., Griffiths, S.G., Lovely, J.E. and Lynch, W.H. (1997). PCR and probe-PCR assays to monitor broodstock Atlantic salmon (*Salmo salar* L.) ovarian fluid and kidney tissue for presence of DNA of the fish pathogen *Renibacterium salmoninarum*. J Clin. Microbiol. 35: 1322-1326

Myata, M., Inglis, V. and Aoki, T. (1996). Rapid indentification of *Aeromonas salmonicida* subsp. *salmonicida* by the polymerase chain reaction. Aquaculture 141: 13-24.

Newton, J.C., Wolfe, L.G., Grizzle, J.M. and Plumb, J.A. (1989). Pathology of experimental enteric septicaemia in channel catfish, *Ictalurus punctatus* (Rafinesque), following immersion exposure to *Edwardsiella ictaluri*. J. Fish Dis. 12: 335-347.

Nishibushi, M. (2006) Molecular identification. In: The Biology of Vibrios. F.L. Thompson, B. Austin and J. Swings (eds). American Society for Microbiology Press, Washington, DC, USA.

Notomi, T., Okayama, H., Masubuchi, H., Yonekawa, T., Watanabe, K., Amino, N. and Hase, T. (2000). Loop mediated isothermal amplification of DNA. Nucleic Acids Res. 15: E63.

O'Brien, D., Mooney, J., Ryan, D., Powell, E., Hiney, M., Smith, P.R. and Powell, R. (1994). Detection of *Aeromonas salmonicida*, causal agent of furunculosis in salmonid fish, from the tank effluent of hatchery-reared Atlantic salmon smolts. Appl. Environ. Microbiol. 60: 3874-3877.

OIE (2006). Manual of Diagnostic Tests for Aquatic Animals 2006 (www.oie.int)

Osorio, C.R., Toranzo, A.E., Romalde, J.L. and Barja, J.L. (2000). Multiplex PCR assay for ureC and 16S rDNA genes clearly discriminates between both subspecies of *Photobacterium damselae*. Dis. Aquat. Organ. 40: 177-183.

Osorio, C.R., Collins, M.D., Romalde, J.L. and Toranzo, A.E. (2004). Characterization of the 23S and SS intergenic spacer region (ITS2) of *Photobacterium damselae*. Dis. Aquat. Organ. 61: 33-39.

Osorio, C.R., Collins, M.D., Romalde, J.L. and Toranzo, A.E. (2005). Variation of the 16S-23S rRNA intergenic spacer regions in *Photobacterium damselae*: a mosaic-like structure. Appl. Environ. Microbiol. 71: 636-645.

Padros, F., Zarza, C., Dopazo, L., Cuadrado, M. and Crespo, S. (2006). Pathology of *Edwardsiella tarda* infection in turbot *Scophthalmus maximus* (L.). J. Fish Dis. 29: 87-94.

Panangala, V.S., van Santen, V.L., Shoemaker, C.A. and Klesius, P.H. (2005). Analysis of 16S-23S intergenic spacer regions of the rRNA operons in *Edwardsiella ictaluri* and *Edwardsiella tarda* isolates from fish. J. Appl. Microbiol. 99: 657–669.

Panangala, V.S., Shoemaker, C.A., McNulty, S.T., Arias, C.R. and Klesius, P.H. (2006). Intra- and interspecific phenotypic characteristics of fish-pathogenic *Edwardsiella ictaluri* and *E. tarda*. Aquaculture Res. 37: 49-60.

Panangala, V.S., Shoemaker, C.A., van Santen, V.L., Dybvig, K. and Klesius, P.H. (2007). Multiplex-PCR for simultaneous detection of 3 bacterial fish pathogens, *Flavobacterium columnare*, *Edwardsiella ictaluri* and *Aeromonas hydrophila*. Dis. Aquat. Organ. 74: 199-208.

Pascho, R.J., Elliott, D.G., Mallet, R.W. and Mulcahy, D. (1987). Comparison of five techniques for the detection of *Renibacterium salmoninarum* in coho salmon. Trans. Am. Fish. Soc. 11: 882-890.

Pascho, R.J., Elliott, D.G. and Streufert, J.M. (1991). Brood stock segregation of chinook salmon *Oncorhynchus tshawytscha* by use of the enzyme-linked immunosorbent assay (ELISA) and the fluorescent antibody technique (FAT) affects the prevalence and levels of *Renibacterium salmoninarum* infection in progeny. Dis. Aquat. Organ. 12: 25-40.

Pascho, J.R., Chase, D. and McKibben, C.L. (1998). Comparison of the membrane filtration fluorescent antibody test, the enzyme-linked immunosorbent assay, and the polymerase chain reaction to detect *Renibacterium salmoninarum* in salmonid ovarian fluid. J. Vet. Invest. 10: 60-66.

Pate, M., Jencic, V., Zolnir-Dove, M. and Ocepek, M. (2006). Detection of mycobacteria in aquarium fish in Slovenia by culture and molecular methods. Dis. Aquat. Organ. 64: 29-35.

Pazos, F., Santos, Y., Macias, A.R., Nuñez, S. and Toranzo, A.E. (1996). Evaluation of media for the successful culture of *Flexibacter maritimus*. J. Fish Dis. 19: 193-197.

Plumb, J.A. (1999). Edwardsiella septicaemias. In: Fish Diseases and Disorders. Volume 3: Viral, Bacterial and Fungal Infections. P.T.K.Woo and D.W. Bruno (eds). CABI Publishing, New York, USA, pp. 479-490.

Powell, M., Overturf, K., Hogge, C. and Johnson, K. (2005). Detection of *Renibacterium salmoninarum* in chinook salmon, *Oncorhynchus tshawytscha* (Walbaum) using quantitative PCR. J. Fish Dis. 28: 615-622.

Rajan, P.R., Lin, J.H.Y., Ho, M.S. and Yang, H.L. (2003). Simple and rapid detection of *Photobacterium damselae* subsp. *piscicida* by a PCR technique and plating method. J. Appl. Microbiol. 95: 1375-1380.

Rhodes, L.D., Nilsson, W.B. and Strom, M.S. (1998). Sensitive detection of *Renibacterium salmoninarum* in whole fry, blood, and other tissues of Pacific salmon by reverse transcription-polymerase chain reaction. Mol. Mar. Biol. Biotechnol. 7: 270-279.

Rhodes, L.D., Durkin, C., Nance, S.L. and Rice, C.A. (2006). Prevalence and analysis of *Renibacterium salmoninarum* infection among juvenile Chinook salmon *Oncorhynchus tshawytscha* in north puget sound. Dis. Aquat. Organ. 71: 179-190.

Rodriguez-Lazaro, D., D'Agostino, M., Herrewegh, A., Pla, M., Cook, N. and Ikonomopoulos, J. (2005). Real time PCR-based methods for detection of *Mycobacterium avium* subsp. *paratuberculosis* in water and milk. International J. Food Microbiol. 10: 93-104.

Rogers, W.A. (1981). Serological detection of two species of *Edwardsiella* infecting catfish. In: International Symposium on Fish Biologics: Serodiagnostics and Vaccines. Dev. Biol. Std. 49: 169-217.

Romalde, J.L., Magariños, B., Lores, F. and Toranzo, A.E. (1999). Assessment of a magnetic bead-EIA based kit for rapid diagnosis of fish pasteurellosis. J. Microbiol. Methods 38: 147-154.

Salati, F., Tassi, P., Cau, A. and Kusuda, R. (1997). Microbiological study on Gram-negative pathogenic bacteria isolated from diseased sea bass, *Dicentrarchus labrax* Linneo 1758, cultured in Italy. Bollettino Societa Italiana di Patologia Ittica 9: 9-22.

Savan, R., Igarashi, A., Matsuoka, S. and Sakai, M. (2004). Sensitive and rapid detection of edwardsiellosis in fish by a loop-mediated isothermal amplification method. Appl. Environ. Microbiol. 70: 621-624.

Shotts, E.B., Blazer, V.S. and Waltman, W.D. (1986). Pathogenesis of experimental *Edwardsiella ictaluri* infection in channel catfish (*Ictalurus punctatus*). Can. J. Fish. Aquat. Sci. 43: 36-42.

Shotts, E.B. and Teska, J.D. (1989). Bacterial of aquatic vertebrates. In: Methods for Microbiological Examination of Fish and Shellfish. B. Austin and D.A. Austin (eds), Ellis Horwood, Chichester, UK, pp. 164-186.

Smith, I.W. (1964). The occurrence and pathology of Dee disease. Freshwater and Salmon Fish. Res. 34: 1-12.

Sorum, H., Holstad, G., Lunder, T. and Hastein, T. (2000). Grouping by plasmid profiles of atypical *Aeromonas salmonicida* isolated from fish, with special reference to salmonid fish. Dis. Aquat. Organ. 41: 159-171.

Starliper, C.E., Smith, D.R. and Shatzer, T. (1997). Virulence of *Renibacterium salmoninarum* to salmonids. J. Aquat. Anim. Health 9: 1-7.

Suzuki, K. and Sakai, D.K. (2007). Real time PCR for quantification of viable *Renibacterium salmoninarum* in chum salmon *Oncorhynchus keta*. Dis. Aquat. Organ. 74: 209-213.

Takahashi, H., Hara-Kudo, Y., Miyasaka, J., Kumagai, S. and Konuma, H. (2005). Development of a quantitative real-time polymerase chain reaction targeted to the *toxR* for detection of *Vibrio vulnificus*. J. Microbiol. Methods 61: 77-85.

Toranzo, A.E. and Barja, J.L. (1992). First report of furunculosis in turbot reared in floating cages in northwest of Spain. Bull. Eur. Assoc. Fish Pathol. 12: 147-149.

Toranzo, A.E., Barreiro, S., Casal, J.F., Figueras, A., Magariños, B. and Barja, J.L. (1991). Pasteurellosis in cultured guilthead seabream (*Sparus aurata*): first report in Spain. Aquaculture 61: 81-97.

Toranzo, A.E., Magariños, B. and Romalde, J.L. (2005). A review of the main bacterial fish diseases in mariculture systems. Aquaculture 246: 27-61.

Toyama, T., Kita-Tsukamoto, K. and Wakabayashi, H. (1996). Identification of *Flexibacter maritimus, Flavobacterium branchiophilum* and *Cytophaga columnaris* by PCR targeted 16S ribosomal DNA. Fish Pathol. 31: 25-31.

Traxler, G.S. and Bell, G.R. (1988). Pathogens associated with impounded Pacific herring *Clupea harengus pallasi*, with emphasis on viral erythrocytic necrosis (VEN) and atypical *Aeromonas salmonicida*. Dis. Aquat. Organ. 5: 93-100.

Waltman, W.D., Shotts, E.B. and Blazer, V.S. (1985). Recovery of *Edwardsiella ictaluri* from Danio (*Danio devario*). Aquaculture 46: 63-66.

Waltman, W.D., Shotts, E.B. and Hsu, T.C. (1986). Biochemical characteristics of *Edwardsiella ictaluri*. Appl. Environ. Microbiol. 51: 101-104

Walton, D.T., Abbot, S.L. and Janda, J.M. (1993). Sucrose positive *Edwardsiella tarda* mimicking biogroup 1 strain isolated from a patjent with cholelithiasis. J. Clin. Microbiol. 31: 155-156.

Warsen, A.E., Krug, M.J., LaFrentz, S., Staneck, D.R., Loge, F.J. and Call, D.R. (2004). Simultaneous discrimination between 15 fish pathogens by using 16S ribosomal DNA PCR and DNA microarrays. Appl. Environ. Microbiol. 70: 4216-4221.

Weinstein, M., Litt, M., Kertesz, D., Wyper, P., Rose, D., Coulter, M., McGreer, A., Facklam, R., Ostach, C., Wiley, B., Borczyk, A. and Low, D. (1997). Invasive infections due to a fish pathogen *Streptococcus iniae*. New Eng. J. Med. 9: 589-594.

Whaltman, W.D. and Shotts, E.B. (1986). Antimicrobial susceptibility of *Edwardsiella tarda* from the United States and Taiwan. Vet. Microbiol. 12: 277-282.

Wiklund, T. and Dalsgaard, I. (1998). Occurrence and significance of atypical *Aeromonas salmonicida* in non-salmonid fish species: a review. Dis. Aquat. Organ. 32: 49-69.

Wiklund, T., Madsen, L., Bruun, M.S. and Dalsgaard, I. (2000) Detection of *Flavobacterium psychrophilum* from fish tissue and water samples by PCR amplification. J. Appl. Microbiol. 88: 299-307.

Wilson, T., Carson, J. and Bowman, J. (2002). Optimisation of one-tube PCR-ELISA to detect femtogram amounts of genomic DNA. J. Microbiol. Methods 51: 163-170.

Wilson, T. and Carson, J. (2003). Development of sensitive, high–throughput one tube RT-PCR-enzyme hybridisation assay to detect selected bacterial fish pathogens. Dis. Aquat. Organ. 54: 127-134.

Wise, D.J. and Johnson, M.R. (1998). Effect of feeding frequency and Romet-medicated feed on survival, antibody response and weight gain of fingerling channel catfish (*Ictalurus punctatus*) after natural exposure to *Edwardsiella icaluri*. J. World Aquacul. Soc. 29: 170-176.

Yeh, H-Y., Shoemaker, C.A. and Klesius, P.H. (2005) Evaluation of a loop-mediated isothermal amplification method for rapid detection of channel catfish *Ictalurus punctatus* important bacterial pathogen *Edwardsiella ictaluri*. J. Microbiol. Methods 63: 36-44.

Yu, L., Yuan, L., Feng, H. and Li, S.F.Y. (2004). Determination of the bacterial pathogen *Edwardsiella tarda* in fish species by capillary electrophoresis with blue light-emitting diode induced fluorescence. Electrophoresis 25: 3139-3144.

Zapulli, V., Patarnello, T., Patarnello, P., Frassineti, F., Franch, R., Manfrin, A., Castagnaro, M. and Bargelloni, L. (2005). Direct identification of *Photobacterium damselae* subsp. *piscicida* by PCR-RFLP analysis. Dis. Aquat. Organ. 65: 53-61.

Zlotkin, A., Hershko, H. and Eldar, A. (1998). Possible transmission of *Streptococcus iniae* from wild fish to cultured marine fish. Appl. Environ. Microbiol. 64: 4065-4067.

Zorrilla, I., Babelona, M.C., Moriñigo, M.A., Sarasquete, C. and Borrego, J.J. (1999). Isolation and characterisation of the causative agent of pasteurellosis, *Photobacterium damselae* subsp. *piscicida*, from *Solea senegalensis* (Kaup). J. Fish Dis. 22: 167-171.

4

Shrimp Diseases and Molecular Diagnostic Methods

Sonia A. Soto-Rodríguez[1], Bruno Gomez-Gil[1] and Ana Roque[2]*

INTRODUCTION

The contribution of aquaculture to global supplies of fish, crustaceans, mollusks and other aquatic animals continues to grow, increasing from 3.9% of total production by weight in 1970 to 32.4% in 2004 (FAO, 2007). Aquaculture continues to grow more rapidly than all other animal food-producing sectors. Worldwide, the sector has grown at an average rate of 8.8% per year since 1970, compared with only 1.2% for capture fisheries systems. From the global production, 87.4% of penaeid shrimps come from Asia and the period 2000-2004 has seen a strong growth (19.2%) in production of farmed shrimps worldwide (FAO, 2007). However, disease outbreaks cause serious economic losses in several countries. The pathogens are often exotic viruses, suspected to have spread through a variety of pathways including international trade in aquaculture products. The most important diseases of cultured penaeid shrimps have infectious etiologies: viral, rickettsial and bacterial etiologies, but a few important diseases have fungal and protozoan agents as their cause.

Diagnostic methods for these causative agents include morphological pathology, wet mount preparations, histopathology, traditional micro-biology, bioassay methods, electron microscopy, and the application of serological techniques. Diagnostic methods play two significant roles in aquatic animal health management and disease control. Some techniques

[1]*CIAD, A.C. Mazatlan Unit for Aquaculture and Environmental Management, A.P. 711 Mazatlan, Sinaloa 82010, Mexico*
[2]*IRTA-Sant Carles de la Rápita, Carr al Poblenou SN km 5.5 Sant Carles de la Rapita. Spain*
[*]Corresponding author: Tel.: +52 669 989 8700; Fax: +52 669 989 8701; E-mail: ssoto@ciad.mx

are used to screen healthy animals (surveillance studies); and others are used to determine the cause of unfavorable health (diagnosis). Diagnostic methods are also divided, depending of the technique used, in presumptive and confirmative.

Molecular methods, such as one-step and nested PCR/RT-PCR (Polymerase chain reaction/reverse transcriptase PCR) and real-time PCR, are nowadays commonly used by shrimp disease diagnostic laboratories for surveillance and confirmatory diagnosis. Some of these methods are also being used by the shrimp industry to improve farm-level aquatic animal health management. This chapter discusses the most devastating infectious diseases for commercial shrimp production and the molecular diagnostic methods used to detect causative agents.

VIRAL DISEASES

White Spot Disease (WSD)

WSD is a lethal disease, mainly of juveniles shrimp, characterized by a rapid mortality caused by the white spot syndrome virus (WSSV). The virus belongs to genus *Whispovirus* with a large circular double-stranded DNA genome of 292,967 bp (van Hulten et al., 2001); however, different genome sizes have been reported from diverse virus isolates, which also show differences in virulence (Lo et al., 1999).

Outbreaks of WSD were first reported in Taiwan between 1991 and 1992, and the virus was first described from Japan where an initial outbreak occurred with *Penaeus japonicus* in 1993 (Takahashi et al., 1994). After WSSV was detected in the USA and Latin America in 1999, with disastrous effects to the aquaculture industries of the Pacific White Shrimp *Penaeus vannamei* and the Pacific Blue Shrimp *Penaeus stylirostris*. Natural infections have subsequently been observed in *P. chinensis*, *P. indicus*, *P. merguiensis*, the Black Tiger Shrimp *P. monodon* and *P. setiferus*. Now the geographical distribution of WSSV is throughout Asia and America which is currently the most serious viral pathogen of shrimp. This has led to an increase in the import requirements of shrimp broodstock in various countries, and even the closing of the border for the import of animals coming from areas where the virus is present.

The gross signs observed in acutely affected shrimp with WSD are a rapid reduction in food consumption, lethargy, having a loose cuticle with white spots of 0.5 to 2.0 mm in diameter, which are most apparent under the carapace. In many cases, i.e. *P. vannamei* from America, moribund shrimp with WSD display a pink to reddish-brown coloration, due to expansion of the chromatophores and few, if any, white spots. WSD can infect cells of mesodermal and ectodermal origin, such as the subcuticular epithelium.

This virus infects a wide range of aquatic crustaceans by vertical and horizontal transmission, with different mortality results. The spread of infection between regions may be through cannibalism of dead, infected shrimp (Lotz and Soto, 2002) and water-borne routes. Organisms exposed to contaminated effluent discharge from shrimp packing plants may transfer the pathogen to areas not contaminated with the virus (Sanchez-Martinez et al., 2007). Imported WSSV-bait shrimp originated from China was introduced into the Gulf of Mexico and Texas (Hasson et al., 2006); this represents a potential threat to marine crustacean fisheries. There are many host vectors of WSSV like rotifers, bivalves, polychaete worms, Brine Shrimp *Artemia salina*, copepods and Euphydradae insect larvae (Vijayan et al., 2005). All these organisms can accumulate high concentrations of viable WSSV, although there is no evidence of virus replication (Lo et al., 1996a; Li et al., 2003; Yan et al., 2004).

Diagnostic Methods for WSD

Antibodies (Abs)-based test are being tested for the detection of WSSV. A reverse passive latex agglutination assay (RPLA), which detects WSSV from stomach tissue homogenate has been proposed as a useful method for virus detection on site (Okumura et al., 2004). Histological methods using Hematoxiline and Eosine stain (H&E, Lightner, 1996) have been used as routine diagnostic technique of WSSV by demonstration of eosinophilic to pale basophilic, intranuclear inclusion bodies in hypertrophied nuclei of the cuticular epithelial cells and connective tissue, and less frequently, in other tissues.

Immuno-histochemistry uses WSSV-specific Abs with histological sections or wet mounts to demonstrate the presence of WSSV antigen in infected cells. Dot-blot nitrocellulose enzyme immunoassay (Nadala and Loh, 2000) and western blotting (Magbanua et al., 2000) have also been applied to detect WSSV in shrimp and carrier species.

PCR has been extensively used to assess the prevalence and geographic distribution of WSD among cultured shrimps and many methods have been described for the detection of WSSV (Lo et al., 1996b; Vaseeharan and Ramasamy, 2003; Hossain et al., 2004). Furthermore, PCR protocols have the added advantage of being applicable to nonlethal testing of valuable broodstock shrimp. A haemolymph sample may be taken with an insulin syringe, or a pleopod may be biopsied and used as the sample. Detection of WSSV has been also tested in multiplex reactions with other shrimp virus (Khawsak et al., 2008). At present, there are commercial kits available for the detection of many shrimp virus including WSSV using PCR-based techniques i.e. IQ2000^{TM} for Farming IntelliGene Tech. Corp.

The PCR standard protocol recommended by World Organization for Animal Health (OIE) for all situations where WSD diagnosis is required according to Lo et al. (1996b, 1997). A positive result in the first step of this protocol implies a serious WSSV infection; when a positive result is obtained in the second amplification step only, a latent or carrier-state infection is indicated (OIE, 2006).

Additionally for confirmation of suspected new hosts of WSSV or DNA amplification troubles among different protocols, the DNA fragment amplified from the two-step nested diagnostic PCR should be cloned and sequenced.

Methods for quantification of WSSV (Tang and Lightner, 2000), real-time PCR (Dhar et al., 2001) and isothermal DNA amplification (Kono et al., 2004) have also been described. Other molecular methods like *in situ* hybridization (ISH) uses WSSV-specific DNA probes with histological sections to demonstrate the presence of WSSV nuclei acid in infected cells (Nunan and Lightner, 1997; Wang et al., 1998). A mini-array method allows one-step multiple detection of WSSV by hybridization of a PCR product, increased pathogen detection considerably (Quere et al., 2002).

PCR is the best method according to OIE (2006) currently available for surveillance and diagnosis of WSSV in post-larvae (PL), juvenile and adult shrimp for reasons of availability, utility, and diagnostic specificity and sensitivity.

Taura Syndrome (TS)

TS can cause cumulative mortalities ranged from 40 to 90% involving stocks of *P. vannamei*. The causative agent of TS is the Taura syndrome virus (TSV), a 32 nm, non-enveloped icosahedral virus containing a single-stranded positive-sense RNA genome of 10,205 nucleotides (Bonami et al., 1997), classified in the family Dicistroiridae. ORF 2 contains the sequences for TSV structural proteins, including the three major capsid proteins VP1, VP2 and VP3 (Bonami et al., 1997; Mari et al., 2002). The virus replicates in the cytoplasm of host cells and at least three genotypic groups of TSV have been identified based on the sequence of the VP1: (1) the Americas group; (2) the South-East Asian group; and (3) the Belize group (Nilsen et al., 2005; Tang and Lightner, 2005). At least two distinct antigenic variants have been demonstrated that react to monoclonal antibody1 (MAb) 1A1 and, the MAb 1A1 non-reactors were also subdivided into Types B (TSV 98 Sinaloa, Mexico) and Type C (TSV 02 Belize) based on host species and virulence (Poulos et al., 1999; Mari et al., 2002).

TS was reported as a new disease in 1992 in commercial shrimp farms located on River Taura on the Gulf of Guayaquil, Ecuador (Jimenez, 1992) but its viral etiology was not established until 1994. Since TS discovery, this disease has spread, principally through the regional and international transfer of live PL and broodstocks, to most shrimp growing regions along the Atlantic and Pacific coasts of USA, Mexico, Latin America, including Hawaii and South-East Asia (Hasson et al., 1999a; Lightner, 2003; Nielsen et al., 2005).

The most susceptible specie to TSV is *P. vannamei* reported in all life stages (PL, juveniles and adults) except in eggs and zygotes, although *P. stylirostris* and *P. setiferus* can also be infected. PL and juvenile's *P. schmittii, P. aztecus, P. duorarum, P. chinensis, P. monodon,* and *P. japonicus* have been infected experimentally (Bondad-Reantaso et al., 2001; OIE, 2006).

TS causes three disease phases in infected shrimp. The peracute/acute phase of the disease is characterized by moribund shrimp displaying a pale reddish coloration of the tail fan and pleopods, soft shells, empty gut and usually dies during the process of moulting. If the shrimp survive the recovery phase begins with multifocal melanized cuticular lesions. In the chronic phase of TS infection, infected shrimp appear and behave normally, but remain persistently infected maintaining a subclinical infection. These shrimp may transmit the virus horizontally to other susceptible shrimp through cannibalization of infected moribund and dead shrimp. Vertical transmission is suspected, but this has not been conclusively demonstrated.

Aquatic insects like the water boatman (*Trichocorixa reticulate*) have been shown to serve as a vector of TSV (Bondad-Reantaso et al., 2001). Additionally, TSV has been demonstrated to remain infectious for up to 48 h in the feces passed by wild or captive sea gulls (*Larus atricilla*) and chickens (*Gallus domesticus,* Garza et al., 1997). Viable TSV has also been found in frozen shrimp products in the USA that originated in Latin America and South-East Asia (Lightner, 1996).

Diagnostic Methods for TS

Presumptive diagnosis of TS by wet-mount preparations of uropod, antennal scale, pleopods and gills may show necrosis of the cuticular epithelium where lesions will contain numerous spherical structures. Confirmatory diagnosis may be accomplished with routine histological techniques. During the acute phase there are multifocal areas of necrosis in the cuticular epithelium including appendages, gills, esophagus, and the stomach. Sub-cuticular lesions are characterized by large numbers of spherical, eosinophilic to densely basophilic inclusions, which are pyknotic and karyorrhectic nuclei and give the tissue a kind of "buck shot" appearance.

However, during the recovery and chronic phases, TSV-type lesions, hemocytic infiltration or other signs of a host inflammatory response are absent (Lightner, 1996). Histologically, numerous lymphoid organ spheroids (LOS) are observed which may detach and become ectopic

(Lightner, 1996). When LOS are assayed by ISH with a cDNA probe for TSV some cells give positive reactions for the virus, while no other target tissues react (Hasson et al., 1999b). Abs-based detection methods are available for detection of TSV; they may be used to assay samples of hemolymph, tissue homogenates or Davidson's AFA fixed tissue sections from shrimp (Poulos et al., 1999). MAb 1A1 may be used to distinguish some variants of TSV from other strains or screen possible TSV carriers by indirect fluorescent antibody or immunohistochemistry tests (Erickson et al., 2005), but any positive result should be cross-checked with another confirmatory technique.

Molecular methods of ISH and RT-PCR for TSV have been developed, a nonradioactive, DIG (digoxigenin)-labelled cDNA probes for TSV may be produced in the laboratory or obtained from commercial sources. The ISH method provides greater diagnostic sensitivity than do traditional methods for TSV diagnosis. False-negative ISH results may occur if tissues are left in Davidson's fixative for more than 24-48 h. The low pH of fixative might cause acid hydrolysis of the TSV RNA genome. This artefact can be avoided through the use of an "RNA friendly" fixative developed for shrimp (Hasson et al., 1997).

The method of RT-PCR for TSV recommended by OIE (2006) follows the protocol used by Nunan et al. (1998). The fragment amplified is from a conserved sequence located in the intergenic region and ORF 2 of TSV. Tissue samples (hemolymph, pleopods and whole small shrimp) may be assayed using this method. Real-time RT-PCR methods have the advantages of speed, specificity and sensitivity of almost 100 copies of the target sequence from the TSV genome (Dhar et al., 2002; Tang et al., 2004). A protocol of reverse transcription loop-mediated isothermal amplification (RT-LAMP) has been developed for detecting TSV. The protocol allows amplification of DNA under isothermal conditions using a set of four specially designed primers that recognize six distinct sequences of the target sequence (Kiatpathomchaia et al., 2007).

Infectious Hypodermal and Hematopoietic Necrosis (IHHN)

IHHN is an acute disease causing high mortalities in cultured juveniles of *P. stylirostris*. The disease is caused by Infectious Hypodermal and Hematopoietic Necrosis Virus (IHHNV), a parvovirus averaging 20-22 nm in diameter, with a density of 1.40 g/mL in CsCl. The virus is a linear single strand DNA with an estimated size of 4.1 Kbp, and a capsid that has four polypeptides with molecular weights of 74, 47, 39, and 37.5 kD. IHHN was first detected in *P. stylirostris* from Hawaii in 1981 (Lightner et al., 1983a), now it has been reported from stocks around the world including the Americas, Oceania and Asia (Lightner 1996; OIE, 2006).

Four IHHNV genotypes have been documented from the: (1) Americas and East Asia; (2) South-East Asia; (3) East Africa; and (4) western Indo-Pacific region including Madagascar and Mauritius (Tang and Lightner, 2002). The first two genotypes are infectious to the representative penaeids, *Penaeus vannamei* and *P. monodon*, while the latter two genetic variants may not be infectious to these species (Tang et al., 2003). Some Australian and African shrimp samples positive for IHHNV carry an IHHNV DNA fragment in their genome (Tang and Lightner, 2006).

Although IHHNV occurs widely in several species of shrimp, it causes acute mass mortality only with juveniles and sub-adults of *P. stylirostris*. The virus has been detected in cultured shrimp from Guam, French Polynesia, Hawaii, Israel and New Caledonia; an IHHN-like virus has also been reported from Australia (Bondad-Reantaso, 2001). IHHNV is presumed to be enzootic in wild penaeid populations in Indo-Pacific, Ecuador, western Panama and western Mexico (Lightner, 1996).

IHHN causes the chronic disease 'runt-deformity syndrome' (RDS) in *P. vannamei* juvenile, which display bent rostrums, wrinkled antennal flagella, cuticular roughness, wide size variations, rather than mortalities (Kalagayan et al., 1991). Similar RDS signs have been observed in cultured *P. stylirostris*. Co-infection of IHHNV with other shrimp virus is commonly found in cultured *P. vannamei* around world (Bonnichon et al., 2006; Zhang et al., 2007). *P. stylirostris* juveniles with acute IHHN show a marked reduction in food consumption, followed by changes in behavior and appearance. Shrimp may have white colored spots in the cuticular epidermis, giving such shrimp a mottled appearance. IHHN in *P. monodon* is usually subclinical, but reduced growth rates and culture performance has been reported (Primavera and Quinitio, 2000; Flegel, 2006). The target organs of the virus are tissues of ectodermal and mesodermal origin including gills, cuticular epithelium, connective and hematopoietic tissues, lymphoid organ (LO), antennal gland, and the ventral nerve cord.

Transmission mechanisms of IHHNV can be vertical and horizontal; *P. stylirostris* and *P. vannamei* that survive IHHNV epizootics may carry the virus for life and pass it on to their progeny *via* infected eggs (Motte et al., 2003). IHHNV has been detected in all life stages of *P. vannamei*. Eggs produced by IHHNV-infected females generally fail to develop and hatch (OIE, 2006). Vertically infected larvae of *P. stylirostris* do not become diseased; above PL-35 only gross signs of the disease may be observed. IHHNV can be also transmitted horizontally by cannibalism or contaminated water where the incubation period and severity of the disease are age dependent, young juveniles always being the most severely affected No vectors have been found in natural infections.

Diagnostic Methods for IHHN

IHHN can be diagnozed by histological (H&E) demonstration of the presence of intranuclear, Cowdry type A inclusions, these occur in hypertrophied nuclei of cells as eosinophilic, often haloed inclusions surrounded by marginated chromatin in tissues of ectodermal and mesodermal origin (Lightner, 1996). Gene probe and PCR methods provide greater diagnostic sensitivity and specificity. There is a number of Dot-blot and ISH tests for IHHNV, so commercial products are readily available. OIE (2006) recommend a protocol that uses a DIG-labelled DNA probe for IHHNV and follows generally the methods outlined in Mari et al. (1993) and Lightner (1996). Since Cowdry type A inclusions are rarely seen in *P. monodon*, ISH assays with a specific IHHNV probe may be necessary to obtain a definitive diagnosis (Flegel, 2006). Dot-blot hybridization and ISH test assays were recently used to detect IHHNV in *P. vannamei* from rearing ponds (Yang et al., 2007). No Abs-based antigen detection methods have been successfully developed for IHHNV detection.

There are multiple geographical variants of IHHNV, some of which are not detected by all of the available methods. PCR methods for IHHNV detection in hemolymph, ovaries and sperm are available, by single step (Tang et al., 2003; Tang and Lightner, 2006,) or nested PCR (Motte et al., 2003). Commercial PCR kits have been already developed. According to OIE (2006), a pair of primers appears to be the most suitable for detecting all the known genetic variants of IHHNV. Hence, confirmation of unexpected positive or negative PCR results for IHHNV with a second primer set, or use of another diagnostic method (i.e. real time PCR, DNA probe) is recommended. The protocol follows the method of Tang et al. (2000) and uses two primers sets, 392F and 392R and 389F and 389R, from the non-structural protein-coding region [Open Reading Frame (ORF) 1] of the IHHNV genome. The real-time PCR method recommended by OIE (2006) using TaqMan chemistry for IHHNV follows generally the method used in Tang and Lightner (2001) offers extraordinary sensitivity that can detect a single copy of the target sequence from the virus genome.

Yellowhead Disease (YHD)

YHD is a serious penaeid disease caused by yellowhead virus (YHV) (genotype 1). A closely related strain of YHV, named Gill-Associated virus (GAV genotype 2), has been reported from Australian shrimp farms. YHV is a positive-sense, single stranded RNA, rod shaped, enveloped, cytoplasmic virus (genus *Okavirus*) in a new family Roniviridae (OIE, 2006). YHD was first described in 1991 as an epizootic in Thai shrimp farms (Flegel, 2006). At present, YHD is a dangerous disease of *P. monodon*

(PL15 and older appears particularly susceptible) in intensive culture systems in Asia, among susceptible species, GAV (genotype 2) are known to occur commonly in healthy *P. monodon*, which appears to be the natural host. However, YHV (genotype 1) infections are usually detected only during an epizootic event, and do not in healthy *P. monodon* (OIE, 2006).

YHV has been shown to experimentally infect and cause high mortality in representative cultured and wild penaeid species from the Americas *P. vannamei*, *P. stylirostris*, *P. setiferus*, *P. aztecus*, and *P. duorarum*; where PL stages were found to be relatively resistant to challenge (Lightner, 2003; Pantoja and Lightner, 2003). Presumptive infections were observed in pond-reared *P. setiferus* juveniles from South Texas (Lightner, 1996).

YHD can cause up to 100% mortality in infected *P. monodon* ponds within 3 d of the first clinical signs. Diseased shrimp aggregate at the edges of the ponds or near the surface and show a light yellow coloration of the dorsal cephalothorax area (Chantanachookin et al., 1993). The virus targets tissues of ectodermal and mesodermal origin including LO, hemocytes, hematopoietic and connective tissue, gills, gonads, nerve tracts, and ganglia.

YHV can be transmitted horizontally by ingestion of infected tissue, co-habitation with infected shrimp; or by injection of Wild Rice Shrimp *Acetes* sp. extracts (Lightner, 2003; Flegel, 2006). There are no known vectors of YHV, only suspected carriers, including the Brackish Water Shrimp, *Palaemon styliferus* and *Acetes* sp. (Flegel, 2006). Survivors of YHD infection can maintain chronic sub-clinical infections and vertical transmission is suspected with such individuals.

Diagnostic Methods for YHD

A rapid presumptive diagnosis of YHD is hemolymph smears that show moderate to high numbers of blood cells with pycnotic and karyorrhexic nuclei with no evidence of bacteria. It is important since these can produce similar hemocyte nucleus changes (Flegel, 2006). Histologically, shrimp tissue shows a generalized multifocal to diffuse severe necrosis with the presence of densely basophilic spherical, cytoplasmic inclusions particularly in gill sections (Flegel, 2006). Moribund shrimp show necrosis of gill and stomach sub-cuticular cells. Inflammation response occurs, but is diffused and not conspicuous unless a secondary bacterial infection is concurrent with YHD (Lightner, 1996).

Monoclonal antibodies (MAbs) against YHV have been produced for field use and to assist in histopathological diagnosis. A Dot-blot assays that use the scFv (single chain variable fragment) antibody detect YHV-infected shrimp at 24 h post-infection and do not cross-react with WSSV and TSV proteins, therefore might find application in rapid and sensitive diagnostic tests to farmed shrimp (Intorasoot et al., 2007).

RT-PCR and gene probe are the best methods for YHV detection; however, problems with RNA degradation have arisen with field samples. To solve this problem, hemolymph samples from shrimp can be spotted on a commercial filter paper (ISOCODE®) and then a semi-nested RT-PCR can be assayed for YHV (Kiatpathomchai et al., 2004).

Three RT-PCR methods are suggested by OIE (2006). The first one-step protocol, adapted from Wongteerasupaya et al. (1997) will detect only YHV genotype and can be used for confirmation of YHV in shrimp collected from suspected YHD outbreaks. The second one is a more sensitive multiplex RT-nested PCR (mRT-PCR) used for differential detection of YHV and GAV in disease outbreaks of shrimp or for screening of healthy carriers. This test is available in a suitably modified form from commercial sources. The third one is also a sensitive nested mRT-PCR procedure that can be used for screening healthy shrimp for viruses in the yellowhead complex, including GAV, but it will not discriminate between genotypes. Assignment of genotype can be achieved by nucleotide sequence analysis of the RT-PCR product.

The mRT-PCR was proven for simultaneously detection of the main major shrimp viruses including YHV, WSSV, TSV, IHHNV (Khawsak et al., 2008). The primer sets could amplify viral nucleic acids resulting in PCR products with different sizes. They are highly specific and do not cross-hybridize with other viral or shrimp nucleic acids.

Recently, real-time PCR analysis has been also applied to detect the presence of YHV-positive LOS and samples with a low viral load (Anantasomboon et al., 2008). RT-LAMP (reverse transcription loop-mediated isothermal amplification) assay was developed for detecting the structural glycoprotein gene of YHV. The assay is a rapid, cost-effective, sensitive and specific procedure used to detect YHV in the heart and gill from infected shrimp (Mekata et al., 2006) and has a potential usefulness for rapid diagnosis of YHD.

There are several ISH methods for YHV detection, the protocol recommended by OIE (2006) is in agreement with Tang and Lightner (1999) and can be suitable for detection of YHV or GAV of shrimp samples preserved with RF-fixative (Hasson et al., 1997).

Monodon Baculovirus Disease (MBV)

A spherical baculovirosis or MBV disease is caused by a PmSNPV (singly enveloped nuclear polyhedrosis virus from *P. monodon*). MBV is the first reported virus of *P. monodon* and the second virus of all penaeid shrimp. It is a double stranded circular DNA virus of 80-100 × 106 Da of the family Baculoviridae. As with all NPVs, it is a polygonal viral inclusion body (called occlusion body) composed of a protein matrix called polyhedrin that contains embedded viral particles (Flegel, 2006).

MBV was implicated in the collapse of the shrimp industry in Taiwan in mid 1980, consequently the virus was considered one of the major diseases of Indo-Pacific and East Asian shrimp.

Based on the wide geographical and host species of MBV, the existence of different strains of MBV is likely. MBV infections have been reported in the shrimp genera: *Penaeus, Metapenaeus, Fenneropenaeus* and *Melicertus* (Lightner, 1996; Spann and Lester, 1996). MBV is enzootic in wild penaeids in Asia, India, Middle East, Australia, Indonesia, New Caledonia, East and West Africa, Madagascar and Italy (Lightner, 1996; OIE, 2006). MBV has been also reported in imported penaeids from the Mediterranean, Africa, Tahiti, North and South America and the Caribbean (Lightner, 1996; Bondad-Reantaso, 2001).

Likewise, despite the consequent direct exposure of *P. vannamei, P. stylirostris* and *F. californiensis* to MBV, these have not been associated with severe pathology (Lightner et al., 1983b). MBV have not caused shrimp mortality when rearing conditions have been good, shrimp survive high levels of virus load without effects and no visible resistance response (Flegel, 2006).

MBV disease is typically most severe in the PL stages, when mortalities may reach over 90% of affected populations. Severity of infection typically decreases over PL-25, and may become undetectable in pond-reared juveniles; hence may be present with slight mortalities (Lightner, 1986a). Gross signs of MBV infection may include reduced feeding and growth rates and increased fouling. Severely affected larvae and PL may exhibit a white midgut line through the abdomen. In juveniles, especially reared in high density, MBV infection prevalence and severity may be over 70%. The disease course may be subacute or chronic, but usually with low mortality rates (Karunasagar et al., 1998). Adults can be infected with no evident signs. MBV is strictly enteric infecting epithelial cells of the HP tubules and the anterior midgut and all life stages of the host (OIE, 2006).

Transmission of MBV is horizontal by ingestion of infected tissue, feces, virus-contaminated detritus or water of the spawned egg masses (Lightner, 1996). Even though MBV is not considered a serious pathogen for *P. monodon*, it should be eliminated from the farming system because infections could slow growth in shrimp intensive cultures (Flegel, 2006).

Diagnostic Methods for MBV

MBV can be easily recognized in wet mounts preparations of *P. monodon* PL, hepatopancreas (HP) or midgut tissue, stained with 0.05% solution of malachite green. The virus appears as single or multiple spherical polygonal occlusion bodies located in enlarged nuclei (Flegel, 2006). Wet mounts of fecal strands is perhaps most useful as a nonlethal method for screening valuable broodstock for MBV.

Histology may be used as a definitive diagnosis of MBV, occlusions bodies appear as single or multiple; eosinophilic spherical bodies within the hypertrophied nuclei of HP tubule or midgut epithelial cells (Lightner, 1996). Early in any infection process (or in light infections) occlusion bodies may not be easily demonstrated.

Dot-blot methods using specific DNA probes for MBV are not recommended for most routine diagnostic applications. Pigments present in the HP leave a colored spot on the hybridization membrane that can result in the masking of a positive test or in the false interpretation of a negative test.

Non-radioactive DIG-labeled gene probes to MBV only work well with histological sections because there are substances present in the HP and feces that provide wrong results (Poulos et al., 1994). Extraction of DNA from the HP or feces prior to blotting is recommended. The first report of MBV detection by PCR came from Australia after which, several PCR methods have been developed and applied for prevalence studies of MBV in wild and farm shrimp including the prawn *Macrobrachium rosenbergii* from different geographical regions (Uma et al., 2005; Mishra and Shekhar, 2006). However PCR tests designed for East and South-East Asian isolates of MBV have shown to give false negative results for MBV infected *P. monodon* from Africa (OIE, 2006). In fact, substances in the HP and feces of shrimp inhibit the DNA polymerase used in the PCR assay. Therefore, DNA extraction is required before the assay can be successfully applied (Hsu et al., 2000). Commercial PCR kits for MBV detection are available for shrimp farmers.

Simultaneous detection of YHV, TSV, WSSV, HPV, MBV and IHHNV using highly specific primers have been developed (Xie et al., 2007; Khawsak et al., 2008). These PCR techniques offer an efficient and rapid tool to be used in the field for screening viral infection in cultured shrimps. Nested PCR methods have been also tested (Umesha et al., 2006) although OIE recommend the protocol of Belcher and Young (1998) which is capable of detecting down to eight viral genome equivalents of MBV. OIE also recommend an alternative single-step PCR method (Surachetpong et al., 2005) used by the OIE Reference Laboratory at the University of Arizona which is less prone to contamination.

Infectious Myonecrosis (IMN)

IMN is an emerging disease of cultured shrimp caused by the Infectious Myonecrosis Virus (IMNV) which is a non-enveloped icosahedral virus of the family Totiviridae (Nibert, 2007) with a diameter of 40 nm and a double-stranded RNA genome of 7,560 bp containing two open reading frames, designated as ORF1 and ORF2 (Poulos et al., 2006). ORF1 encodes a putative RNA-binding protein and a capsid protein and, ORF2 codes for a putative RNA dependent RNA polymerase.

In September 2002, unusual mortality in farmed *P. vannamei* was observed in Piaui in northeastern Brazil, but IMNV was first reported from Brazil in 2004. IMN causes significant mortalities in juvenile and subadult pond-reared stocks of *P. vannamei*. Mortality rates ranged from 35 to 55% in 12 g shrimp with increased food conversion ratios (Pinheiro Gouveia et al., 2007). Outbreaks of the disease seemed to be associated with certain types of environment and physical stresses (i.e. extremes in salinity and temperature), and possibly with the use of low quality feeds (Pinheiro Gouveia et al., 2007). A high prevalence of the virus was found in *P. vannamei* reared in six countries in the Northeast of Brazil (Pinheiro et al., 2007) and also in Indonesian farms stocked with *P. vannamei* (Senapin et al., 2007) considering the need for establishing monitoring programs at the national level.

Apart from *P. vannamei*, experimentally susceptible species to IMNV are *P. monodon* and *P. stylirostris* (Tang et al., 2005); hence no mortalities were observed in these two species. IMN is presented as a disease in *P. vannamei* with an acute onset of gross signs and elevated mortalities, but it progresses with a more chronic course accompanied by persistent low level mortalities reducing the population by 50-70% (Lightner, 2003) Affected animals might present lethargy, opacity or reddish coloration in the last abdominal segment and totally necrotic tails.

Histopathological analyses of shrimp in acute phase of MNV present coagulative muscle necrosis. In shrimp recovering from acute disease, the myonecrosis of striated abdominal muscles appears to progress to liquefactive necrosis, accompanied with hemocytic infiltration and fibrosis (Lightner, 2003). Severe formation and accumulation of LOS was also among the main lesions and, ectopic LOS are often found in the hemocoel and connective tissues of the heart and antennal gland (Pinheiro Gouveia et al., 2007). Perinuclear pale basophilic to darkly basophilic inclusion bodies are sometime evident in muscle, connective tissue, hemocytes and LOS (Lightner, 2003).

Diagnostic Methods for IMN

An ISH procedure using a specific molecular probe labelled with digoxigenin specifically detected IMNV in infected tissues including the skeletal muscle, lymphoid organ, hindgut, and phagocytic cells within the HP and heart (Tang et al., 2005). Myonecrosis lesions are strongly positive by ISH in Davidson's fixed tissues (Andrade et al., 2008) which INMV RNA is not labile to the fixative.

One-step and nested RT-PCR have been used in infected tissue with IMNV, nested step detected until 10 copies of INMV (Poulos and Lightner, 2006). The primers shown to be specific for IMNV and no amplicons were detected using RNA extracted from shrimp infected with TSV, YHV, IHHNV and WSSV. Commercial RT-PCR kits target the capsid gene, can be purchased to detect IMNV in shrimp samples. However to solve some difficulty with very weak false positive results, an alternative, nested RT-PCR detection method with specific primers to target the viral RdRp region was tested with better results (Senapin et al., 2007). Andrade et al. (2007) developed a real-time RT-PCR diagnostic method to detect IMNV in shrimp using primers and TaqMan probe from ORF1 region of the IMNV. The assay does not cross-react with other viruses, including YHV, TSV, IHHNV, and WSSV and the necrotizing hepatopancreatic bacterium (NHPB).

BACTERIAL DISEASES

Necrotizing Hepatopancreatic Bacterium (NHPB)

NHPB is a severe bacterial disease of cultured shrimp. The causative agent is a Gram-negative, pleomorphic, obligate intracellular pathogen, member of the á-Proteobacteria. There is a single bacterium present in two distinct forms, the predominant form is rod-shaped (0.25 × 0.9 μm), rickettsia-like, lacks flagella and occasionally exhibits a transverse constricted zone indicative of replication by binary fission. The helical form (0.25 × 2-3.5 μm) possesses eight flagella on the basal apex and one or possibly two flagella on the helix crest.

NHPB was first identified in cultured *P. vannamei* from Texas (USA) in 1985. In 1993, a disease with similar clinical and histopathologic features to NHPB was diagnozed in *P. vannamei* from Peru. Later, both NHPB diagnoses were confirmed by PCR and restriction fragment length polymorphism (RFLP) analysis (Loy et al., 1996a). Transmission electronic microscopy showed that NHP from Peru and Texas are morphologically very similar, if not identical (Lightner, 1996).

NHPB has been associated with mass mortality and serious epizootic events in cultured *P. vannamei* in Peru, Ecuador, Venezuela, Brazil, Panama, and Costa Rica (Lightner, 1996) and Mexico. In 2006, an outbreak of NHPB was reported from a shrimp farm from the southern coast of the Gulf of Mexico, although with low shrimp mortalities (del Río-Rodríguez et al., 2006). Although NHPB is capable of causing severe mortality in cultured shrimps, it is not considered among the list of notifiable diseases of shrimp by the OIE. At present penaeid species affected by NHPB are *P. vannamei, P. setiferus, P. stylirostris, F. aztecus* and *F. californiensis.*

Gross signs displayed by affected shrimp may include reduced feed intake, anorexia, empty guts, elevated food conversion ratios, slow growth, soft shells and flaccid bodies, expansion of chromatophores, especially in pleopods and uropods and heavy surface fouling. Sometimes bacterial shell disease is present (melanized and necrotic lesions of the cuticle), lethargy with atrophied HP (with a pale, black streaks or watery center).

Mortality rates can approach 90% within 30 d of the onset of clinical signs. It was thought that environmental factors such as temperature and salinity could greatly influence the occurrence of NHPB disease; outbreaks of NHPB in cultured *P. vannamei* were observed in Texas and Peru following temperatures from 29 to 35°C with salinities of 20 to 40%. Similar environmental conditions were associated with outbreaks of NHPB in Venezuela, Brazil, Ecuador, Costa Rica and Panama (Lightner, 1996). However in Southern Mexico the recorded temperature and salinity levels were lower than in other regions (del Río-Rodríguez et al., 2006). Transmission rates of NHPB are highest at the intermediate salinities of 20 and 30%, probably these salinities are optimal for transmission (Vincent and Lotz, 2007).

The disease in *P. vannamei* can be transmitted horizontally through the ingestion of infectious material, i.e. cannibalism of dead or dying infected shrimp (Vincent et al., 2004).

Diagnostic Methods for NHPB

Wet-mount prepared from the HP of shrimp with developing or advanced NHPB may show reduced or absent lipid droplets with a few to multiple melanized or roughness HP tubules. The presence of these signs may provide a tentative diagnosis of NHPB in farming regions, with a prior history of NHPB.

Histological techniques demonstrate moderate to extreme atrophy of the HP tubule epithelial cells typically reduced to cuboidal in morphology, contain little or no stored lipid vacuoles (in R-cells), and reduced secretory vacuoles (in B-cells). Multifocal granulomatous lesions may contain masses of pale basophilic, Gram negative, rickettsial-like bacteria free in the cytoplasm. Such cells may display normal or pyknotic nuclei. *Vibrio*-like bacteria are often present as presumed secondary infections in the HP tubule lumens in severely affected shrimp.

MAbs were used for detecting NHPB in HP samples from moribund *P. vannamei* (Bradley-Dunlop et al., 2004). The development of diagnostic tools for NHPB is progressing rapidly, in particular the nucleic acid probes. Methods using gene probes specific for NHPB detection are available. A procedure of ISH was developed for labelling a single-stranded DNA probe with digoxigenin, the probe includes the V1 and V2 variable regions of the 16S ribosomal RNA (rRNA) gene designed to hybridize to complementary sequences of the 16S rRNA of the NHP bacterium (Loy and Frelier, 1996). However, hybridization assays using HP and feces of shrimp can provide false positive results (Lightner, 1996). Several protocols using amplification by PCR with primers specific for 16S rRNA gene sequences of HP or fecal samples have been used for NHPB diagnosis (Brinez et al., 2003). The PCR protocol commonly used follows the method by Loy et al. (1996b). Additionally commercial kits can be purchased which are commonly used by diagnostic laboratories and shrimp facilities around the world. Quantification of NHPB in HP and feces samples by real-time PCR allows an evaluation of time course of infection and estimates the bacterial load (Vincent and Lotz, 2005). There are many diagnostic molecular methods for shrimp diseases, but OIE (2006) recommend only those which sequence or primers had been proved and validated (Table 4.1 and Table 4.2).

Vibriosis

Vibriosis is the only bacterial disease caused by Gram negative free-living bacteria described, but perhaps not well understood. This bacterial

Table 4.1 Molecular methods recommended by OIE (2006) for surveillance and confirmatory diagnosis of penaeid shrimp pathogen

Pathogen	Surveillance	Confirmatory	Reference
WSSV	Gene amplification	Gene amplification, DNA probes, sequence	Lo et al. (1996b, 1997), Nunan and Lightner (1997), Claydon et al. (2004)
TSV	Gene amplification, DNA probes	Gene amplification, DNA probes, sequence	Nunan et al. (1998), Hasson et al. (1999a, 1999b), Lightner (1996)
IHHNV	Gene amplification, DNA probes	Gene amplification, DNA probes	Nunan et al. (2000), Lightner (2001), Mari et al. (1993)
YHV	Gene amplification	Gene amplification, DNA probes, sequence	Wongteerasupaya et al. (1997), Cowley et al. (2004), Tang and Lightner (1999)
MBV	Gene amplification	Gene amplification, DNA probes, sequence	Belcher and Young (1998), Surachetpong et al. (2005)
*IMNV	Gene amplification	Gene amplification	Poulos and Lightner (2006), Senapin et al. (2007)
*NHPB	Gene amplification	Gene amplification, DNA probes	Loy et al. (1996b), Vincent and Lotz (2005).

Gene amplification: different methods of PCR and RT-PCR; DNA probes: dot blot and *in-situ* hybridization; *: no OIE listed.

Table 4.2 Primer sequences recommended by OIE (2006) for detection of penaeid shrimp pathogens

Pathogen	Primers (5´→3´)	Amplicon size (bp)	Comments	Reference
WSSV	146F1 ACT ACT AAC TTC AGC CTA TCTAG 146R1 TAA TGC GGG TGT AAT GTT-CTT ACG A	1447	Single-step PCR	Lo et al. (1996b, 1997)
	146F2 GTA ACT GCC CCT TCC ATC TCC A 146R2 TAC GGC AGC TGC TGC ACC TTG T	941	Nested PCR	Lo et al. (1996b, 1997)
TSV	9992F AAG TAG ACA GCC GCG CTT 9195R	231	RT-PCR; the fragment amplified is from a conserved sequence located in the intergenic region and ORF	Nunan et al. (1998)
	TCA ATG AGA GCT TGG TCC TSV1004F TTG GGC ACC AAA CGA CAT T TSV1075R GGG AGC TTA AAC TGG ACA CAC TGT		Real-time PCR uses TaqMan chemistry; primers were selected from the ORF1 region of the TSV genomic sequence	Tang et al. (2004)
IHHNV	389F CGG AC ACA ACC CGA CTT TA 389R GGC CAA GAC CAA AAT ACG AA	389	Single-step PCR; both primers are from the nonstructural proteins-coding region (ORF 1) and are the most suitable for detecting of all the known genetic variants	Tang et al (2000)
	392F GGG CGA ACC AGA ATC ACT TA 392R	392		GenBank AF218266

Table 4.2 contd...

Target	Primer name and sequence	Size (bp)	Description	Reference
	ATC CGG AGG AAT CTG ATG TG IHHNV1608F; TAC TCC GGA CAC CCA ACC A IHHNV1688R; GGC TCT GGC AGC AAA GGT AA		Real-time RPC, primers encodes nonstructural proteins	Tang and Lightner (2001)
YHV	10F CCG CTA ATT TCA AAA ACT ACG 144R; AAG GTG TTA TGT CGA GGA AGT Y	135	Single-step RT-PCR can be used for confirmation of YHV	Cowley et al. (2004)
	ACG CTC TGT GAC AAG CAT GAA GTT G; GTA GTA GAG ACG AGT GAC ACC TAT	277	Nested mRT-PCR; sequence primer is specific for YHV	Cowley et al. (2004)
		406	Nested mRT-PCR; sequence primer is specific for GAV	
MBV	MBV1.4F CGA TTC CAT ATC GGC CGA ATA MBV1.4r; TTG GCA TGC ACT CCC TGA GAT	533	First step of nested PCR	Belcher and Young (1998)
	MBV1.4NF TCC AAT CGC GTC TGC GAT ACT MBV1.4NR; CGC TAA TGG GGC ACA AGT CTC	361	Second step of nested PCR	
*IMNV	IMNV412F GGA CCT ATC ATA CAT AGC GTT TGC IMNV545R; AAC CCA TAT CTA TTG TCG CTGGAT	134	RT-PCR uses TaqMan probe from ORF1 region of the IMNV	Andrade et al. (2007)
*NHPB	F GCA AGC TTA GAG TTT GAT CCT GGC TCA R; GCG AAT TCA CGG CTA CCT TGT TAC GAC TT		The primers are designed to target the 16S rDNA with near-full-length replication of the 16S rDNA	Loy et al. (1996b)

disease may cause mass mortalities of cultured shrimp. This name describes a disease caused by several strains of at least three species of the *Vibrio* genus. Vibrios are ubiquitously distributed in estuarine and marine environments, and currently over 70 species of *Vibrio* are described.

Some differences have been observed in vibriosis affecting larval stages, juveniles and broodstocks. During outbreaks in larval and postlarval shrimp rearing, luminescent *Vibrio harveyi* (Lavilla-Pitogo et al. 1990; Karunasagar et al., 1994; Alavandi et al., 2006) and *V. campbellii* (Soto-Rodriguez et al., 2006a, 2006b) have been isolated and proven as true pathogens. During the shrimp grow-out culture, many species have been reported as responsible for vibriosis, but only a few have actually been proven to be pathogens, all others are only members of the normal microbiota of the shrimp and the environment. Species, where some strains have been proven to be pathogenic for shrimp are *V. parahaemolyticus* (Alapide-Tendencia and Dureza 1997; Roque et al., 1998), *V. penaeicida* (Ishimaru et al., 1995; Saulnier et al., 2000), *V. nigripulchritudo* (Goarant et al., 2006), *V. alginolyticus* (Jayasree et al., 2006; Hsieh et al., 2007), and *V. harveyi* (Alapide-Tendencia and Dureza 1997; de la Peña et al., 2001; Jayasree et al., 2006), although caution has to be exercized with *V. harveyi* since problems in its identification have been described. Identification of vibrios, and especially between the species closely related belonging to the so-called "core group" or harveyi clade (Sawabe et al., 2007) is difficult. *V. harveyi* has been misidentified as *V. campbellii*, with only molecular methods being able to clearly differentiate among these "sister species" (Gomez-Gil et al., 2004; Thompson et al., 2007). *V. parahaemolyticus* and *V. alginolyticus* represent another conflict group because of their molecular similarity, but phenotypically the first forms green colonies in TCBS (Thiosulfate Citrate Bile Salts Sucrose) agar while the latter forms yellow. In this review, molecular methods are presented that pay attention and differentiate among these closely related species unambiguously.

Shrimp affected by vibriosis might present the following gross signs depending on the severity of the infection; lethargic swimming at the water surface or motionless at the pond bottom, darkened cuticular color, soft shell, fouling, opaque muscle, body reddening; specially the appendices, and melanized erosions on the cuticle. Histologically, vibriosis can be diagnozed by hemocytic infiltration, focal or multifocal hemocytic nodules in HP and connective tissues, mainly in antennal gland, heart, gills and LO. Presumptive diagnosis of vibriosis has traditionally been done by isolating the causative, or presumptive causative, bacteria from diseased shrimp. These diagnosis has the disadvantage that, since a huge diversity of vibrios are part also of the normal microbiota of shrimp (Gomez-Gil et al., 1998), it is difficult to differentiate true pathogens present in the sample. Not only are the species present important (Aguirre-Guzman et al., 2001), but most of all, if the strain is pathogenic and present in high densities (Monsalud et al., 2003; Soto-Rodriguez et al., 2006b). The virulence factors of some pathogenic strains of *Vibrio* that affect shrimps are described (Heydreyda and Rañoa, 2007), thus making possible the molecular diagnosis of pathogenic strains.

Diagnostic Methods for Vibriosis

Molecular methods to diagnoze vibriosis are based on PCR analysis of suspected strains isolated from diseased organisms. Species-specific primers are available to correctly identify the *Vibrio* species that affect shrimp (Table 4.3), but much less information is available to identify pathogenic strains of these species.

CONCLUSION AND FUTURE PERSPECTIVES

Shrimp diseases, mainly of viral origin, have been a limiting factor in the development of a sustainable industry from a global perspective. Research has focused on providing new high-tech PCR-based diagnostic tests, such as quantitative, multiplex, or isothermal methods to detect shrimp pathogens with a high specificity and sensibility. However infectious diseases have been a recurrent problem in farmed shrimps, mainly in the intensive culture systems. Shrimp farmers have used a vast diversity of chemicals to control bacterial outbreaks for many years. To avoid the abuse of antibiotics, fertilizers, and chemicals, alternative models of sustainable shrimp production are being established around the world. Organic shrimp farming is now developed in several countries of Latin America, the USA, and Asia. These practices are less harmful to the environment and reduce risks to the consumers. Other shrimp farmers use a closed-system, sometimes situated inland rather than in sensitive coastal areas. These farms recycle much of their water and drastically reduce pollution, spread of disease, and habitat destruction. A growing number of shrimp producers are also developing Better Management Practices (BMP) for their operations. The BMP target environment protection, improved shrimp health, food safety, socio-economic sustainability, and profitable farming. BMP include also compliance to systems of surveillance and contingency, planning and traceability records.

Table 4.3 Molecular markers useful to identify species of vibrios pathogenic for penaeid shrimp

Species	Gene	Primers (5′ → 3′)	Amplicon size (bp)	Comments	Reference
V. alginolyticus	B subunit DNA gyrase (*gyrB*)	F-*gyrB* ATT GAG AAC CCG ACA GAA GCG AAG R-*gyrB* CCT AAT GCG GTG ATC AGT GTT ACT	340	Useful also for quantification with SYBR Green real-time PCR. Neg. results for 6 other *Vibrio* spp. (3 strains of *V. parahaemolyticus* included).	Zhou et al. (2007)
	Collagenase	VA-F CGA GTA CAG TCA CTT GAA AGC VA-R CAC AAC AGA ACT CGC GTT ACC	737	mPCR with *V. parahaemolyticus* and *V. cholerae* primers. Neg. amplification of other 8 *Vibrio* spp.	Di Pinto et al. (2005)
	Heat shock protein 40 (*dnaJ*)	VM-F CAG GTT TGY TGC ACG GCG AAG A V.al2-MmR GAT CGA AGT RCC RAC ACT MGG A	144	mPCR with *V. parahaemolyticus* and others. Tested only with 11 clinical species.	Nhung et al. (2007)
V. campbellii	Transmembrane transcriptor regulator (*toxR*)	VcatoxR-F CCG CTT TCT GCT GAC TCT ACC VcatoxR-R GGC TTA GTC AAC ATC AGT ACA CAG	245	mPCR with *V. harveyi*. Not tested with other *Vibrio* spp.	Castroverde et al. (2006)
V. harveyi	*toxR*	VhtoxR-F TTC TGA AGC AGC ACT CAC VhtoxR-R TCG ACT GGT GAA GAC TCA	390	mPCR with *V. campbellii*. Neg. amplification of other 7 *Vibrio* spp.	Conejero and Hedreyda (2003), Castroverde et al. (2006)
	Ornithine decarboxylase gene (*odc*)	SpeFf1 AAC TGA CGG ATG GTG TAG CC SpeFR GGC ACT TAC GAC GGT ACG AT	900	Neg. amplification of *V. campbellii* and *V. parahaemolyticus*	Heydreyda and Rañoa (2007)

Table 4.3 contd...

Species	Target	Primer and sequences	Size (bp)	Notes	Reference
V. nigripulchritudo	*gyrB*	VngF2 CCC GAA CGA AGC GAA A VngR2 ACC TTT CAG TGG CAA GAT G	258	Useful also for quantification with SYBR Green real-time PCR. Neg. results for 10 other *Vibrio* spp.	Goarant et al. (2007)
V. parahaemolyticus	Thermolabile hemolysin (*tlh*)	L-TLH AAA GCG GAT TAT GCA GAA GCA CTG R-TLH GCT ACT TTC TAG CAT TTT CTC TGC	450	Cross-reactions with some *V. alginolyticus* strains	Taniguchi et al. (1986)
	pR72H	VP33 TGC GAA TTC GAT AGG GTG TTA ACC VP32 CGA ATC CTT GAA CAT ACG CAG C	387	Only tested with *V. alginolyticus* (Neg.) and *V. parahaemolyticus* (Pos.)	Lee et al. (1995)
	gyrB	VP-1 CGG CGT GGG TGT TTC GGT AGT VP-2r TCC GCT TCG CGC TCA TCA ATA	285	Neg. amplification of >30 *Vibrio* spp.	Venkateswaran et al. (1998)
	Collagenase	VP-F Gaa agt tga aca tca tca gca cga VP-R GGT CAG AAT CAA ACG CCG	271	mPCR with *V. alginolyticus* and *V. cholerae* primers. Neg. amplification of other 8 *Vibrio* spp.	Di Pinto et al. (2005)
	dnaJ	VM-F CAG GTT TGY TGC ACG GCG AAG A VP-MmR TGC GAA GAA AGG CTC ATC AGA G	96	mPCR with *V. alginolyticus* and others. Tested only with 11 clinical species.	Nhung et al. (2007)
V. penaeicida	16S rRNA	VpF GTG TGA AGT TAA TAG CTT CAT ATC VR CGC ATC TGA GTG TCA GTA TCT	310	Useful also for quantification with SYBR Green real-time PCR. Neg. amplification of 10 other *Vibrio* spp.	Saulnier et al. (2000), Goarant and Merien (2006)

ABBREVIATIONS

Abs	Antibodies
DIG	Digoxigenin
GAV	Gill-associated virus
H&E	Hematoxiline and Eosine stain
HP	Hepatopancreas
IMNV	Infectious myonecrosis virus
ISH	In-situ hybridization
IHHN	Infectious hypodermal and hematopoietic necrosis
IHHNV	Infectious hypodermal and hematopoietic necrosis virus
LOS	Lymphoid organ spheroids
MBV	Monodon baculovirus disease
mRT-PCR	Multiplex RT-PCR
NHPB	Necrotizing hepatopancreatic bacterium
OIE	World Organization for Animal Health
PCR	Polymerase chain reaction
PmSNPV	Singly enveloped nuclear polyhedrosis virus from *P. monodon*
RPLA	Reverse passive latex agglutination assay
RFLP	Restriction fragment length polymorphism
RT-PCR	Reverse transcription PCR
RT-LAMP	Reverse transcription loop-mediated isothermal amplification
TS	Taura syndrome
TSV	Taura syndrome virus
YHD	Yellowhead disease
YHV	Yellowhead virus
WSD	White spot disease
WSSV	White spot syndrome virus

REFERENCES

Aguirre-Guzman, G., Vazquez-Juarez, R. and Ascencio, F. (2001). Differences in the susceptibility of American white shrimp larval substages (*Litopenaeus vannamei*) to four *Vibrio* species. J. Invertebr. Pathol. 78: 215-219.

Alapide-Tendencia, E.V. and Dureza, L.A. (1997). Isolation of *Vibrio* spp. from *Penaeus monodon* (Fabricius) with red disease syndrome. Aquaculture 154: 107-114.

Alavandi, S.V., Manoranjita, V., Vijayan, K.K., Kalaimani, N. and Santiago, T.C. (2006). Phenotypic and molecular typing of *Vibrio harveyi* isolates and their pathogenicity. Lett. Appl. Microbiol. 43: 566-570.

Anantasomboon G., Poonkhumc, R., Sittidilokratna, N., Flegel, T.W. and Withyachumnarnkula, B. (2008). Low viral loads and lymphoid organ spheroids are associated with yellow head virus (YHV) tolerance in whiteleg shrimp *Penaeus vannamei.* Dev. Comp. Immunol. 32: 613-626.

Andrade, T.P.D., Srisuvan, T., Tang, K.F.J. and Lightner, D.V. (2007). Real-time reverse transcription polymerase chain reaction assay using TaqMan probe for detection and quantification of infectious myonecrosis virus (INMV). Aquaculture 264: 9-15.

Andrade, T.P.D., Redman, R. and Lightner, D.V. (2008). Evaluation of the preservation of shrimp samples with Davidson's AFA fixative for infectious myonecrosis virus (IMNV) *in situ* hybridization. Aquaculture 278: 179-183.

Belcher, C.R. and Young, P.R. (1998). Colorimetric PCR-based detection of monodon baculovirus in whole *Penaeus monodon* post-larvae. J. Virol. Methods 74: 21-29.

Bonami, J.R., Hasson, K.W., Mari, J., Poulos, B.T. and Lightner, D.V. (1997). Taura syndrome of marine penaeid shrimp: characterization of the viral agent. J. Gen. Virol. 78: 313-319.

Bondad-Reantaso, M.G., McGladdery, S.E., East, I. and Subasinghe, R.P. (2001). Asia Diagnostic Guide to Aquatic Animal Diseases. FAO Fisheries Technical Paper 402, Supplement 2. Rome, Italy.

Bonnichon, V., Lightner, D.V. and Bonami, J.R. (2006). Viral interference between infectious hypodermal and hematopoietic necrosis virus and white spot syndrome virus in *Litopenaeus vannamei.* Dis. Aquat. Organ. 72: 179-184.

Bradley-Dunlop, D.J., Pantoja, C. and Lightner, D.V. (2004). Development of monoclonal antibodies for detection of necrotizing hepatopancreatitis in penaeid shrimp. Dis. Aquat. Organ. 60: 233-240.

Brinez, B., Aranguren, F. and Salazar, M. (2003). Fecal samples as DNA source for the diagnosis of Necrotizing Hepatopancreatitis (NHP) in *Penaeus vannamei* broodstock. Dis. Aquat. Organ. 55: 69-72.

Castroverde, C.D.M., San Luis, B.B., Monsalud, R.G. and Hedreyda, C.T. (2006). Differential detection of vibrios pathogenic to shrimp by multiplex PCR. J. Gen. Appl. Microbiol. 52: 273-280.

Conejero, M.J. and Hedreyda, C.T. (2003). Isolation of partial toxR gene of *Vibrio harveyi* and design of toxR-targeted PCR primers for species detection. J. Appl. Microbiol. 95: 602-611.

Chantanachookin, C., Boonyaratanapalin, S., Kasornchandra, J., Direkbusarakom, S., Ekpanithanpong, U., Supamataya, K., Siurairatana, S. and Flegel, T.W. (1993). Histology and ultrastructure reveal a new granulosis-like virus in *Penaeus monodon* affected by "yellow-head" disease. Dis. Aquat. Organ. 17: 145-157.

Claydon, K., Cullen, B. and Owens, L. (2004). OIE white spot syndrome virus PCR gives false-positive results in *Cherax quadricarinatus.* Dis. Aquat. Organ. 62: 265-268.

Cowley, J.A., Cadogan, L.C., Wongteerasupaya, C., Hodgson, R.A.J., Spann, K.M., Boonsaeng, V. and Walker, P.J. (2004). Differential detection of gill-associated virus (GAV) from Australia and yellow head virus (YHV) from Thailand by multiplex RT-nested PCR. J. Virol. Methods 117: 49-59.

de la Peña, L.D., Lavilla-Pitogo, C.R. and Paner, M.G. (2001). Luminescent vibrios associated with mortality in pond-cultured shrimp *Penaeus monodon* in the Philippines: Species composition. Fish Pathol. 36: 133-138.

del Río-Rodríguez, R.E., Soto-Rodríguez, S., Lara-Flores, M., Cu-Escamilla, A.D. and Gomez-Solano, M.I. (2006). A necrotizing hepatopancreatitis (NHP) outbreak in a shrimp farm in Campeche, Mexico: A first case report. Aquaculture 255: 606-609.

Dhar, A.K., Roux, M.M. and Klimpel, K.R. (2001). Detection and quantification of infectious hypodermal and hematopoietic necrosis virus and white spot virus in shrimp using real-time quantitative PCR and SYBR green chemistry. J. Clin. Microbiol. 39: 2835-2845.

Dhar, A.K., Roux, M.M. and Klimpel, K.R. (2002). Quantitative assay for measuring the Taura syndrome virus and yellow head virus load in shrimp by real-time RT-PCR using SYBR Green Chemistry. J. Virol. Methods 104: 69-82.

Di Pinto, A., Ciccarese, G., Tantillo, G., Catalano, D. and Forte, V.T. (2005). A collagenase-targeted multiplex PCR assay for identification of *Vibrio alginolyticus, Vibrio cholerae,* and *Vibrio parahaemolyticus.* J. Food Prot. 68: 150-153.

Erickson, H.S., Poulos, B.T., Tang, K.F.J., Bradley-Dunlop, D. and Lightner, D.V. (2005). Taura Syndrome Virus from Belize represents a unique variant. Dis. Aquat. Organ. 64: 91-98.

FAO (2007). The State of World Fisheries and Aquaculture 2006. Food and Agriculture Organization of the United Nations. FAO Fisheries and Aquaculture Department. Rome, Italy.

Flegel, T.W. (2006). Detection of major penaeid shrimp viruses in Asia, a historical perspective with emphasis on Thailand. Aquaculture 258: 1-33.

Garza, J.R., Hasson, K.W., Poulos, B.T., Redman, R.M., White, B.L. and Lightner, D.V. (1997). Demonstration of infectious taura syndrome virus in the feces of sea gulls collected during an epizootic in Texas. J. Aquat. Anim. Health 9: 156-159.

Goarant, C. and Merien, F. (2006). Quantification of *Vibrio penaeicida,* the etiological agent of Syndrome 93 in New Caledonia shrimp, by real-time PCR using SYBR Green I chemistry. J. Microbiol. Methods. 67: 27-35.

Goarant, C., Reynaud, Y., Ansquer, D., de Decker, S., Saulnier, D. and le Roux, F. (2006). Molecular epidemiology of *Vibrio nigripulchritudo,* a pathogen of cultured penaeid shrimp (*Litopenaeus stylirostris*) in New Caledonia. System. Appl. Microbiol. 29: 570-580.

Goarant, C., Reynaud, Y., Ansquer, D., de Decker, S. and Merien, F. (2007). Sequence polymorphism-based identification and quantification of *Vibrio nigripulchritudo* at the species and subspecies level targeting an emerging pathogen for cultured shrimp in New Caledonia. J. Microbiol. Methods 70: 30-38.

Gomez-Gil, B., Tron-Mayen, L., Roque, A., Turnbull, J.F., Inglis, V. and Guerra-Flores, A.L. (1998). Species of *Vibrio* isolated from hepatopancreas, haemolymph and digestive tract of a population of healthy juvenile *Penaeus vannamei.* Aquaculture 163: 1-9.

Gomez-Gil, B., Soto-Rodriguez, S., Garcia-Gasca, A., Roque, A., Vazquez-Juarez, R., Thompson, F.L. and Swings, J. (2004). Molecular characterization of *V. harveyi* related isolates associated with diseased aquatic organisms. Microbiology 150: 1769-1777.

Hasson, K.W., Hasson, J., Aubert, H., Redman, R.M. and Lightner, D.V. (1997). A new RNA-friendly fixative for the preservation of penaeid shrimp samples for virological detection using cDNA genomic probes. J. Virol. Methods 66: 227-236.

Hasson, K.W., Lightner, D.V., Mohney, L.L., Redman, R.M., Poulos, B.T., Mari, J. and Bonami, J.R. (1999a). The geographic distribution of Taura Syndrome Virus (TSV) in the Americas: determination by histology and *in situ* hybridization using TSV-specific cDNA probes. Aquaculture 171: 13-26.

Hasson, K.W., Lightner, D.V., Mohney, L.L., Redman, R.M., Poulos, B.T. and White, B.L. (1999b). Taura syndrome virus (TSV) lesion development and the disease cycle in the Pacific white shrimp *Penaeus vannamei.* Dis. Aquat. Organ. 36: 81-93.

Hasson, K.W., Fan, Y., Reisinger, T., Venuti, J. and Varner, P.W. (2006). White-spot syndrome virus (WSSV) introduction into the Gulf of Mexico and Texas freshwater systems through imported, frozen bait-shrimp. Dis. Aquat. Organ. 71: 91-100.

Heydreyda, C.T. and Rañoa, D.R. (2007). Sequence analysis of the *Vibrio harveyi* ornithine decarboxylase (*odc*) gene and detection of a gene homologue in *Vibrio campbellii.* J. Gen. Appl. Microbiol. 53: 353-356.

Hossain, M.S., Otta, S.K., Chakraborty, A., Sanath Kumar, H., Karunasagar, I. and Karunasagar, I. (2004). Detection of WSSV in cultured shrimps, captured brooders, shrimp postlarvae and water samples in Bangladesh by PCR using different primers. Aquaculture 237: 59-71.

Hsieh, S.L., Ruan, Y.H., Li, Y.C., Hsieh, P.S., Hu, C.H. and Kuo, C.M. (2007). Immune and physiological responses in Pacific white shrimp (*Penaeus vannamei*) to *Vibrio alginolyticus*. Aquaculture 275: 335-341.

Hsu, Y.L., Wang, K.H., Yang, Y.H., Tung, M.C., Hu, C.H., Lo, C.F., Wang, C.H. and Hsu, T. (2000). Diagnosis of *Penaeus monodon*-type baculovirus by PCR and by ELISA of occlusion bodies. Dis. Aquat. Organ. 40: 93-99.

Intorasoot, S., Tanaka, H., Shoyama, Y. and Leelamanit, W. (2007). Characterization and diagnostic use of a recombinant single-chain antibody specific for the gp116 envelope glycoprotein of Yellow head virus. J. Virol. Methods 143: 186-193.

Ishimaru, K., Akagawa-Matsushita, M. and Muroga, K. (1995). *Vibrio penaeicida* sp. nov., a pathogen of kuruma prawns (*Penaeus japonicus*). Intl. J. System. Bacteriol. 45: 134-138.

Jayasree, L., Janakiram, P. and Madhavi, R. (2006). Characterization of *Vibrio* spp. associated with diseased shrimp from culture ponds of Andhra Pradesh (India). J. World Aquacult. Soc. 37: 523-532.

Jimenez, R. (1992). Sindrome de Taura (Resumen). Acuacultura del Ecuador 1: 1-16.

Kalagayan, G., Godin, D., Kanna, R., Hagino, G., Sweeney, J., Wyban, J. and Brock, J. (1991). IHHN virus as an etiological factor in runt-deformity syndrome of juvenile *Penaeus vannamei* cultured in Hawaii. J. World Aquaculture Soc. 22: 235-243.

Karunasagar, I., Pai, R. and Malathi, G.R. (1994). Mass mortality of *Penaeus monodon* larvae due to antibiotic-resistant *Vibrio harveyi* infection. Aquaculture 128: 203-209.

Karunasagar, I., Otta, S.K. and Karunasagar, I. (1998). Disease problems affecting cultured penaeid shrimp in India. Fish Pathol. 33: 413-419.

Khawsak, P., Deesukon, W., Chaivisuthangkura, P. and Sukhumsirichart, W. (2008). Multiplex RT-PCR assay for simultaneous detection of six viruses of penaeid shrimp. Mol. Cell. Probes 22: 177-183.

Kiatpathomchai, W., Jitrapakdee, S., Panyim, S. and Boonsaeng, V. (2004). RT-PCR detection of yellow head virus (YHV) infection in *Penaeus monodon* using dried haemolymph spots. J. Virol. Methods 119: 1-5.

Kiatpathomchaia, W., Jareonramb, W., Jitrapakdeec, S. and Flegel, T.W. (2007). Rapid and sensitive detection of Taura syndrome virus by reverse transcription loop-mediated isothermal amplification. J. Virol. Methods 146: 125-128.

Kono, T., Savan, R., Sakai, M. and Itami, T. (2004). Detection of white spot syndrome virus in shrimp by loop-mediated isothermal amplification. J. Virol. Methods 115: 59-65.

Lavilla-Pitogo, C.R., Baticados, M.C.L., Cruz-Lacierda, E.R. and Pena, L.D. (1990). Occurrence of luminous bacterial disease of *Penaeus monodon* larvae in the Philippines. Aquaculture 91: 1-13.

Lee, C.Y., Pan, S.F. and Chen, C.H. (1995). Sequence of a cloned pR72H fragment and its use for detection of *Vibrio parahaemolyticus* in shellfish with the PCR. Appl. Environ. Microbiol. 61: 1311-1317.

Li, Q., Zhang, J., Chen, Y. and Yang, F. (2003). White spot syndrome virus (WSSV) infectivity for *Artemia* at different developmental stages. Dis. Aquat. Organ. 57: 261-264.

Lightner, D.V. (1996). A Handbook of Shrimp Pathology and Diagnostic Procedures for Diseases of Penaeid Shrimp. World Aquaculture Soc., Baton Rouge, Louisiana, USA.

Lightner, D.V. (2003). The Penaeid Shrimp Viral Pandemics due to IHHNV, WSSV, TSV and YHV: History in the Americas and Current Status. Aquaculture and Pathobiology of Crustacean and Other Species. Proc. 32nd Meeting UJNR Aquaculture Panel Symposium, Davis and Santa Barbara, California, U.S.A., November 17-18th and 20th, 2003, pp. 1-20. http://www.lib.noaa.gov/japan/aquaculture/proceedings/report32/lightner_corrected.pdf

Lightner, D.V., Redman, R.M. and Bell, T.A. (1983a). Detection of IHHN virus in *Penaeus stylirostris* and *P. vannamei* imported into Hawaii. J. World Maric. Soc. 14: 212–225.

Lightner, D.V., Redman, R.M. and Bell, T.A. (1983b). Observations on the geographic distribution, pathogenesis and morphology of the baculovirus from *Penaeus monodon* Fabricius. Aquaculture 32: 209-233.

Lo, C.F., Ho, C.H., Peng, S.E., Chen, C.H., Hsu, H.C., Chiu, Y.L., Chang, C.F., Liu, K.F., Su, M.S., Wang, C.H. and Kou, G.H. (1996a). White spot syndrome baculovirus (WSBV) detected in cultured and captured shrimp, crabs and other arthropods. Dis. Aquat. Organ. 27: 215-225.

Lo, C.F., Leu, J.H., Ho, C.H., Chen, C.H., Peng, S.E., Chen, Y.T., Chou, C.M., Yeh, P.Y., Huang, C.J., Chou, H.Y., Wang, C.H. and Kou, G.H. (1996b). Detection of baculovirus associated with white spot syndrome (WSBV) in penaeid shrimps using polymerase chain reaction. Dis. Aquat. Organ. 25: 133-141.

Lo, C.F., Ho, C.H., Chen, C.H., Liu, K.F., Chiu, Y.L., Yeh, P.Y., Peng, S.E., Hsu, H.C., Liu, H.C., Chang, C.F., Su, M.S., Wang, C.H. and Kou, G.H. (1997). Detection and tissue tropism of white spot syndrome baculovirus (WSBV) in captured brooders of *Penaeus mondon* with a special emphasis on reproductive organs. Dis. Aquat. Organ. 30: 53-72.

Lo, C.F., Hsu, H.C., Tsai, M.F., Ho, C.H., Peng, S.E., Kou, G.H. and Lightner, D.V. (1999). Specific genomic fragment analysis of different geographical clinical samples of shrimp white spot syndrome virus. Dis. Aquat. Organ. 35: 175-185.

Lotz, J.M. and Soto, M.A. (2002). Model of white spot syndrome virus (WSSV) epidemics in *Litopenaeus vannamei*. Dis. Aquat. Organ. 50: 199-209.

Loy, J.K. and Frelier, P.F. (1996). Specific, nonradioactive detection of the NHP bacterium in *Penaeus vannamei* by *in situ* hybridization. J. Vet. Diag. Inv. 8: 324-331.

Loy, J.K., Frelier, P.F., Varner, P. and Templeton, J.W. (1996a). Detection of the etiologic agent of necrotizing hepatopancreatitis in cultured *Penaeus vannamei* from Texas and Peru by polymerase chain reaction. Dis. Aquat. Organ. 25: 117-122.

Loy, J.K., Dewhirst, F.E., Weber, W., Frelier, P.F., Garbar, T.L., Tasca, S.I. and Templeton, J.W. (1996b). Molecular phylogeny and *in situ* detection of the etiologic agent of necrotizing hepatopancreatitis in shrimp. Appl. Environ. Microbiol. 62: 3439-3445.

Magbanua, F.O., Natividad, K.T., Migo, V.P., Alfafara, C.G., de la Peña, F.O., Miranda, R.O., Albaladejo, J.D., Nadala, E.C.B. Jr, Loh, P.C. and Tapay, L.M. (2000). White spot syndrome virus (WSSV) in cultured *Penaeus monodon* in the Philippines. Dis. Aquat. Organ. 42: 77-82.

Mari, J., Bonami, J.R. and Lightner, D.V. (1993). Partial cloning of the genome of infectious hypodermal and hematopoietic necrosis virus, an unusual parvovirus pathogenic for penaeid shrimps; diagnosis of the disease using a specific probe. J. Gen. Virol. 74: 2637-2643.

Mari, J., Poulos, B.P., Lightner, D.V. and Bonami, J.R. (2002). Shrimp Taura syndrome virus, genomic characterization and similarity with member of the genus Cricket paralysis-like viruses. J. Gen. Virol. 83: 915-926.

Mekata, T., Kono, T., Savan, R., Sakai, M., Kasornchandra, J., Yoshida, T. and Itami, T. (2006). Detection of yellow head virus in shrimp by loop-mediated isothermal amplification (LAMP). J. Virol. Methods 135: 151-156.

Mishra, S.S. and Shekhar, M.S. (2006). Surveillance and detection of white spot syndrome virus (WSSV) and, monodon baculovirus (BMV) in different prawn farms in Orissa and West Bengal, using PCR technique and pathological studies. Indian J. Animal Sci. 76: 174-181.

Monsalud, R.G., Magbanua, F.O., Tapay, L.M., Hedreyda, C.T., Olympia, M.S., Migo, V.P., Kurahashi, M. and Yokota, A. (2003). Identification of pathogenic and non-pathogenic *Vibrio* strains from shrimp and shrimp farms in the Philippines. J. Gen. Appl. Microbiol. 49: 309-314.

Motte, E., Yugcha, E., Luzardo, J., Castro F., Leclercq, G., Rodríguez, J., Miranda, P., Borja, O., Serrano, J., Terreros, M., Montalvo, K., Narváez, A., Tenorio, N., Cedeño, V., Mialhe, E. and Boulo, V. (2003). Prevention of IHHNV vertical transmission in the white shrimp *Litopenaeus vannamei*. Aquaculture 219: 57-70.

Nhung, P.H., Ohkusu, K., Miyasaka, J., Sun, X.S. and Ezaki, T. (2007). Rapid and specific identification of 5 human pathogenic *Vibrio* species by multiplex polymerase chain reaction targeted to *dnaJ* gene. Diagn. Microbiol. Infect. Dis. 59: 271-275.

Nibert, M.L. (2007). "2A-like" and "shifty heptamer" motifs in penaeid shrimp infectious myonecrosis virus, a monosegmented double-stranded RNA virus. J. Gen. Virol. 88: 1315-1318.

Nielsen, L., Sang-oum, W., Cheevadhanarak, S. and Flegel, T.W. (2005). Taura syndrome virus (TSV) in Thailand and its relationship to TSV in China and the Americas. Dis Aquat. Organ. 63: 101-106.

Nunan, L.M. and Lightner, D.V. (1997). Development of a non-radioactive gene probe by PCR for detection of white spot syndrome virus (WSSV). J. Virol. Methods 63: 193-201.

Nunan, L.M., Poulos, B.T. and Lightner, D.V. (1998). Reverse transcription polymerase chain reaction (RT-PCR) used for the detection of Taura Syndrome Virus (TSV) in experimentally infected shrimp. Dis. Aquat. Organ. 34: 87-91.

Nunan, L.M., Poulos, B.T. and Lightner, D.V. (2000). Use of polymerase chain reaction (PCR) for the detection of infectious hypodermal and hematopoietic necrosis virus (IHHNV) in penaeid shrimp. Mar. Biotechnol. 2: 319-328.

OIE (2006). Manual of Diagnostic Tests for Aquatic Animals. World Organization for Animal Health. http://www.oie.int/eng/en_index.htm.

Okumura, T., Nagai, F., Yamamoto, S., Yamano, K., Oseko, N., Lnouye, K., Oomura, H. and Sawada, H. (2004). Detection of white spot syndrome virus from stomach tissue homogenate of the kuruma shrimp (*Penaeus japonicus*) by reverse passive latex agglutination. J. Virol. Methods 119: 11-16.

Pantoja, C.R. and Lightner, D.V. (2003). Similarity between the histopathology of White Spot Syndrome Virus and Yellow Head Syndrome Virus and its relevance to diagnosis of YHV disease in the Americas. Aquaculture 218: 47-54.

Pinheiro, A.C., Lima, A.P.S., de Souza. M.E., Neto, E.C.L., Adrião, M., Gonçalves, V. and Coimbra, M. (2007). Epidemiological status of Taura syndrome and infectious myonecrosis viruses in *Penaeus vannamei* reared in Pernambuco (Brazil). Aquaculture 262: 17-22.

Pinheiro Gouveia, R., Freitas, M.D.L. and Galli, L. (2007). Southern brown shrimp found susceptible to IMNV. The Advocate. Global Aquaculture Alliance. Feb. 2007.

Poulos, B.T. and Lightner, D.V. (2006). Detection of infectious myonecrosis virus (IMNV) of penaeid shrimp by reverse-transcriptase polymerase chain reaction (RT-PCR). Dis. Aquat. Organ. 73: 69-72.

Poulos, B.T., Mari, J., Bonami, J.R., Redman, R. and Lightner, D.V. (1994). Use of non-radioactively labeled DNA probes for the detection of a baculovirus from *Penaeus monodon* by *in situ* hybridization on fixed tissue. J. Virol. Methods 49: 187-194.

Poulos, B.T., Kibler, R., Bradley-Dunlop, D., Mohney, L.L. and Lightner. D.V. (1999). Production and use of antibodies for the detection of the Taura syndrome virus in penaeid shrimp. Dis. Aquat. Organ. 37: 99-106.

Poulos, B.T., Tang, K.F.J., Pantoja, C.R., Bonami, J.B. and Lightner, D.V. (2006). Purification and characterization of infectious myonecrosis virus of penaeid shrimp. J. Gen. Virol. 87: 987-996.

Primavera, J.H. and Quinitio, E.T. (2000). Runt-deformity syndrome in cultured giant tiger prawn *Penaeus monodon*. J. Crustacean Biol. 20: 796-802.

Quere, R., Commes, T., Marti, J., Bonami, J.R. and Piquemal, D. (2002). White spot syndrome virus and infectious hypodermal and hematopoietic necrosis virus simultaneous diagnosis by miniarray system with colorimetry detection. J. Virol. Methods 105: 189-196.

Roque, A., Turnbull, J.F., Escalante, G., Gomez-Gil, B. and Alday-Sanz, M.V. (1998). Development of a bath challenge for the marine shrimp *Penaeus vannamei* Boone, 1931. Aquaculture 169: 283-290.

Sanchez-Martinez, J.G., Aguirre-Guzman, G. and Mejía-Ruiz, H. (2007). White Spot Syndrome Virus in cultured shrimp: A review. Aquaculture Res. 2007: 1339-1354.

Saulnier, D., Avarre, J.C., Le Moullac, G., Ansquer, D., Levy, P. and Vonau, V. (2000). Rapid and sensitive PCR detection of *Vibrio penaeicida*, the putative etiological agent of syndrome 93 in New Caledonia. Dis. Aquat. Organ. 40: 109-115.

Sawabe, T., Kita-Tsukamoto, K. and Thompson, F.L. (2007). Inferring the evolutionary history of vibrios by means of multilocus sequence analysis. J. Bacteriol. 189: 7932-7936.

Senapin, S., Phewsaiya, K., Briggs, M. and Flegel, T.W. (2007). Outbreaks of infectious myonecrosis virus (IMNV) in Indonesia confirmed by genome sequencing and use of an alternative RT-PCR detection method. Aquaculture 266: 32-38.

Soto-Rodriguez, S., Armenta, M. and Gomez-Gil, B. (2006a). Effects of enrofloxacin and florfenicol on survival and bacterial population in an experimental infection with luminescent *Vibrio campbellii* in shrimp larvae of *Litopenaeus vannamei*. Aquaculture 255: 48-54.

Soto-Rodriguez, S.A., Simoes, N., Roque, A. and Gomez Gil, B. (2006b). Pathogenicity and colonization of *Litopenaeus vannamei* larvae by luminescent vibrios. Aquaculture 258: 109-115.

Spann, K.M. and Lester, R.J.G. (1996). Baculovirus of *Metapenaeus bennettae* from the Moreton Bay region of Australia. Dis. Aquat. Organ. 27: 53-58.

Surachetpong, W., Poulos, B.T., Tang, K.F.J. and Lightner, D.V. (2005). Improvement of PCR method for the detection of monodon baculovirus (MBV) in penaeid shrimp. Aquaculture 249: 69-75.

Takahashi, Y., Itanii., T., Kondo, M., Maeda, M., Fujii, R., Tonionaga, S., Supamattya, K. and Boonyaratpalin, S. (1994). Electron microscope evidence of bacilliforni virus infection in kuruma shrimp (*Penaeus japonicus*). Fish Pathol. 29: 121-125.

Taniguchi, H., Hirano, H., Kubomura, S., Higashi, K. and Mizuguchi, Y. (1986). Comparison of the nucleotide sequences of the genes for the thermostable direct hemolysin and the thermolabile hemolysin from *Vibrio parahaemolyticus*. Microb. Pathogen. 1: 425-432.

Tang, K.F.J. and Lightner, D.V. (1999). A yellow head virus gene probe: nucleotide sequence and application for *in situ* hybridization. Dis. Aquat. Organ. 35: 165-173.

Tang, K.F.J. and Lightner, D.V. (2000). Quantification of white spot syndrome virus DNA through a competitive polymerase chain reaction. Aquaculture 189: 11–21.

Tang, K.F.J. and Lightner, D.V. (2001). Detection and quantification of infectious hypodermal and hematopoietic necrosis virus in penaeid shrimp by real-time PCR. Dis. Aquat. Organ. 44: 79-85.

Tang, K.F.J. and Lightner, D.V. (2002). Low sequence variation among isolates of infectious hypodermal and hematopoietic necrosis virus (IHHNV) originating from Hawaii and the Americas. Dis. Aquat. Organ. 49: 93-97.

Tang, K.F.J. and Lightner, D.V. (2005). Phylogenetic analysis of Taura syndrome virus isolates collected between 1993 and 2004 and virulence comparison between two isolates representing different genetic variants. Virus Res. 112: 69-76.

Tang, K.F.J. and Lightner, D.V. (2006). Infectious hypodermal and hematopoietic necrosis virus (IHHNV)-related sequences in the genome of the black tiger prawn *Penaeus monodon* from Africa and Australia. Virus Res. 118: 185-191.

Tang, K.F.J., Durand, S.V., White, B.L., Redman, R.M., Pantoja, C.R. and Lightner. D.V. (2000). Postlarvae and juveniles of a selected line of *Penaeus stylirostris* are resistant to infectious hypodermal and hematopoietic necrosis virus infection. Aquaculture 190: 203-210.

Tang, K.F.J., Poulos, B.T., Wang, J., Redman, R.M., Shih, H.H. and Lightner, D.V. (2003). Geographic variations among infectious hypodermal and hematopoietic necrosis virus (IHHNV) isolates and characteristics of their infection. Dis. Aquat. Organ. 53: 91-99.

Tang, K.F.J., Wang, J. and Lightner, D.V. (2004). Quantitation of Taura Syndrome Virus by real-time RT-PCR with a TaqMan assay. J. Virol. Methods 115: 109-114.

Tang, K.F.J., Pantoja, C.R., Poulos, B.T. Redman, R.M. and Lightner, D.V. (2005). *In situ* hybridization demonstrates that *Litopenaeus vannamei, L. stylirostris* and *Penaeus monodon* are susceptible to experimental infection with infectious myonecrosis virus (IMNV). Dis. Aquat. Organ. 63: 261-265.

Thompson, F.L., Gomez-Gil, B., Vasconcelos, A.T. and Sawabe, T. (2007). Multilocus sequence analysis reveals that *Vibrio harveyi* and *V. campbellii* are distinct species. Appl. Environ. Microbiol. 73: 4279-85.

Uma, A., Koteeswaran, A., Indrani, K. and Iddya, K. (2005). Prevalence of white spot syndrome virus and monodon baculovirus in *Penaeus monodon* broodstock and postlarvae from hatcheries in southeast coast of India. Curr. Sci. 89: 1619-1622

Umesha, K.R., Dass, B.K.M., Naik, B.M., Venugopal, M.N., Karunasagar, I. and Karunasagar, I. (2006). High prevalence of dual and triple viral infections in black tiger shrimp ponds in India. Aquaculture 258: 91-96.

van Hulten, M.C., Witteveldt, J., Peters, S., Kloosterboer, N., Tarchini, R., Fiers, M., Sandbrink, S., Lankhorst, R.K. and Vlak, J.M. (2001). The white spot syndrome virus DNA genome sequence. Virology 286: 7-22.

Vaseeharan, B.J. and Ramasamy, R. (2003). PCR-based detection of white spot syndrome virus in cultured and captured crustaceans in India. Lett. Appl. Microbiol. 37: 443-447.

Venkateswaran, K., Dohmoto, N. and Harayama, S. (1998). Cloning and nucleotide sequence of the *gyrB* gene of *Vibrio parahaemolyticus* and its application in detection of this pathogen in shrimp. Appl. Environ. Microbiol. 64: 681-687.

Vijayan, K.K., Stalin Raj, V., Balasubramanian, C.P., Alavandi, S.V., Thillai Sekhar, V. and Santiago, T.C. (2005). Polychaete worms—a vector for white spot syndrome virus (WSSV). Dis. Aquat. Organ. 63: 107-111.

Vincent, A.G. and Lotz, J.M. (2005). Time course of necrotizing hepatopancreatitis (NHP) in experimentally infected *Litopenaeus vannamei* and quantification of NHP-bacterium using real-time PCR. Dis. Aquat. Organ. 67: 163-169.

Vincent, A.G. and Lotz, J.M. (2007). Effect of salinity on transmission of necrotizing hepatopancreatitis bacterium (NHPB) to Kona stock *Litopenaeus vannamei*. Dis. Aquat. Organ. 75: 265-268.

Vincent, A.G., Breland, V.M. and Lotz, J.M. (2004). Experimental infection of Pacific white shrimp *Litopenaeus vannamei* with necrotizing hepatopancreatitis (NHP) bacterium by per os exposure. Dis. Aquat. Organ. 61: 227-233.

Wang, C.S., Tsai, Y.J. and Chen, S.N. (1998). Detection of white spot disease virus (WSDV) infection in shrimp using *in situ* hybridization. J. Invert. Pathol. 72: 170-173.

Wongteerasupaya, C., Boonsaeng, V., Panyim, S., Tassanakajon, A., Withyachumnarnkul, B. and Flegel, T.W. (1997). Detection of yellow-head virus (YHV) of *Penaeus monodon* by RT-PCR amplification. Dis. Aquat. Organ. 31: 181-186.

Xie, Z., Pang, Y., Deng, X., Tang, X., Liu, J., Lu, Z.F. and Khan, M.I. (2007). A multiplex RT-PCR for simultaneous differentiation of three viral pathogens of penaeid shrimp. Dis. Aquat. Organ. 76: 77-80.

Yan, D.C., Dong, S.L., Huang, J., Yu, X.M., Feng, M.Y. and Liu, X.Y. (2004). White spot syndrome virus (WSSV) detected by PCR in rotifers and rotifer resting eggs from shrimp pond sediments. Dis. Aquat. Organ. 59: 69-73.

Yang, B., Song, X.L., Huang, J., Shi, C.Y. and Liu, L. (2007). Evidence of existence of infectious hypodermal and hematopoietic necrosis virus in penaeid shrimp cultured in China. Vet. Microbiol. 120: 63-70.

Zhang, C., Yuan, J. and Shi, Z. (2007). Molecular epidemiological investigation of infectious hypodermal and hematopoietic necrosis virus and Taura syndrome virus in *Penaeus vannamei* cultured in China. Virol. Sinica 22: 380-388.

Zhou, S., Hou, Z., Li, N. and Qin, Q. (2007). Development of a SYBR Green I real-time PCR for quantitative detection of *Vibrio alginolyticus* in seawater and seafood. J. Appl. Microbiol. 103: 1897-1906.

5

Bacterial Antibiotic Resistance in Aquaculture

Sarter Samira[1]* and Benjamin Guichard[2]

INTRODUCTION

Aquaculture, which is defined as the farming of aquatic organisms including fish, molluscs, crustaceans and aquatic plants, is currently one of the fastest growing food production sectors in the world (FAO/ NACA/WHO, 1999). In terms of food fish supply, the aquaculture sector in the world excluding China produced about 15 million tonnes of farmed aquatic products in 2004, compared with about 54 million tonnes from capture fisheries destined for direct human consumption. Corresponding figures reported for China were about 31 million tonnes from aquaculture and 6 million tonnes from capture fisheries—a powerful indication of the dominance of aquaculture in China (FAO, 2006). In 2004, the respective contribution of freshwater, mariculture and brackish-water productions of fish, crustaceans and molluscs represented 56.6, 36.0 and 7.4% by quantity.

The FAO statistics reported for 2006 showed that in global terms:

- 97.5% of cyprinids, 87.4% of penaeids and 93.4% of oysters come from Asia and the Pacific.
- 55.6% of the world's farmed salmonids come from northern countries of Western Europe. However, carps dominate in the Central and Eastern European regions, both in quantity and in value.

[1]*CIRAD, UMR QUALISUD, Montpellier, F-34398 France*
[2]*AFSSA-ANMV, BP 90203, 35302 Fougères, France*
*Corresponding author: E-mail: samira.sarter@cirad.fr

- In North America, the dominating cultured species are channel catfish in the USA, and Atlantic and Pacific salmon in Canada.
- In Latin America and the Caribbean, over the past decade, salmonids have overtaken shrimp as the top aquaculture species group.
- In the sub-Saharan Africa region, where aquaculture is still a minor sector, Nigeria is the leading country with 44,000 tonnes of catfish, tilapia and other freshwater fishes.
- In the Near East and North Africa, Egypt has become the leading country providing 92% of the regional production which make it now the second biggest tilapia producer after China and the world's top producer of mullets.

As aquaculture is becoming an important growing sector around the world, it becomes a more concentrated industry based on large-scale farms in the main producing countries. Among the main hazards threatening this industry, are animal diseases which may cause serious economic and stock losses (FAO/NACA/WHO, 1999). Antibiotics have been used worldwide to treat infections caused by a variety of pathogenic or opportunistic bacteria of fish including *Aeromonas hydrophila*, *Aeromonas salmonicida*, *Edwardsiella tarda*, *Pasteurella piscicida*, *Vibrio anguillarum*, *Yersinia ruckeri*, *Flavobacterium psychrophilum* and others. However, the intensive use of antimicrobial agents in aquaculture has been associated with the increase of bacterial resistance in the exposed microbial ecosystems. As the use of antimicrobials is essential for animal and human health as well, it is well recognized that the issues of antimicrobial use and misuse are of global concern because the occurrence of antibiotic resistances will affect both the economics of aquaculture industries and human health as well. In fact, potential transfer of resistant bacteria or resistant genes from animals to humans may occur through the food chain because reservoirs of antibiotic resistance can interact between different ecological systems (Teuber, 2001; van den Bogaard and Stobberingh, 2000; Witte, 2000). Other sources of contamination have also been reported through fish handling or cross-contamination between foods when other foods are eaten that have been cross-contaminated by bacteria from fish.

According to the Centres for Disease Control and Prevention (USA, Atlanta), which lead several programmes to monitor antimicrobial resistance, resistant strains of three major human pathogens—*Salmonella* sp., *Campylobacter* sp. and *Escherichia coli*—are linked to the use of antibiotics in animals. Currently, antibiotic resistance has been reported in a wide range of human pathogenic or opportunistic bacteria such as *Vibrio cholerae* 01 (Dalsgaard et al., 2001), *Klebsiella pneumoniae* (Carneiro et al., 2003), *Salmonella* sp. from imported seafood (Zhao et al., 2003; Randall et al., 2004), *Pseudomonas aeruginosa* (Ziha-Zarifi et al., 1999) and also in fish pathogens (Aoki, 1988; Inglis et al., 1995; Schmidt et al., 2001; Teuber, 2001).

Antibiotic use can also select antibiotic resistance in non-pathogenic bacteria, the resistance genes of which can be transferred to other pathogens of diverse origins, resulting in antibiotic-resistant infections for animals or humans as well. From this point of view, commensal fish microflora are important reservoirs to be considered because of their broad environmental distribution. This microflora includes bacteria that are natural inhabitants of the environments where aquaculture is practised.

The literature has shown that -

- Once acquired, resistance genes could be maintained even in the absence of the corresponding antibiotic (Chiew et al., 1998).
- Farming practices impact extends beyond the individual farm environment (Schmidt et al., 2000).
- In response to a variable antibiotic pressure, bacteria optimizes its resistance system towards multiple drugs to survive (Baquero et al., 1998). Consequently, the contamination of the environment with bacterial pathogens resistant to antimicrobial agents is a potential threat not only as a source of disease but also as a source from which resistance genes can easily spread to other pathogens of diverse origins.

ANTIBIOTIC USE IN AQUACULTURE

Mode of Action

Antibiotics are drugs of natural or synthetic origin that have the capacity to kill or to inhibit the growth of microorganisms (Table 5.1). The first antibiotic, discovered in 1926 by Alexander Fleming, was penicillin, a substance produced by fungi that appeared to inhibit bacterial growth. For animal husbandry they can be used either to treat infections, or to prevent the spread of a disease at the group level (metaphylactic treatment), or as growth promoters (this latter use does not exist in aquaculture). Antibiotics should not be used for prophylaxis, for which vaccines are the medicines of choice. It may however happen, mostly in the case of high pathogenic pressure, when no efficient vaccine is available.

Table 5.1 Typical modes of action of common antibiotics (FAO, 2005)

Mechanism	Comments	Examples
Damage cell membrane, allowing contents to leak out. Bactericidal.	High toxicity to animals and humans; topical use only.	polymixins.
Inhibitors of bacterial cell wall synthesis.	Animals and humans not affected because their cells do not have walls.	penicillins; aminopenicillins; cephalosporins (cephalexin); bacitracin (topical); vancomycin.
Inhibitors of folic acid synthesis. Folic acid is needed to make RNA and DNA for growth and multiplication, and bacteria must synthesize folic acid. Bacteriostatic.	Animals and humans obtain folic acid from their diets, so they are not affected.	sulphonamides; sulfasalazine; trimethoprim; co-trimoxazole.
Inhibitors of DNA function. DNA is needed for cell growth and division. Most are bactericidal.	Drugs used affect bacterial (or fungal) cells more than animal or human cells.	nalidixic acid; ofloxacin; metronidazole; rifampicin; enrofloxacin; sarafloxacin
Inhibitors of protein synthesis. Proteins are synthesized on cell structures called ribosomes. Bactericidal or bacteriostatic.	High doses can affect animals and humans because some ribosomes are similar to those in bacteria.	tetracyclines; aminoglycosides; chloramphenicol; florfenicol; macrolides; spectinomycin; lincosamides.

Range of Antimicrobial Agents used in World Aquaculture

Europe

Few countries (with the exception of Norway) have compiled an authoritative list of antimicrobial agents used in their aquaculture industries. With respect to the use of antimicrobial agents used in Europe, two sources have been used.

The first was a survey sent at the beginning of 2006 to more than 30 European agencies for Veterinary Medicinal Products (VMP), on antibiotics registered for use in fish in their countries. Twenty-six countries answered the survey, among which six countries did not have any chemical VMP registered for use in fish. Among the 20 countries that have (or will soon have) such products, only nine registered one active substance, four registered two active substances, four registered three active substances, two registered four active substances, and only one country registered five different active substances. At the European level, 8 different active substances and 50 distinct VMP are registered, mostly medicated premixes, designed to be incorporated in medicated feed. Instead, the countries that registered the highest number of active substances and/or products are not the biggest farmed fish producers, but those that have the most ancient and diversified fish production. Oxytetracycline is the most common antibiotic, with a wide range of national products: registered in 12 countries, it is commercialized as 19 different products. In contrast, only five different products with trimethoprim-sulfadiazine are registered in seven countries, with one single product being commercialized in four countries. Oxolinic acid is registered in six countries (six products) and flumequine and florfenicol in five countries (and five products) each. Three antibiotics are uncommon. Amoxicillin is registered in three countries, chlortetracycline and sarafloxacin in one country each. This survey shows that the therapeutic arsenal to treat bacterial fish diseases is very narrow in Europe, a threatening situation for the durable efficacy of available antibiotics (Guichard and Lièek, 2006).

A second source of information was less direct and was derived from a survey of laboratories performing susceptibility tests of bacteria isolated from farmed finfish (Smith, 2006). In this survey, laboratories reported the agent to which the routinely reported susceptibility. The survey of licensed agents is likely to produce an underestimate of the range of agents used as it will not include agents used off-label or agents used that are not licensed. The susceptibility testing survey may on the other hand represent an overestimate of the range of agents used. Table 5.2 presents a comparison of the data obtained from the two sources and it is clear that there is a similarity in the picture that is obtained from both.

The Americas

In North America (the United States and Canada), only three classes of antimicrobials are registered for use in finfish. These include potentiated sulphonamides (ormetoprim-sulphadimethoxine and trimethoprim-sulfadiazine) (Canada only), tetracyclines (oxytetracycline) and chloramphenicols (florfenicol).

In South America, oxolinic acid, flumequine, florfenicol and oxytetracycline are registered for use in Chile (Cárdenas, 2003).

Asia

In general, the range of antimicrobial agents available or in use in Asian countries is larger than in Europe or America. Japan has a very large aquaculture industry. While the extent of use of antimicrobial agents is difficult to determine, it is thought to be considerable. There is published evidence of 29 different antibiotics/antibiotic presentations authorized for use in Japan (Wilder, 1996). Estimates of antimicrobial agent use in other

Table 5.2 Data on the range of antimicrobial agents used in European finfish culture

Agent	Frequency of susceptibility testing in 29 laboratories (Smith, 2006)	Frequency of registration in 20 European countries
Oxytetracycline	28 (97%)	11 (55%)
Trimethoprim-sulfadiazine	26 (90%)	7 (35%)
Oxolinic acid	20 (69%)	6 (30%)
Florfenicol	17 (59%)	5 (25%)
Flumequine	17 (59%)	5 (25%)
Enrofloxacin	12 (41%)	
Amoxycillin	11 (38%)	3 (15%)
Erythromycin	8 (28%)	
Ampicillin	5 (17%)	
Chloramphenicol, Ormetoprim-sulfadiazine	4 (14%)	
Gentamycin	3 (10%)	
Naladixic acid, Furazolidone, Nitrofurantoin	2 (7%)	
Amikacin, Doxycycline Marbofloxacin, Neomycin, Norfloxacin, Novobiocin, Penicillin and Streptomycin	1 (3%)	

Asian countries are difficult to obtain but available information suggests that a wide variety of agents are used. Some data can be obtained from the papers presented at the conference on the use of chemicals in aquaculture in Asia held in the Philippines in 1996 (Arthur et al., 2000). It should be noted that none of the papers differentiated between use on small-scale farms and large production units.

A report from Vietnam recorded 122 antibiotic products being used in shrimp culture in that country. Many of these preparations contained similar agents, for example, 77 contained quinolones, and so the total range of agents is probably much smaller (Van, 2005).

Surveys from Sri Lanka indicated that erythromycin and tetracycline were the most commonly used antibiotics in the shrimp industry. Further, this report observed that these chemotherapeutic agents were used in an *ad hoc* manner where farmers selected arbitrary dosages for the agents, which were applied in bath or mixed with feed (Wijegoonawardena and Siriwardena, 1996).

Reports from the Philippines also listed a significant number of antimicrobial agents used in shrimp culture (tetracycline, rifampicin, chloramphenicol, nitrofuran and erythromycin) and in the rearing of other food or ornamental aquatic animals (oxolinic acid, ormetoprim, sulfadimethoxine, sulfamerazine, sulfisoxazole, trimethoprim/sulfadiazine, streptomycin, erythromycin, furazolidone, gentamycin, kanamycin, neomycin and nyfurpyrinol) (Cruz-Lacierda and de la Peña, 1996).

In India, a survey on the use of chemotherapeutants in carp farms in 1995 showed that in extensive farms, 4% used oxytetracycline and 2% chloramphenicol and in semi-intensive farms, 12% used oxytetracycline, 1% chloramphenicol and 3% other antibiotics. In extensive shrimp farms, 31% used oxytetracycline, and 13% other antibiotics; in semi-intensive shrimp farms 71% used oxytetracycline, 3% chloramphenicol and 2% other antibiotics; in intensive shrimp farms, 67% used oxytetracycline (Pathak et al., 1996).

Estimates of the Amounts of Antimicrobial Agents used in Aquaculture

The data on the amount of antibiotics used in aquaculture are scarce but increasing, even if this question is still considered as highly sensitive because of commercial consequences on international trade. However, more and more countries are now following the early example of Norway that started publishing the amounts of antimicrobials used in salmonids farming at the beginning of the 1990s, and has continued to the present day (www.fiskeridir.no).

There are often difficulties comparing such data, because of the lack of homogeneity in the way they are presented. Some are given in annual weights of commercial products, others in weights of active principles, more often in amounts of active principles per tonne of animal produced. The biomass of treated animals would be an interesting parameter to obtain in order to assess the real therapeutic importance of each antibiotic, but is rarely available. In some cases, usage of therapeutic agents is only expressed as annual expenditures, or as a percentage of the farms using them, which can hardly be compared with any other usage data.

In Norway, the amounts of active substances used in salmon and trout farming increased from 3660 kg in 1980 (81% oxytetracycline, 11% sulfamerazine, 8% trimethoprim-sulfadiazine) to a peak of 48,570 kg in 1987 (56% oxytetracycline, 32% nitrofurazolidone, 8% oxolinic acid, 4% trimethoprim-sulfadiazine) (Grave et al., 1990). The fish production also increased steadily in the same period, but the antibiotic use increased at a higher rate: 5 g/tonne of fish in 1980, 30 g/tonne in 1984 and more than 100 g/tonne in 1987 (Grave et al., 1990).

From 1988 onwards, the use of antibiotics decreased regularly, most of all during the 1990s with the massive development of vaccination (Markestad and Grave, 1997). It should be noted that Hiney and Smith (2000) have argued that improvements in husbandry also played a major role in this decline.

During the 1996-1999 period, the amounts of anti-bacterials used in Norwegian fish farming were approximately one tonne of active substances per year, and slightly less during the 2000-2004 period (Lunestad and Grave, 2005). In spite of a substantial increase in the farmed fish production, the amounts of antibacterial agents used in Norwegian aquaculture have decreased by 98% since 1987.

In 2006, the antibacterials used to treat bacterial fish diseases in Norway were oxolinic acid, florfenicol, flumequine and oxytetracycline, for a total amount of 1428 kg.

In Swedish fish farming, oxytetracycline was the dominant antibacterial drug between 1988 and 1993 (Björnerot et al., 1996). The heaviest usage of oxytetracycline was in 1990, and there was a marked decrease in the following years that resulted in a decrease of the total usage of antibacterial drugs in fish farming. The usage of antibacterials increased from 5 g/tonne in 1988 to 14 g/tonne in 1990, and decreased to 4 g/tonne in 1993. The amount of antibiotics used in Swedish fish farming decreased at the beginning of the 1990s to stabilize around 50 kg of active substance per year between 1998 and 2003 (Swedish Veterinary Antimicrobial Resistance Monitoring, 2003). The use of antibiotics in 2003 is estimated to less than 2 g/ton of fish produced. The use of antibiotics for Arctic char (*Salvelinus alpinus*) is however 15 times higher than for rainbow trout.

During an environmental review of the British Columbia salmon farming industry, it was shown that in 1997 it required 156 g of antibiotics to produce one metric tonne of Atlantic salmon (Fraser et al., 2004). Oxytetracycline accounted for 90% of drug use in British Columbia, prescribed at the dose of 75 to 100 mg/kg of body weight, thus partially explaining this amount of drugs per tonne of fish.

In France, The National Agency for Veterinary Medicinal Products (AFSSA-ANMV) publishes since 1999 a yearly survey on sales of antibiotics for veterinary use, as declared by pharmaceutical companies (Chevance and Moulin, 2007). For farmed fish, the estimated weights of active principles sold were of 6.40 tonnes in 1999, 5.87 tonnes in 2002 and 5.55 tonnes in 2006. A recent survey on prescriptions in French fish farms showed that antibiotics mostly used in 2005 were trimethoprim-sulfadiazine (36%), oxolinic acid (22%), flumequine (19%), oxytetracycline (15%) and florfenicol (8%) (Guichard, 2005).

In Vietnam, imports of drugs for aquatic use increased from approximately 640 tonnes in 2001 to 85,500 tonnes in 2004, from Thailand, the USA, India, Taiwan, Germany, Hong-Kong, Indonesia, France and China (Van, 2005). Marine fish farmers use at least eight antibiotics for disease treatment. Shrimp hatcheries use 5-19 different antibiotics during each production cycle.

No mention of amounts of antibiotics used for aquaculture in other Asian countries could be found (Arthur et al., 2000). There are comparisons (Table 5.3) however of estimated expenditures (in $ US/ha/yr) on chemicals, excluding fertilizers, but including antiparasitics and disinfectants, in several countries.

Table 5.3 Cost of chemotherapy ($ US/ha/yr) in various Asian countries

		Bangladesh	Cambodia	Nepal	Pakistan	Sri Lanka	Vietnam
Carp	Extensive	0.7		12.6	0.7		0.1
	Semi-extensive	7.3	0.95	10.9	7.8		2.9
Shrimps	Extensive	4.5	0			15.3	2.3
	Semi-extensive	477.7				235.7	44.6
	Intensive		791.9			276.6	

DEFINITION AND MECHANISMS OF RESISTANCE

Smith et al. (1994) defined that a strain or a species is termed "resistant" if it has the ability to function, survive, or persist in the presence of higher concentrations of an antimicrobial agent than the members of the parental population from which it emerged, or from other species respectively. Cross-resistance refers to the fact that resistance to one antimicrobial compound within a class of antimicrobials often confers resistance to other members of the same class (EUCAST, 2000).

Besides this microbiological definition, others definitions are given by Acar and Röstel (2001) depending on the scientific discipline and the goals involved.

- Clinical definition: the bacteria survive an adequate treatment with an antibiotic
- Pharmacological definition: the bacteria survive a range of concentrations expressing the various amounts of an antibiotic present in the different compartments of the body when the antibiotic is administered at the recommended dose
- Epidemiological definition: any group of bacterial strains, which can be distinguished from the normal (Gauss) distribution of minimum inhibitory concentrations to an antibiotic

The development of antibiotic resistance in bacteria is mainly based on the selective pressure exerted by use of antibiotics and the presence of resistance genes (White and McDermott, 2001).

The bacterial resistance is based on different mechanisms, achieving either:

- Antibiotic inactivation or destruction by enzymes, e.g. beta-lactamase (beta-lactams), esterase (macrolides), phosphorylase (aminoside, macrolides), acetyltransferase (chloramphenicol)
- Active antibiotic exclusion out of the cell (tetracyclines, macrolides, phenicols, quinolones, beta-lactams)
- Change of the permeability of the cell wall which limits the entry of the antibiotic into the cell (tetracyclines, phenicols, beta-lactams)
- Expression of a variant of the molecule targeted by the antibiotic having a low binding affinity with the antibiotic (sulphonamides, tetracyclines, macrolides, beta-lactams, fluoroquinolones)
- Creation of an altered enzymatic pathway instead of those targeted by the antimicrobial
- All these mechanisms aim to limit the access of the particular antibiotic to its target site in the bacteria.

The resistance phenotype could be:

- Inherent to a genus and called intrinsic resistance, in the case of the absence of affinity between the antimicrobial agent and its target site in the cell, or the inability of the antibiotic to enter the cell, or the absence of the target site in the bacteria. This is a stable genetic property encoded in the chromosomal genes, or
- Acquired if a susceptible organism to a particular antibiotic acquires the genetic material encoding an alternative antibiotic-resistant target molecule. This occurs through:
  - The mutation in the chromosomal genes encoding the targets. For full resistance to develop, all genes involved must have mutations, which lower the frequency of such an event in occurring. But, it could occur in combination with the other following mechanisms. The transfer of this resistance will persist down the generations as cells divide
  - The capture of a new gene on plasmids (conjugation), or from a bacteriophage (transduction), or by the introduction of naked DNA (transformation)
  - The intragenic recombination of the chromosomal gene encoding the sensitive target molecule with related genes, resulting in novel encoded resistant proteins (mosaic genes) (Maiden, 1998)

Plasmid is self-replicating extrachromosomal DNA. Integrons are mobile DNA elements, which are able to incorporate single, or groups of antibiotic resistance genes contained within cassette structures by site-specific recombination and are found in both chromosomal and extrachromosomal DNA. Among the four known classes, the class 1 integrons have been reported to be important for mobilization of antibiotic resistance. The gene cassettes of integrons constitute the smallest mobile elements known and they can be shared between the various integrons. Multiple gene cassettes can be inserted into a single integron. Class 1 integrons may be carried on R plasmids or in transposons which are DNA elements that are able to move between various genetic structures whether it is within the same bacteria or bacteria in another taxa. They can be transferred through conjugation, transformation or transduction mentioned earlier (Roe and Pillai, 2003).

These mechanisms of mobility participate to the horizontal spread of resistance genes between cells and confer a multiple antimicrobial resistant phenotype to the bacteria. Several resistance determinants in fish bacteria have been reported to be carried by transferable genetic elements such as plasmids, transposons or integrons (Sandvang et al., 1997; Adams et al., 1998; Rosser and Young, 1999; Yoo et al., 2003; Barlow et al., 2004). Mobile DNA elements-encoded resistance determinants were often found in fish or water-associated bacteria as for tetracyclines, chloramphenicol, sulphonamide, trimethoprime or β-lactams resistance (Rosser and Young, 1999; Chopra and Roberts, 2001; Schmidt et al., 2001; Sorum and L'Abee-Lund, 2002; Barlow et al., 2004). Integron-associated antibiotic resistance genes were detected in several Gram-negative bacteria such as *Escherichia coli, Proteus* sp., *Aeromonas* sp., *Morganella morganii, Shewanella* spp.

ANTIBIOTIC RESISTANCE IN AQUACULTURE

Resistance to a variety of antibiotics has been largely reported in bacteria isolated from aquaculture environments such as the water body, sediments and fish (Table 5.4). Antibiotic resistance among *Vibrio* and *Aeromonas* strains has been found higher in the shrimp hatcheries than in the *Penaeus monodon* culture ponds, suggesting use of antibiotics in the hatcheries rather than in the farms (Vaseeharan et al., 2005). In many studies, the bacterial resistance levels were correlated to the pattern of antimicrobial use in the farms (McPhearson et al., 1991; Spanggaard et al., 1993; Guardabassi et al., 2000; Tendencia and Leobert, 2002). The analysis of resistance susceptibility of 123 bacterial isolates from water, sediment and different fish farms (catfish, tilapia, common carp and gouramy) in five provinces of the Mekong River showed that 90% of the isolates were resistant to tetracycline, 76% to ampicillin, 100% to chloramphenicol, 65% to nitrofurantoin and 89% to trimethoprim-sulfamethoxazole (Phuong et al., 2005). High resistance to oxytetracycline was found among 103 bacterial strains isolated from freshwater Chilean salmon farms at four different locations since this antibiotic is used for curative purposes to treat a variety of salmon diseases, but also as a preventive treatment

Table 5.4   Reported antibiotic resistant bacteria isolated from aquaculture environments

Multiple antibiotic resistance patterns (a-b) Individual antibiotic resistance (a,b)	Bacterial species	Aquaculture origin	Reference
Amx- Am- Ctx-C- Ffc- Te- Otc- -E-Fx-Sxt*.	Gram negative bacteria	Salmon farms/ Chile	Miranda and Zemelman, 2002b
Otc-Ak-Am-Cb-Cf. Otc-Ak-Am-C-Ge-Net-Nf-Pef.	*Vibrio* spp.	Penaeid shrimps/ Mexico	Molina-Aja et al., 2002
Am-C-Na-Ge-Te-Nf-To.	*Salmonella typhimurium*	Finfish/India	Ruiz et al., 1999
Am-C-E-Km-Sxt-Te.	*Aeromonas hydrophila*	Tilapia/Malaysia	Son et al., 1997
Am-C-Nf-Otc-Sxt-Na.	*Enterobacter gergoviae*	Catfish/Vietnam	Sarter et al., 2007
Oa-Sxt-Otc-Amx.	*Flavobacterium psychrophilum*, Aeromonads	Rainbow trout/ Denmark	Schmidt et al., 2000
Am-Te-Sxt.	*Escherichia coli*	Seabob shrimps (*Xiphopenaues kroyeri*)/Brazil	Teophilo et al., 2002
Am, Amx, Ar, C, E, Enr, Ffc, Ge, Km, Oa, Otc, P, Sm, Sxt, Te.	*Photobacterium damselae* subsp. *piscicida*	Fish/Japan	Thyssen and Ollevier, 2001
Am,Cht,Te,Km,Na, Nm.	*Vibrio* spp., *Aeromonas* spp.	Water *Penaeus monodon* hatcheries and ponds/India	Vaseeharan et al., 2005
Te-C-Km-Sa-Sm. C-Te-Oa-Sm.	*Escherichia coli*	Fish/Korea	Yoo et al., 2003

Ak: amikacin, Am: ampicillin, Amx: amoxicillin, Ar: flumequine, Cb: carbenicillin, Cf: cephalotin, Ctx: Cefotaxine, C: chloramphenicol, Cht: Chlortetracycline, Enr: enrofloxacin, E: erythromycin, Ffc: florfenicol, Fx: Furazolidone, Ge: gentamicin, Km: kanamycin, Net: netilmicin, Nf: nitrofurantoin, Oa: oxolinic acid, Otc: oxytetracycline, Pef: pefloxacine, P: penicillin, Sa: sulphonamide, Sm: streptomycin, Sxt: trimethoprim-sulfamethoxazole, Sxt: cotrimoxazole, Te: tetracycline, To: tobramycin

before the smoltification process (Miranda and Zemelman, 2002a). Other studies have demonstrated rises in the frequencies of resistance to agents that were not those used on the farm (Austin, 1985). However, the correlation between resistance incidence and the antibiotic concentration is not that evident because high frequencies of resistant microflora have also been observed in the water body which has had no recent contact with antimicrobial agents. For instance, resistance to oxytetracycline has been reported in situations where this antibiotic was not present in the aquatic environment. It seems that non-specific mechanisms might be involved and which are related to fish feed accumulation on the sediments as a result of over-feeding (Kapetanaki et al., 1995; Vaughan et al., 1996). High incidence of oxytetracycline-resistant were found in aeromonads (69%) and in flavobacteria (72%) populations, although this antibiotic has been very rarely used in the Danish aquaculture during the past five years prior to the assays (Schmidt et al., 2000). Unusual high incidence of oxytetracycline-resistant (72%), trimethoprim-sulfadiazine resistant (44%) and multiresistant (50%) aeromonads was observed at the inlet of one of the tested farms, as this inlet is located near the outlet of a previous fish farm. This result shows the importance of the environmental impact of the antibiotic resistance in aquaculture.

The high level of multi-resistance in these bacteria may be due to an adaptation to the "fluctuating antibiotic environment" as proposed by Baquero et al. (1998). Such an environment could be easily achieved by intensive farming practices, or by the use and abuse of different drug families or broad spectrum molecules. In response to this variable antibiotic pressure, the selected bacterial strains are those, which possess multipurpose or multiple mechanisms to survive. Multipurpose mechanism is well illustrated by the multidrug efflux leading to the simultaneous pumping out of different antibiotic molecules. The drug efflux by membrane transporters is the only mechanism identified for multidrug resistance in bacteria (George, 1996). AcrAB efflux system for instance, is responsible for multiple antibiotic resistance phenotype in an *Escherichia coli* mutant which became resistant to several antimicrobials including tetracycline, chloramphenicol, ampicillin, nalidixic acid and rifampicin (Okusu et al., 1996). Several efflux pumps associated with increasing levels of quinolone resistance have been characterized in *E. coli* and others Gram-negative bacteria as well (Ruiz, 2003).

Associated-resistance has been reported for streptomycin and trimethoprim-sulfamethoxazole in Enterobacteriaceae (Chiew et al., 1998); for oxolinic acid and oxytetracycline in *Aeromonas salmonicida* (Barnes et al., 1990); for ampicillin and oxytetracycline in Gram-negative bacteria from cultured catfish (McPhearson et al., 1991) and for oxytetracycline and trimethoprim-sulfamethoxazole in *Aeromonas salmonicida* (Schmidt et al., 2001). Oxolinic acid resistant *Acinetobacter* spp. isolates from a freshwater trout farm were more resistant to oxytetracycline when compared to oxolinic acid sensitive strains (Guardabassi et al., 2000).

Although the oxolinic acid and oxytetracycline drugs have different mechanisms of action, it has been shown that the alteration in outer membrane proteins of oxolinic acid-resistant mutants could benefit to tetracyclines, as well as chloramphenicol and some β-lactams (Baquero, 1990; Sanders et al., 1984). On the other hand, the expression of an outer membrane protein (OMP54) in *Stenotrophomonas maltophila* was associated with an increase of the minimum inhibitory concentration (MIC) for tetracycline, chloramphenicol and quinolones, but not for β-lactams (Alonso and Martinez, 1997).

Such associated-resistance is very important to consider in intensive farming conditions because it could lead to particular resistance patterns without a direct use of the corresponding drugs by the farmer (EUCAST, 2000), as mentioned earlier.

The second mechanism that may be involved under the antibiotic pressure is the multiple mechanism of resistance, for example the collection of several β-lactamases in the cell or in a single plasmid (Kotra and Mobashery, 1998) to provide the bacteria with resistance to different β-lactams. This group is not stable to β-lactamases of either Gram-positive or Gram-negative bacteria. β-lactams inhibits the cell wall synthesis by preferentially binding to specific penicillin-binding proteins (PBP) that are located inside the bacterial cell wall. Considering that these PBP vary among different bacterial species, the intrinsic activity of β-lactams will depend on their ability to bind the necessary PBP. So depending on availability and significant quantity of β-lactamase and/or PBP, particular organisms could be intrinsically resistant or sensitive. Among Enterobacteriaceae, such diversity exists, and it has been argued that each species of enterobacteria contains its own chromosomally encoded β-lactamase (Livermore, 1998; Stock and Wiedemann, 2001). However, Stock and Wiedemann (2001) did not detect any α-lactamase activity in *Edwardsiella hoshinae* and *E. ictaluri*. Regarding Vibrionaceae species, such diversity has been reported as well for fish and shrimps bacteria (Castro-Escarpulli et al., 2003; Dalsgaard et al., 1994; Molina-Aja et al., 2002). These observations underline the importance of several factors including the fish species, disease history within the farm, farming practices, water quality, animal density that should be considered before comparing different fish farming.

TRANSFER OF THE RESISTANCE TO HUMAN PATHOGENS

Pathogen agents from aquacultured fish through food ingestion or direct contact can contaminate humans (Table 5.5). The resistance of human pathogens is a serious threat for public health since bacterial resistance can compromise the efficiency of targeted antibiotic treatment. Actually, several studies have reported transfer of antibiotic genes between enteric bacteria or clinical strains and fish microflora (Andersen and Sandaa, 1994; Goni-Urriza et al., 2000; Rhodes et al., 2000; Furushita et al., 2003). Kruse and Sorum (1994) demonstrated that conjugation and transfer of R plasmids can occur between bacterial strains of human, animal, and fish origins that are unrelated either evolutionarily or ecologically even in the absence of antibiotics. In 1991, an epidemic of *Vibrio cholerae* 01 infections

Table 5.5 Bacteria of significance as human pathogens isolated from fish or their immediate environment and the preferred antimicrobial agents for the treatment of the infections that they cause (Smith et al., 1994)

Pathogen	Disease	Preferred treatment
Pathogens primarily entering the host via the mouth		
Salmonella spp.	Food poisoning	Ampicillin, amoxicillin, trimethoprim-sulfamethoxazole
Vibrio parahaemolyticus	Food poisoning	No mention
Campylobacter jejeuni	Gastroenteritis	Erythromycin
Aeromonas hydrophila	Diarrhoea	Ciprofloxacin, norfloxacin, trimethoprim-sulfamethoxazole
Aeromonas hydrophila	Septicaemia	Cephalosporins
Plesiomonas shigelloides	Gastroenteritis	Trimethoprim-sulfamethoxazole, tetracycline, ciprofloxacin
Edwardsiella tarda	Diarrhoea	Ampicillin
Pathogens primarily entering the host via the skin		
Pseudomonas aeruginosa	Wound infection	Aminoglycoside + antipesudomonad penicillin
Pseudomonas fluorescens	Wound infection	No mention
Mycobacterium fortuitum	Mycobacteriosis	Amikacin + cefoxitin
Mycobacterium marinum	Mycobacteriosis	Rifampicin + ethambutol, trimethoprim-sulfamethoxazole
Erysipelothrix rhusiopathiae	Erysipelioid	Penicillin
Leptospira interrogans	Leptospirosis	Penicillin G, ampicillin

affected Latin America; the epidemic strain in Latin America was susceptible to the 12 antimicrobial agents tested, but in coastal Ecuador, the epidemic strain had become multiresistant. The cholera epidemic in Ecuador began among persons working on shrimp farms. The multi-resistant phenotype was present in non-cholera *Vibrio* infections that were pathogenic to the shrimp. The resistance may have been transferred to *V. cholerae* 01 from other vibrios (Weber et al., 1994).

Evidence has been given that the aquaculture and human compartments of the environment are interactive (Rhodes et al., 2000). Tetracycline resistance-encoding plasmid has been transferred for instance between *Aeromonas* spp. and *E. coli* and between human and aquaculture environments in distinct geographical locations. Another study showed that an increase of both tetracycline and beta-lactams resistance levels have been detected in Enterobacteriaceae isolates from respectively, 12.5% to 24.3% and 0% to 20.5% in a stream before and after the stream passed a wastewater discharge station (Goni-Urriza et al.,

2000). Demersal and pelagic fish captured in water close to urban sewage carried an important proportion of gills and intestines bacteria that were resistant to ampicillin, tetracycline, streptomycin, chloramphenicol, nalidixic acid and nitrofurantoin (Miranda and Zemelman, 2001). Antibiotic-resistant bacteria have been isolated from the carcasses of catfish from the retail market and from different sources of imported frozen seafood (DePaola and Roberts, 1995; Zhao et al., 2003) A strain of *Salmonella* Derby isolated from those frozen anchovies imported from Cambodia was multi-resistant to ampicillin, amoxicillin-clavulanic acid, sulfamethoxazole, chloramphenicol, tetracycline and trimethoprim-sulfamethoxazole. Such bacteria can be transferred during food preparation at home or by handling in the market.

Since the prohibition of chloramphenicol, a broad spectrum antimicrobial agent, in animal use in most countries in the mid-1990s, fluorinated derivative, florfenicol had become an important drug in aquaculture. It has been reported, among aquatic animal bacteria (Taiwan 1995 till 1998), that florfenicol-resistant isolates could also resist to chloramphenicol, but that chloramphenicol-resistant isolates did not always resist florfenicol (Ho et al., 2000). Florfenicol resistance gene, *floR*, has been reported in Japan in marine fish pathogen *Pasteurella piscicida* (Kim and Aoki, 1996). A similar gene, flo_{St} mediating resistance to chloramphenicol and which also confers resistance to florfenicol, has been found in a major human pathogen *Salmonella enterica* serotype *typhimurium* DT104 (R-type ACSSuT), a resistant strain to five drugs including ampicillin, chloramphenicol, streptomycin, sulfonamides, and tetracycline (Bolton et al., 1999). Molecular studies based on the nucleotide sequence analysis suggested a potential horizontal transfer of some antimicrobial resistance determinants to *S. typhimurium* DT 104 R-type ACSSuT from other bacteria (Cloeckaert et al., 2000; Lawson et al., 2002).

LABORATORY METHODS FOR THE ANTIBIOTIC SUSCEPTIBILITY TESTING

Isolates subjected to antimicrobial susceptibility testing must be pure, and correctly identified to the genus or the species level. Different methods are available for laboratory testing which are mainly based on *disc diffusion*, *broth dilution* (micro or macro-dilutions referring to the culture volume, either 2 ml or using micro-titration plates respectively) and *agar dilution*. All these methods need standardization of the assay parameters such as the composition, volume, pH, and humidity of the medium, the temperature of incubation, the preparation of the inocula.

The *disc diffusion technique* consists in the diffusion of a particular antibiotic of a specified concentration from disks, tablets or strips into a solid culture plate inoculated with a standard bacterial culture. This method is mostly used for routine measurement of bacterial resistance, because it is easy to perform for screening different antibiotics and does not require expensive equipment. Depending on the diameter of the zone of inhibition, the strains will be classified as sensitive, intermediate or resistant, once breakpoints are established.

The *dilution methods* are more suitable to determine the Minimum Inhibitory Concentration (MIC: the lowest concentration of the antibiotic at which the microorganism does not demonstrate visible growth). The *broth dilution* uses standardized liquid medium inoculated with a specified concentration of a pure culture, which is tested against different concentrations of the antibiotic. This technique allows the determination of the Minimum Bactericidal Concentration (MBC: the lowest concentration of the antibiotic at which no bacterial growth occurs on the agar plates) as well. In *agar dilution*, the antibiotic is included in the solid medium in the geometrical progression of concentrations, and a specified culture is then applied onto the surface of the plate. This last technique is recommended for fastidious species. Numerous media have been published for antimicrobial susceptibility testing, however the Clinical and Laboratory Standards Institute (CLSI, 2005a, b) and *Comité de l'Antibiogramme de la Société Française de Microbiologie* (SFM, 2007) recommend the use of Mueller-Hinton medium. CLSI has proposed specific guidelines for bacteria isolated from aquatic animals. Dalsgaard (2001) has reviewed the different media that have been reported for *agar disc diffusion method* and MIC for susceptibility testing of fish pathogens, highlighting the need for harmonization in this field. Some studies added sodium chloride or seawater to the medium for a better growth of halophilic strains, but some others have reported that the ions present in seawater, particularly Mg^{2+} and Ca^{2+}, inhibit the antimicrobial activity of the antibiotic (Smith, 1998). Methodological problems rise for setting breakpoint concentrations taking into account the *in vitro* MIC data, the pharmacokinetics of the agents and the clinical efficacy in the treatment of fish diseases.

A recent study (Smith, 2006) which reported a summary of the breakpoints being used in the 25 surveyed laboratories using disc diffusion protocols to assess the susceptibility of Gram-negative, Group 1 organisms (CLSI, 2005b) isolated from finfish, showed important variations in the breakpoints currently in use (Table 5.6).

The *ad hoc* group on antimicrobial resistance of the World Organization for Animal Health (OIE) has developed a guideline on the standardization and harmonization of laboratory methodologies used for the detection and quantification of antimicrobial resistance (White et al., 2001). The standardization of techniques is an essential issue to achieve the

Table 5.6 Frequency distribution of breakpoints (mm) currently in use in responding laboratories for the nine most commonly tested antimicrobial agents (Smith, 2006)

Zone (mm)	AMX 25 µg		ENR 5 µg		ERY 15 µg		FLO 30 µg		FLU 30 µg		OXA 2 µg		OTC 30 µg		SFO 25 µg		SFT 25 µg	
	S*	R**	S	R	S	R	S	R	S	R	S	R	S	R	S	R	S	R
6–7																1		
8–9																		3
10–11			1				1			2	1	6		4		3		7
12–13		3	2			3	5		3		2		1	3			1	4
14–15							1		1		2	1	10					3
16–17	1		1	5	2		1	3	1	2	1		2	1	3		9	1
18–19	1		2	1	3		5	1	3		1	1	12	2			6	
20–21			1				4	2	2	3			2	1			1	
22–23	1		5		1		1						1		1			2
24–25							2	1			1	1					1	3
26–27									2		1							
28–29			1										1				3	1
30–31			1				1		2				1					
32–33							1		1				1				1	
34–35													1				2	
36–37																		
38–39													1					
40–41																		
42–43																		
44–45									1	1								

AMX: amoxycillin, ENR: enrofloxacin, ERY: erythromycin, FLO: florfenicol, FLU: flumequine, OTC: oxytetracycline, OXA: oxolinic acid, SFO: ormetoprim/sulfadimethoxine, SFT: trimethoprim/sulfmethoxazole

S* indicates breakpoints used to determine sensitivity.

R** indicates breakpoints used to determine resistance.

reproducibility of the results and to allow data comparison among different countries.

CONCLUSION AND RECOMMENDATIONS

Reducing the selective pressure exerted on the microbial communities by restricting the use of antibiotics seems essential to preserve the efficiency of antibiotics and keep the production costs low for farmers. But, it may be speculated that such an approach might be not sufficient in eliminating antibiotic resistance as much as (i) resistance might result from gratuitous selection by others agents and (ii) resistance might be carried on mobile genetic elements. For this reason, alternatives strategies for disease control such as application of probiotic lactic acid bacteria and vaccines should be developed as well. For instance, the Norwegian aquaculture has registered a significant reduction in the use of antibiotics, up to 98% from 1987 to 2004, following the introduction of effective vaccines which was concomitant with better production management as well as general improvement of the hygiene (Lillehaug et al., 2003; Lunestad and Grave, 2005). At the same time, salmon and trout production has risen more than 10 folds.

Several international consultations and reports (GESAMP, 1997; FAO, 2005; Alday et al., 2006) have addressed rich and accurate recommendations encouraging better management practices of antibiotics in aquaculture at the different levels of responsibilities (governance, farmers, chemical industries, scientists). Antibiotics in aquaculture should be restricted to therapeutic purposes only, and prophylactic use avoided. OIE established a series of hygiene measures available for aquaculture farmers. Implementation of codes of practices environmentally sound for a responsible aquaculture, taking into account these guidelines, is essential to prevent and control resistance of bacteria.

In many countries of the world, among which many are important aquaculture-producing countries, no veterinary medicinal product is registered for use in aquatic animals, resulting in an uncontrolled, off-label use of antibiotics. Authorities should encourage the registration of products specifically designed for use in aquatic animals, and evaluate their quality, safety, efficacy and environmental impact before delivering marketing authorization. In this goal for instance, the European Medicines Evaluation Agency is currently revising its guidelines on the efficacy of veterinary medicinal products for use in aquatic species, which is due to be released in 2009.

ABBREVIATIONS

CLSI   Clinical and Laboratory Standards Institute
MBC   Minimum bacterial concentration
MIC   Minimum inhibitory concentration
OIE   World Organization for Animal Health
OMP   Outer membrane protein
PBP   Penicillin binding protein
VMP   Veterinary Medicinal Products

REFERENCES

Acar, J. and Röstel, B. (2001). Antimicrobial resistance: an overview. Revue Scientifique et. Technique de l' Office International des Epizooties 20: 797-810.

Adams, C.A., Austin, B., Meaden, P.G. and McIntosh, D. (1998). Molecular characterization of plasmid-mediated oxytetracycline resistance in *Aeromonas salmonicida*. Appl. Environ. Microbiol. 64(11): 4194-4201.

Alday, V., Guichard, B., Smith, P. and Uhland, C. (2006). Towards a risk analysis of antimicrobial use in aquaculture, Joint FAO/WHO/OIE Expert Consultation on Antimicrobial Use in Aquaculture and Antimicrobial Resistance. Seoul, South Korea, June 13-16, 2006.

Alonso, A. and Martinez, J.L. (1997). Multiple antibiotic resistance in *Stenotrophomonas maltophilia*. Antimicrob. Agents Chemother. 41(5): 1140-1142.

Andersen, S.R. and Sandaa, R.A. (1994). Distribution of tetracycline resistance determinants among Gram-negative bacteria isolated from polluted and unpolluted marine sediments. Appl. Environ. Microbiol. 60(3): 908-912.

Aoki, T. (1988). Drug-resistant plasmids from fish pathogens. Microbiol. Sci. 5(7): 219-223.

Arthur, J.R., Lavilla-Pitogo, C.R. and Subasinghe, R.P. (2000). Use of chemicals in aquaculture in Asia, Proceedings of the Meeting on the Use of Chemicals in Aquaculture in Asia, 20-22 May, 1996, Tigbauan, Iloilo, Philippines.

Austin, B. (1985). Antibiotic pollution from fish farms: effects on aquatic microflora. Microbiol. Sci. 2(4): 113-117.

Baquero, F. (1990). Resistance to quinolones in Gram-negative microorganisms: mechanisms and prevention. Eur. Urol. 17: 3-12.

Baquero, F., Negri, M.-C., Morosini, M.-I. and Blazquez, J. (1998). Antibiotic-selective environments. Clin. Infect. Dis. 27(Suppl 1): S5-11.

Barlow, R.S., Pemberton, J.M., Desmarchelier, P.M. and Gobius, K.S. (2004). Isolation and characterisation of integron-containing bacteria without antibiotic selection. Antimicrob. Agents Chemother. 48(3): 838-842.

Barnes, A.C., Lewin, C.S., Hastings, T.S. and Amyes, S.G.B. (1990). Cross resistance between oxytetracycline and oxolinic acid in *Aeromonas salmonicida* associated with alterations in outer membrane proteins. FEMS Microbiol. Lett. 72(3): 337-339.

Björnerot, L., Franklin, A. and Tysén, E. (1996). Usage of antimicrobial and antiparasitic drugs in animals in Sweden between 1988 and 1993. Vet. Rec. 139: 282-286.

Bolton, L.F., Kelley, L.C., Lee, M.D., Fedorka-Cray, P.J. and Maurer, J.J. (1999). Detection of multidrug-resistant *Salmonella enterica* Serotype typhimurium DT104 based on a gene which confers cross-resistance to florfenicol and chloramphenicol. J. Clin. Microbiol. 37(5): 1348-1351.

Cárdenas, F.C. (2003). Antibioticos y acuicultura – Un análisis de sus potenciales impactos para el medio ambiante, la salud humana y animal en Chile. Análisis de Políticas Públicas 17: 1-16

Carneiro, L.A.M., Silva, A.P.S., Merquior, V.L.C. and Queiroz, M.L.P. (2003). Antimicrobial resistance in Gram-negative bacilli isolated from infant formulas. FEMS Microbiol. Lett. 228: 175-179.

Castro-Escarpulli, G., Figueras, M.J., Aguilera-Arreola, G., Soler, L., Fernandez-Rendon, E., Aparicio, G.O., Guarro, J. and Chacon, M.R. (2003). Characterisation of *Aeromonas* spp. isolated from frozen fish intended for human consumption in Mexico. Int. J. Food Microbiol. 84(1): 41-49.

Chevance, A. and Moulin, G. (2007). Suivi des ventes de médicaments vétérinaires contenant des antibiotiques en France en 2006. Report of the National Agency for Veterinary Medicinal Products, 38 p. www.anmv.afssa.fr/antibioresistance/Rapport_ATB_final%202006.pdf

Chiew, Y.-F., Yeo, S.-F., Hall, L.M.C. and Livermore, D.M. (1998). Can susceptibility to an antimicrobial be restored by halting its use? The case of streptomycin versus Enterobacteriaceae. J. Antimicrob. Chemother. 41: 247-251.

Chopra, I. and Roberts, M. (2001). Tetracycline antiobiotics: mode of action, applications, molecular biology, and epidemiology of bacterial resistance. Microbiol. Mol. Biol. Rev. 65(2): 232-260.

Cloeckaert, A., Sidi Boumedine, K., Flaujac, G., Imberechts, H., D'Hooghe, I. and Chaslus-Dancla, E. (2000). Occurrence of a *Salmonella enterica* serovar Typhimurium DT 104-like antibiotic resistance gene cluster including the *flo*R gene in *S. enterica* serovar Agona. Antimicrob. Agents Chemother. 44: 1359-1361.

CLSI (Clinical Laboratory Standards Institute) (2005a). Clinical and Laboratory Standards Institute. Methods for broth dilution susceptibility testing of bacteria isolated from aquatic animals; Proposed guideline. CLSI document M49-P.

CLSI (Clinical Laboratory Standards Institute) (2005b). Clinical and Laboratory Standards Institute. Methods for antimicrobial disk susceptibility testing of bacteria isolated from aquatic animals; Proposed guideline. CLSI document M42-P.

Cruz-Lacierda, E.R. and de la Peña L.D. (1996). The use of chemicals in aquaculture in the Philippines. In: Use of chemicals in Aquaculture in Asia: Proceedings of the Meeting on the Use of Chemicals in Aquaculture Asia. 20-22 May 1996, Tigbauan, Iloilo, Philippines.

Dalsgaard, I. (2001). Selection of media for antimicrobial susceptibility testing of fish pathogenic bacteria. Aquaculture 196(3-4): 267-275.

Dalsgaard, I., Nielson, B. and Larsen, J.L. (1994) Characterization of *Aeromonas salmonicida* subsp. *salmonicida*: a comparative study of strains of different geographic origins. J. Appl. Bacteriol. 77(1): 21-30 .

Dalsgaard, A., Forslund, A., Sandvang, D., Arntzen, L. and Keddy, K. (2001). *Vibrio cholerae* O1 outbreak isolates in Mozambique and South Africa in 1998 are multiple-drug resistant, contain the SXT element and the *aadA2* gene located on class 1 integrons. J. Antimicrob. Chemother. 48: 827-838.

DePaola, A. and Roberts, M.C. (1995). Class D and E tetracycline resistance determinants in gram-negative bacteria from catfish ponds. Mol. Cell. Probes 9(5): 311-313.

EUCAST. (2000). European Committee for Antimicrobial Susceptibility Testing (EUCAST) of the European Society of Clinical Microbiology and Infectious Diseases (ESCMID). Terminology relating to methods for the determination of susceptibility of bacteria to antimicrobial agents. Clin. Microbiol. Infect. 6(9): 503-508.

FAO. (2005). Fisheries Technical Paper No. 469: 97p. Rome, Italy.

FAO. (2006). State of World Aquaculture 2006, by R. Subasinghe. FAO Fisheries Technical Paper No. 500. Rome, Italy.

FAO/NACA/WHO. (1999). Joint Study Group. Food safety issues associated with products from aquaculture: WHO Technical Report Series N° 883.

Fraser, E., Stephen, C., Bowie, W.R. and Wetzstein, M. (2004). Availability and estimates of veterinary antimicrobial use in British Columbia. Can. Vet. J. 45: 39-311.

Furushita, M., Shiba, T., Maeda, T., Yahata, M., Kaneoka, A., Takahashi, Y., Torii, K., Hasegawa, T. and Ohta, M. (2003). Similarity of tetracycline resistance genes isolated from fish farm bacteria to those from clinical isolates. Appl. Environ. Microbiol. 69(9): 5336-5342.

George, A.M. (1996). Multidrug resistance in enteric and other Gram-negative bacteria. FEMS Microbiol. Lett. 139: 1-10.

GESAMP (Joint group of experts on the scientific aspects of marine environmental protection) (1997). Towards safe and effective use of chemicals in coastal aquaculture. Reports and studies No 65: 40 p. Rome: FAO.

Goni-Urriza, M., Capdepuy, M., Arpin, C., Raymond, N., Caumette, P. and Quentin, C. (2000). Impact of an urban effluent on antibiotic resistance of riverine Enterobacteriaceae and *Aeromonas* spp. Appl. Environ. Microbiol. 66(1): 125-132.

Grave, K., Engelstad, M., Søli, N. E. and Håstein, T. (1990). Utilization of antibacterial drugs in salmonid farming in Norway during 1980-1988. Aquaculture 86: 347-358.

Guardabassi, L., Dalsgaard, A., Raffatellu, M. and Olsen, J. E. (2000). Increase in the prevalence of oxolinic resistant *Acinetobacter* spp. observed in a stream receiving the effluent from a freshwater trout farm following the treatment with oxolinic acid-medicated feed. Aquaculture 188: 205-218.

Guichard, B. (2005). A survey on fish pathology activity in France. Poster, 12th International Conference of the European Association of Fish Pathologists, Copenhagen, Denmark, 11-16 September 2005.

Guichard, B. and Lièek, E. (2006). A comparative study of antibiotics registered for use in farmed fish in European countries. Poster, OIE Global Conference on Aquatic Animal Health, Bergen, Norway, 9-12 October 2006.

Hiney, M. and Smith, P. (2000). Oil-adjuvanted furunculosis vaccines in commercial fish farms: a preliminary epizootiological investigation. Aquaculture 190: 1-9.

Ho, S.P., Hsu, T.Y., Che, M.H. and Wang, W.S. (2000). Antibacterial effect of chloramphenicol, thiamphenicol and florfenicol against aquatic animal bacteria. J. Vet. Med. Sci. 62(5): 479-485.

Inglis, V., Cafini, M. and Yoshida, T. (1995). The interaction of trimethoprim and quinolones against gram-negative fish pathogens. J. Appl. Bacteriol. 79(2): 135-140.

Kapetanaki, M., Kerry, J., Hiney, M., O'Brien, C., Coyne, R. and Smith, P. (1995). Emergence, in oxytetracycline-free marine mesocosm, of organisms capable of colony formation on oxytetracycline-containing media. Aquaculture 134: 227-236.

Kim, E. and Aoki, T. (1996). Sequence analysis of the florfenicol resistance gene encoded in the transferable R-plasmid of a fish pathogen, *Pasteurella piscicida*. Microbiol. Immunol. 40(9): 665-669.

Kotra, L.P. and Mobashery, S. (1998). ß-Lactam antibiotics, ß-lactamases and bacterial resistance. Bull. Inst. Pasteur 96: 139-150.

Kruse, H. and Sorum, H. (1994). Transfer of multiple drug resistance plasmids between bacteria of diverse origins in natural microenvironments. Appl. Environ. Microbiol. 60(11): 4015-4021.

Lawson, A.J., Dassama, M.U., Ward, L.R. and Threlfall, E.J. (2002). Multiple resistant *Salmonella enterica* serovar Typhimurium DT 12 and DT 120: A case of MR DT 104 in disguise? Emerg. Infect. Dis. 8(4): 434-436.

Lillehaug, A., Lunestad, B.T. and Grave, K. (2003). Epidemiology of bacterial diseases in Norwegian aquaculture—a description based on antibiotic prescription data for the ten-year period 1991 to 2000. Dis. Aquat. Organ. 53: 115-125.

Livermore, D.M. (1998). ß-Lactamase-mediated resistance and opportunities for its control. J. Antimicrob. Chemotherapy 41(Suppl. D): 25-41.

Lunestad, B.T. and Grave, K. (2005). Therapeutic agents in Norwegian aquaculture from 2000 to 2004: Usage and residue control. Bull. Eur. Assoc. Fish Pathologists 25(6): 284-290.

Maiden, M.C.J. (1998). Horizontal genetic exchange, evolution, and spread of antibiotic resistance in bacteria. Clin. Infect. Dis. 27 (Suppl 1): 12-20.

Markestad, A. and Grave, K. (1997). Reduction of antibacterial drug use in Norwegian fish farming due to vaccination. Develop. Biol. Std. 90: 365-369.

McPhearson, R.M., DePaola, A., Zywno, S.R., Motes, M.L.J. and Guarino, A. (1991). Antibiotic resistance in Gram-negative bacteria from cultured catfish and aquaculture ponds. Aquaculture 99: 203-211.

Miranda, C.D. and Zemelman, R. (2001). Antibiotic resistant bacteria in fish from the Concepcion Bay, Chile. Mar. Pollut. Bull. 42(11): 1096-1102.

Miranda, C.D. and Zemelman, R. (2002a). Bacterial resistance to oxytetracycline in Chilean salmon farming. Aquaculture 212: 31-47.

Miranda, C.D. and Zemelman, R. (2002b). Antimicrobial multiresistance in bacteria isolated from freshwater Chilean salmon farms. Sci. Total Environ. 293: 207-218.

Molina-Aja, A., Garcia-Gasca, A., Abreu-Grobois, A., Bolan-Mejia, C., Roque, A. and Gomez-Gil, B. (2002). Plasmid profiling and antibiotic resistance of *Vibrio* strains isolated from cultured penaeid shrimp. FEMS Microbiol. Lett. 213(1): 7-12.

Okusu, H., Ma, D. and Nikaido, H. (1996). AcrAB efflux pump plays a major role in the antibiotic resistance phenotype of *Escherichia coli* multiple-antibiotic-resistance (Mar) mutants. J. Bacteriol. 178(1): 306-308.

Pathak, S.C., Ghosh, S.K. and Palanisamy K. (1996). The use of chemicals in aquaculture in India. In: Use of Chemicals in aquaculture in Asia: Proceedings of the Meeting on the Use of Chemicals in Asia. 20-22 May 1996, Tigbauan, Iloilo, Philippines.

Phuong, N.T., Oanh, D.T.H., Dung, T.T. and Sinh, L.X. (2005). Bacterial resistance to antimicrobials use in shrimps and fish farms in the Mekong Delta, Vietnam. Paper presented at the Proceedings of the International Workshop on "Antibiotic Resistance in Asian Aquaculture Environments", Chiang May, Thaïland.

Randall, L.P., Cooles, S.W., Osborn, M.K., Piddock, L.J.V. and Woodward, M.J. (2004). Antibiotic resistance genes, integrons and multiple antibiotic resistance in thirty-five serotypes of *Salmonella enterica* isolated from humans and animals in the UK. J. Antimicrob. Chemother. 53: 208-216.

Rhodes, G., Huys, G., Swings, J., McGann, P., Hiney, M., Smith, P. and Pickup, R. W. (2000). Distribution of oxytetracycline resistance plasmids between aeromonads in hospital and aquaculture environments: implication of Tn*1721* in dissemination of the tetracycline resistance determinant Tet A. Appl. Environ. Microbiol. 66(9): 3883-3890.

Roe, M.T. and Pillai, S.D. (2003). Monitoring and identifying antibiotic resistance mechanisms in bacteria. Poultry Sci. 82: 622-626.

Rosser, S.J. and Young, H.K. (1999). Identification and characterisation of class 1 integrons in bacteria from an aquatic environment. J. Antimicrob. Chemother. 44: 11-18.

Ruiz, J. (2003). Mechanism of resistance to quinolones: target alterations, decreased accumulation and DNA gyrase protection. J. Antimicrob. Chemother. 51: 1109-1117.

Ruiz, J., Capitano, L., Nunez, L., Castro, D., Sierra, J. M., Hatha, M., Borrego, J.J. and Vila, J. (1999). Mechanisms of resistance to ampicillin, chloramphenicol and quinolones in multiresistant *Salmonella typhimurium* strains isolated from fish. J. Antimicrob. Chemother. 43: 699-702.

Sanders, C.C., Sanders, W.E.J., Goering, R.V. and Werner, V. (1984). Selection of multiple antibiotic resistance by quinolones, beta-lactams, and aminoglycosides with special reference to cross-resistance between unrelated drug classes. Antimicrob. Agents Chemother. 26(6): 797-801.

Sandvang, D., Aarestrup, F. and Jensen, L.B. (1997). Characterisation of integrons and antibiotic resistance genes in Danish multiresistant *Salmonella enterica* Typhimurium DT104. FEMS Microbiol. Lett. 157: 177-181.

Sarter, S., Nguyen, H.N.K., Hung, L.T., Lazard, J. and Montet, D. (2007). Antibiotic resistance in Gram-negative bacteria isolated from farmed catfish. Food Control 18(11): 1391-1396.

Schmidt, A.S., Bruun, M.S., Dalsgaard, I., Pedersen, K. and Larsen, J.L. (2000). Occurrence of antimicrobial resistance in fish-pathogenic and environmental bacteria associated with four Danish rainbow trout farms. Appl. Environ. Microbiol. 66(11): 4908-4915.

Schmidt, A.S., Bruun, M.S., Larsen, J.L. and Dalsgaard, I. (2001). Characterization of class 1 integrons associated with R-plasmids in clinical *Aeromonas salmonicida* isolates from various geographical areas. J. Antimicrob. Chemother. 47: 735-743.

SFM (Comité de l'Antibiogramme de la Société Française de Microbiologie) (2007). Comité de l'antibiogramme de la Société Française de Microbiologie. Communiqué 2007 (http://www.sfm.asso.fr/)

Smith, P. (1998). Towards the establishment of a breakpoint concentration for the determination of resistance to oxolinic acid in marine microflora. Aquaculture 166(3-4): 229-239

Smith, P. (2006). Breakpoints for disc diffusion susceptibility testing of bacteria associated with fish diseases: A review of current practice. Aquaculture 261(4): 1113-1121.

Smith, P., Hiney, M.P. and Samuelson, O.B. (1994). Bacterial resistance to antimicrobial agents used in fish farming: a critical evaluation of method and meaning. Annu. Rev. Fish Dis. 4: 273-313.

Son, R., Rusul, G., Sahilah, A.M., Zainuri, A., Raha, A.R. and Salmah, I. (1997). Antibiotic resistance and plasmid profile of *Aeromonas hydrophila* isolates from cultured fish, Telapia (*Telapia mossambica*). Lett. Appl. Microbiol. 24(6): 479-482.

Sorum, H. and L'Abee-Lund, T.M. (2002). Antibiotic resistance in food-related bacteria—a result of interfering with the global web of bacterial genetics. Int. J. Food Microbiol. 78: 43-56.

Spanggaard, B., Jorgensen, F., Gram, L. and Huss, H.H. (1993). Antibiotic resistance in bacteria isolated from three freshwater fish farms and an unpolluted stream in Denmark. Aquaculture 115: 195-207.

Stock, I. and Wiedemann, B. (2001). Natural antibiotic suceptibility of *Edwardsiella tarda, E. ictaluri,* and *E. hoshinae.* Antimicrob. Agents Chemother. 45(8): 2245-2255.

Swedish Veterinary Antimicrobial Resistance Monitoring. The National Veterinary Institute (SVA), Uppsala, Sweden, 2003. www.sva.se, ISSN 1650-6332.

Tendencia, A.A. and Leobert, D.d.P. (2002). Level and percentage recovery of resistance to oxytetracycline and oxolinic acid of bacteria from shrimp ponds. Aquaculture 213(1-4): 1-13.

Teophilo, G.N.D., dos Fernandes Vieira, R.H.S., dos Prazeres Rodrigues, D. and Menezes, F.G. (2002). *Escherichia coli* isolated from seafood: toxicity and plasmid profiles. Int. Microbiol. 5: 11-14.

Teuber, M. (2001). Veterinary use and antibiotic resistance. Curr. Opin. Microbiol. 4: 493-499.

Thyssen, A. and Ollevier, F. (2001). In vitro antimicrobial susceptibility of *Photobacterium damselae* subsp. *piscicida* to 15 different antimicrobial agents. Aquaculture 200: 259-269.

van den Bogaard, A.E. and Stobberingh, E.E. (2000). Epidemiology of resistance to antibiotics, links between animals and humans. Int. J. Antimicrob. Agents 14: 327-335.

Van, P.T. (2005). Current status of aquatic veterinary drugs usage in aquaculture in Vietnam. Paper presented at the Antibiotic Resistance in Asian Aquaculture Environnment, Proceedings of the International Workshop held in Chiang Mai, Thailand.

Vaseeharan, B., Ramasamya, P., Muruganc, T. and Chenb, J.C. (2005). In vitro susceptibility of antibiotics against *Vibrio* spp. and *Aeromonas* spp. isolated from *Penaeus monodon* hatcheries and ponds. Int. J. Antimicrob. Agents 26: 285-291.

Vaughan, S., Coyne, R. and Smith, P. (1996). The critical importance of sample site in the determination of the frequency of oxytetracycline resistance in the effluent microflora of a fresh water fish farm. Aquaculture 139: 47-54.

Weber, J.T., Mintz, E.D., Canizares, R., Semiglia, A., Gomez, I., Sempertegui, R., Davila, A., Greene, K.D., Puhr, N.D., Cameron, D.N., et al. (1994). Epidemic cholera in Ecuador: multidrug-resistance and transmission by water and seafood. Epidemiol. Infect. 112(1): 1-11.

White, D.G. and McDermott, P.F. (2001). Emergence and transfer of antimicrobial resistance. Journal of Dairy Science 84(E.Suppl.): 151-155

White, D.G., Acar, J., Anthony, F., Franklin, A., Gupta, R., Nicholls, T., Tamura, Y., Thompson, S., Threlfall, E.J., Vose, D., van Vuuren, M., Wegener, H.C. and Costarrica, M.L. (2001). Antimicrobial resistance: standardisation and harmonisation of laboratory methodologies for the detection and quantification of antimicrobial resistance. Revue Scientifique et. Technique de l' Office International des Epizooties 20(3): 849-858.

Wijegoonawardena, P.K.M. and Siriwardena, P.P.G.S.N. (1996). The use of chemotherapeutic agents in shrimp hatcheries in Sri Lanka. In: Use of Chemicals in Aquaculture in Asia: Proceedings of the Meeting on the Use of Chemicals in Asia. 20-22 May 1996, Tigbauan, Iloilo, Philippines.

Wilder, M.N. (1996). Government regulation concerning the use of chemicals in aquaculture in Japan. In: Use of Chemicals in Aquaculture in Asia: Proceedings of the Meeting on the Use of Chemicals in Asia. 20-22 May 1996, Tigbauan, Iloilo, Philippines.

Witte, W. (2000). Ecological impact of antibiotics use in animals on different complex microflora: environment. Int. J. Antimicrob. Agents 14: 321-325.

Yoo, M.H., Huh, M.D., Kim, E.H., Lee, H.H. and Jeong, H.D. (2003). Characterisation of chloramphenicol acetyltransferase gene by multiplex polymerase chain reaction in multidrug-resistant strains isolated from aquatic environments. Aquaculture 217: 11-21.

Zhao, S., Datta, A.R., Ayers, S., Friedman, S., Walker, R.D. and White, D.G. (2003). Antimicrobial-resistant *Salmonella* serovars isolated from imported foods. Int. J. Food Microbiol. 84(1): 87-92.

Ziha-Zarifi, I., Llanes, C., Köhler, T., Pechere, J.-C. and Plesiat, P. (1999). In vivo emergence of multidrug-resistant mutants of *Pseudomonas aeruginosa* overexpressing the active efflux system MexA-MexB-OprM. Antimicrob. Agents Chemother. 43(2): 287-291.

6

DNA Vaccines Application in Aquaculture: Prospects and Constraints

Frøydis Gillund[1], Tom Christian Tonheim[2], Anne Ingeborg Myhr[3] and Roy Dalmo[4]

INTRODUCTION

Aquaculture is currently the fastest growing food producing industry in the world. The sector has grown at an average rate of 8.8% per year since 1970, compared with only 1.2% for captured fish and 2.8% for terrestrial farmed meat production systems over the same period (FAO, 2007). Thus, farmed fish is expected to play an important role in future food supplies, especially as the capture fishing industry has declined and wild stocks are diminishing. Several factors, including access to high quality feed supply and diseases caused by pathogens such as bacteria, viruses and parasites, do however represent constraints for further expansion of the aquaculture industry. The effectiveness of pathogen transportation in water in combination with high stock density of fish in aquaculture creates favourable living conditions for various pathogens. It has been estimated that 10% of farmed fish die due to infectious diseases (Leong and Fryer,

[1]*Genøk – Center for Biosafety, The Science Park in Breivika, Po Box 6418, 9294 Tromsø, Norway. E-mail: froydis.gillund@genok.org*
[2]*Department of Marine Biotechnology, Norwegian College of Fishery Science, University of Tromsø, N-9037 Tromsø, Norway. E-mail: tom-cto@fom-as.no*
[3]*Genøk – Center for Biosafety, The Science Park in Breivika, Po Box 6418, 9294 Tromsø, Norway. E-mail: anne.myhr@genok.org*
[4]*Department of Marine Biotechnology, Norwegian College of Fishery Science, University of Tromsø, N-9037 Tromsø, Norway. E-mail: royd@nfh.uit.no*

1993). This has promoted extensive research in disease protection of fish. During the 1980s antibiotics represented the most common practice for protection against bacterial diseases (Sommerset et al., 2005). Increase in antibiotic resistance among bacteria associated with animal production during the 1990s and its possible public health implications led to an intensified surveillance of bacterial resistance in many countries (Chapter 5, this volume). Within aquaculture, it was acknowledged that widespread use of antibiotics could cause both therapeutic and environmental problems, as antibiotics are released into the surrounding water during treatment of bacterial fish diseases. In order to limit antibiotic resistance within fish pathogens and bacterial populations in areas surrounding fish farms, effort was put into the development of efficient vaccines. The first vaccines against infectious bacterial diseases in farmed fish were introduced into commercial aquaculture in the early 1980s (Lorenzen and LaPatra, 2005). Since the 1990s bacterial vaccines have been routinely used, resulting in significant reduction in the use of antibiotics. There are however few efficient measures to combat diseases caused by viruses or parasites. Despite extensive research, few viral vaccines, based on either live, attenuated virus or vaccines containing recombinant viral antigens are available and there are no commercial vaccines against fish parasites (Lorenzen and LaPatra, 2005).

Deoxyribonucleic acid (DNA) vaccination is considered as a promising strategy to combat viral diseases both in animals and humans. A number of clinical trials with DNA vaccines against human diseases such as acquired immuno deficiency syndrome (AIDS), hepatitis and malaria are currently undertaken. In 2003, the U.S. Center for Disease Control and Prevention allowed for the vaccination of the Californian condor as an attempt to protect this endangered species from becoming infected with the West Nile virus (Bouchie, 2003). Currently, three DNA vaccines have been licensed for animal health applications. The first DNA vaccine to obtain licensure was the IHNV (Infectious Hematopoietic Necrosis Virus) DNA vaccine (Apex-IHN®) against the salmonid rhabdoviruses and infectious pancreatic necrosis virus (IPNV) in domesticated Atlantic salmon. It was developed by Aqua Health Ltd. (an affiliate of Novartis, Charlottetown, Canada) and was approved for marketing by the Canadian Food Inspection Agency in July 2005 (Novartis, 2005). Additionally one DNA vaccine preventing the disease caused by West Nile virus in horses and a therapeutic DNA vaccine to treat melanoma in dogs has been approved by the US Department of Agriculture (Salonius, 2007).

Thus, DNA vaccines are currently considered as one of the most efficient measures to combat viral diseases in fish, promote fish welfare and improve disease control. Despite the attractive characteristics of DNA vaccination, a number of questions remain to be solved relating to potential undesirable biological impacts such as: unintended immune responses, the fate of the DNA construct after injection of the vaccine and the possibility of spread to the aquatic environment. In addition, there are legislative and ethical issues involved: Are DNA vaccinated fish to be treated as a GMO and how will this influence public acceptance?

This chapter intends to illustrate some of the prospects and constraints of DNA vaccination in aquaculture. First by explaining what DNA vaccines are and the working mechanisms of DNA vaccines that stimulates the immune system. Then, some of the advantages will be illustrated and potential limitation of DNA vaccination of fish, before areas are identified with lack of scientific understanding and the importance of these uncertainties. Finally, ethical issues related to the regulation and public acceptance of fish treated with DNA vaccines will be discussed.

DNA VACCINES

Circular plasmid DNA (pDNA) may be used for gene delivery to mammals and fish. The pDNA is double stranded deoxyribonucleic acids (DNA, not different from chromosomal DNA), capable of being replicated autonomously usually in prokaryotes. Plasmid DNA used for gene delivery normally contains promoter- and enhancer sequences, the gene of interest, a poly-adenylation sequence, transcriptional termination sequence, antibiotic resistance gene and origin of replication. To express the gene of interest, pDNA is transcribed and mRNA is translated to protein by the cell's own apparatus. This opens up the applicability of pDNA in two important areas: Gene therapy and DNA vaccination.

The definitions of gene therapy and DNA vaccines are not consistent throughout the literature. Often, gene therapy also encompasses the use of DNA vaccines. The Norwegian Biotechnology Advisory Board defines gene therapy on animals as: "The intentional transfer of genetic material to somatic cells for purposes other than influencing the immune system" (Foss, 2003). Gene therapy often aims to achieve a long-lasting physiologically matched expression of the gene, without activating the immune system. In contrast, DNA vaccination is defined as the intentional transfer of genetic material to somatic cells for the purposes of influencing the immune system (Foss, 2003). For DNA vaccination, a short-term expression is sufficient for evoking an immune response.

WORKING MECHANISMS OF DNA VACCINES

The immune response after DNA vaccination is primarily launched by antigen presenting cells (APC), e.g. dendritic cells (DC) (Restifo et al.,

2000) in concert with T lymphocytes. Professional APC such as macrophages and DC have been shown to contain pDNA after intermuscular administration (Casares et al., 1997; Chattergoon et al., 1998). Following which, the APC at the site of administration may prime immune cells such as naïve T-lymphocytes following antigen presentation (Banchereau and Steinman, 1998). APC containing cytosolic encoding pDNA may transcribe and translate the transgene and thereby produce immunogenic proteins, mimicking an infection of an intracellular (cytosolic) pathogen and allow presentation of the foreign antigen peptide by major histocompatibility complex (MHC) class I on the surface of the APC. Antigen presenting cells can also take up soluble antigens released from another transgene producing cell (e.g. a myocyte), process it and present the peptide on MHC class II molecules at the cell surface. The T cell receptor (TCR) may recognize these peptides presented on the MHC class I and MHC class II, which stimulate a CD 8^+ T cell (cytotoxic T cell) and a CD4$^+$ T cell (helper T cell) response, respectively. In addition, APC can also take up antigen-laden apoptotic bodies of transfected myocytes (or other cells) and present the relevant peptides on MHCs. During certain circumstances, exogenously derived peptides may be presented on both MHC I and II.

This cross-priming stemmed from the observation that professional APCs could present antigen or peptides from the classical exogenous pathway via the MHC class I pathway of antigen presentation (Giri et al., 2004). In contrast to the TCR restricted recognition of MHC class I/II, the B-lymphocyte, the precursor to the antibody secreting cells, may directly recognize antigens by their immunoglobulin B-cell receptor (Banchereau and Steinman, 1998).

Thus, one of the unique features of DNA vaccines is their ability to stimulate both cellular (including cytotoxicity) and humoral immune responses (Weiner and Kennedy, 1999; Whitton et al., 1999; Restifo et al., 2000; Utke et al., 2007), as described in Figure 6.1.

The cellular immune response is represented, for simplicity, by the activated T_{h1} cells that may secrete proinflammatory cytokines, and CD8$^+$ T cells that may kill cells presenting the transgene. The adaptive humoral immune response is, however, represented by activation of B-lymphocytes and antibody production.

Several findings suggest that the fish immune system resembles the mammalian immune system; homologies are found between mammalian and fish MHC class I (Hashimoto, 1999), TCR (T-cell receptors) (Partula et al., 1995; Hordvik et al., 1996; Partula et al., 1996; De Guerra and Charlemagne, 1997; Zhou et al., 1997; Wilson et al., 1998; Scapigliati et al., 2000; Wermenstam and Pilstrom, 2001; Fisher et al., 2002; Kamper and

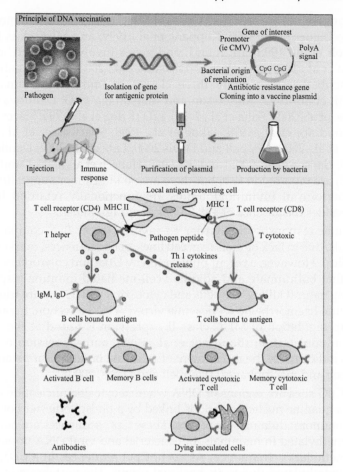

Fig. 6.1 The principle for DNA vaccination, from isolation of the gene encoding the antigenic protein to final immune responses. CD4, cluster of differentiation 4 co-receptor; CD8, cluster of differentiation 8 co-receptor; MHC-I, major histocompatibility complex class I protein; MHC-II, major histocompatibility complex class II protein; IgM, immunoglobulin M; IgD, immunoglobulin D; TCR: T cell receptor; Th1, T helper 1. (From Dafour, 2001)

McKinney, 2002; Nam et al., 2003; Hordvik et al., 2004) and the TCR co-receptor CD8 (Hansen and Strassburger, 2000) suggesting similarities in the antigen presentation process. The lymphocytes detected in fish are equivalent to the mammalian T- and B-cells (Clem et al., 1991). However, different subsets of T cells are not well characterized, but CD8 (Hansen and Strassburger, 2000; Shen et al., 2002; Moore et al., 2005; Somamoto et al., 2005) and CD4 (Suetake et al., 2004; Dijkstra et al., 2006; Edholm et al.,

2007) homologues are reported. Genes of MHC class IIβ (Hashimoto et al., 1990; Rodrigues et al., 1995; Koppang et al., 1999) and MHC class I gene (Grimholt et al., 1993) are found, and genes involved in MHC class I (Hansen et al., 1999) and class II (Dijkstra et al., 2003) loading pathways are described in fish. Only three classes of immunoglobulins: IgM (Hordvik et al., 1992; Andersson and Matsunaga 1993; Danilova et al., 2000; Lee et al., 2001; Saha et al., 2005), IgD (Krieg et al., 1995; Stacey et al., 1996; Boudinot et al., 1998; Jakob et al., 1998; Sparwasser et al., 1998; Hemmi et al., 2000; Heppell and Davis 2000; Latz et al., 2004; Danilova et al., 2005; Hansen et al., 2005) and IgZ/IgT (Hansen et al., 2005; Danilova et al., 2005) are reported in fish, but there are no indications suggesting that this repertoire of immunoglobulins are functionally retarded in any way—besides the absence of e.g. IgG, IgA and IgE in fish.

Immune responses are harder to investigate in fish compared to mammals, due to lack of tools (e.g. cell lines, gene sequences, markers and antibodies). However, a recent publication by Utke and co-workers (2007) shows that both innate and adaptive cell-mediated immune responses involving natural killer (NK) cells and cytotoxic T cell (CTL), are triggered after a viral haemorrhagic septicaemia virus (VHSV) infection. In rainbow trout, up-regulation of MHC class II expression is found at the site of pDNA administration (Boudinot et al., 1998), and expression of these genes is indicative of the recruitment of activated immunocompetent cells (e.g. B-cell and macrophages) (Heppell and Davis, 2000).

The CpG sites are regions of DNA where a cytosine nucleotide occurs next to a guanine nucleotide and is linked by a phosphodiester bond and have immunomodulatory properties. These CpG sequences are normally highly methylated in mammals, but bacterial and viral DNA possesses a lower methylation frequency. As such, CpG sequences in a pDNA are often unmethylated due to its bacterial origin. The vertebrate immune system recognizes unmethylated CpG-sequences as "foreign" and "danger" signal. Cells and animals treated with synthetic CpG oligodeoxynucleotides (ODN) and pDNA may produce cytokines that may induce macrophage activation, B-cell proliferation and immunoglobulin secretion (Krieg et al., 1995; Stacey et al., 1996). The CpG DNA can also directly activate monocytes, macrophages and dendritic cells to secrete T_{h1}-like cytokines (Stacey et al., 1996; Jakob et al., 1998; Sparwasser et al., 1998) to induce a cellular immune response, which is the desired response for combating intracellular pathogens. Leukocyte activation by CpG DNA is not mediated through a cell surface receptor but depends on uptake of the DNA (Krieg et al., 1995) into lysosomal compartments harbouring specific receptors, the so-called Toll-like receptor 9 (TLR9), for recognition (Hemmi et al., 2000; Latz et al., 2004). TLR9 is expressed in dendritic cells (Latz et al., 2004), macrophages (Latz et al., 2004), B-cells (Huggins et al., 2007) and liver endothelial cells (Martin-Armas et al., 2006). TLRs general function are as innate pattern recognition receptors (PRRs) that initiate immune activation in response to detection of pathogen specific molecular structures (Medzhitov and Janeway, 1997; Medzhitov and Janeway, 2000; Underhill and Ozinsky, 2000), such as CpGs. The CpG induced immune response seems to occur without apparent adverse effects (Krieg, 2000). Thus, it may be beneficial for a plasmid vector to also include CpG sequences to increase and direct the immune response. Not all CpG sequences possess the same immune stimulatory activities, and the effects depend on the ODN sequence, length, concentration and the recipient species (Carrington and Secombes, 2006). As TLR9 has been proposed to bind to CpG ODN, there are indications also that CpG containing pDNAs binds to TLR9 (Bauer et al., 2001; Zhao et al., 2004). However, it is not clear whether the immunostimulatory CpG sequences in the backbone of pDNA behave similar to that predicted from studies of synthetic CpG ODNs (Stevenson, 2004).

The TLR9 gene is found in fish (Meijer et al., 2004; Franch et al., 2006) and is expressed in several epithelial and lymphoid tissues (Takano, 2007), but it remains to be elucidated whether pDNA binds to fish TLR9. However, CpG ODN administered to fish, exerts various immune responses similar to that of mammals, such as activation of macrophages (Lee et al., 2003; Meng et al., 2003; Tassakka and Sakai, 2003; Tassakka and Sakai, 2004), proliferation of leukocytes (Jørgensen et al., 2003; Tassakka and Sakai, 2003; Carrington, 2004) and stimulation of cytokine expression (Jørgensen et al., 2001a; Jørgensen et al., 2001b; Tassakka and Sakai, 2004).

IMPACTS OF INTRODUCING DNA VACCINES TO AQUACULTURE

Several factors need to be investigated in order to describe potential beneficial and adverse impacts of introducing DNA vaccines to aquaculture. This includes: (i) Immunological reactions following DNA vaccine injection, (ii) The fate of the DNA vaccines after injection (including distribution, expression and integration) and (iii) spreading to the environment. Few studies have been conducted to investigate these issues (Myhr and Dalmo, 2005). Furthermore, the occurrence of the possible impacts is highly case-specific. Site and level of gene expression has been shown to be dependent on the administration volume (Anderson et al., 1996; Heppell et al., 1998), dose (Hansen et al., 1991; Anderson et al., 1996), age (Hansen et al., 1991) and fish size (Anderson et al., 1996; Heppell et al., 1998; Tonheim et al., 2007). Additionally, differences in stress levels, growth conditions and exposure to other pathogens are other parameters that could affect vaccine efficacy in the field (Lorenzen and LaPatra, 2005)

Advantages of DNA Vaccines

A huge advantage of DNA vaccination is that pDNA, like live or attenuated viruses, effectively induces both humoral and cell-mediated immune responses. Immune responses to antigen present in soluble form, such as for recombinant protein, generally induces only antibody responses. In addition, DNA vaccines overcome the safety concerns compared to live attenuated vaccines where a possible reversion to virulent forms may occur. Another benefit compared to alternative vaccination strategies is that pDNA possesses intrinsic immunostimulatory capacity, due to the presence of CpG sequences (Coombes and Mahony, 2001). Furthermore, it is possible to construct a vector encoding several antigens given in a single administration, and thus create a vaccine for multiple diseases. Immune responses against other pathogens or microbes that were not foreseen when developing the vaccine; e.g. cross-protection against heterologous virus at early time-points after DNA injection is also shown to occur (Lorenzen et al., 2002).

From a practical point of view, DNA vaccines are relatively inexpensive and easy to produce, and can be manufactured using identical production processes. DNA vaccines are very stable as dried powder or in a solution, unlike conventional vaccines that often need to be stored at proper conditions (e.g. cold environments) (Griffiths, 1995; Minor, 1995; Heppell and Davis, 2000). DNA vaccines administered without any additional adjuvants create much less direct tissue damage or/and inflammatory responses compared to conventional oil-adjuvanted vaccines (Lillehaug et al., 1992; Midtlyng, 1996; Babiuk et al., 1999, Garver et al., 2005; Kurath et al., 2006) and with no systemic toxicity (Parker et al., 1999). Perhaps the most important use of DNA vaccine technology is the possibility to create vaccines for targeted diseases where traditional vaccines have scored sub-optimally.

Limitations of DNA Vaccines

DNA vaccination encompases several attractive characteristics but also some limitations. For example, DNA vaccines induce specific immune responses only against the protein component of the pathogens. Thus, DNA vaccination strategies may not induce immune responses against carbohydrates and highly glycosylated proteins. This suggests that DNA vaccination strategies cannot be a substitute for the more traditional polysaccharide containing vaccines in evoking immune responses, against microbes that have an outer membrane made of e.g. lipopolysaccharides (Minor, 1995).

Lack of Scientific Understanding by Use of DNA Vaccines

At present there is a lack of scientific understanding with regard to a number of questions related to potential undesirable biological impacts by DNA vaccination that will be dealt with here subsequently.

Induction of autoimmune responses by antibody response to mammalian or fish DNA may be a disadvantage of DNA vaccination. The pDNA contains CpG sequences that may act as an adjuvant (Krieg et al., 1995; Piesetsky et al., 1995; Klinman et al., 1996), and there is a possibility that the combined adjuvant activity with the presence of DNA may induce antibody responses to the DNA itself. Such a process may be detrimental to the host, as exemplified by an occurrence of glomerulonephritis in mice (Mor, 2001). An additional safety concern associated with the use of DNA vaccines after inter-muscular administration is that myocytes, taking up the pDNA and express the encoded antigen, may become targets for antigen-specific CTLs. This would result in the development of autoimmune myositis (Goebels, 1992; Robertson, 1994).

Other unwanted immunological effects include susceptibility to tolerance. Tolerance seems to mainly be a problem in neonatal individuals, where an immature immune system could recognize the DNA vaccine encoded protein as a "self" protein (Mor et al., 1996; Mor, 1998; Smith and Klinman, 2001). However, during neonatal DNA vaccination, tolerance is an exception rather than the rule (Bot, 2000). The susceptibility to tolerance induction was shown to wane within one wk after birth in mice and to increase with increasing pDNA dose ((Ichino et al., 1999).

There is at present a need for research with focus on the stability and persistence of the DNA construct after injection. It is highly relevant here if the pDNA can integrate into the chromosomal DNA. Integration of pDNA could potentially lead to mutations, genomic instability and abnormalities (Gregoriadis, 1998; Taubes, 2000; Smith and Klinman, 2001). Available evidence indicates that genetic vaccines currently being tested clinically rarely integrate (Parker et al., 1999; Ledwith et al., 2000; Manam et al., 2000) and have a low probability of creating a tumourigenic event (Moelling, 1997). However, attempts to enhance vaccine efficacy by modifying the vector, co-administration of vaccine with agents that increase cellular uptake or altering the site or method of delivery, may increase the potential for integration (Smith and Klinman, 2001; Wang et al., 2004). Homologous recombination could occur when the foreign DNA possesses a sequence that is very similar to a sequence of the genomic

DNA and the two sequences "cross over", although random insertion is a more frequent event (Nicols et al., 1995). Environmental release of pDNA may occur by several ways, e.g. leakage of DNA vaccine from the administration site after vaccination (mainly in fish), by consumption of pDNA residues in the meat of DNA vaccinated animals, by spills or waste of DNA vaccine and from the production process. After vaccination, pDNA may find its way to the intestine (Hohlweg and Doerfler, 2001) where pDNA may be taken up by intestinal bacteria or secreted by the faeces. Antibiotic resistant genes may then be spread to various bacterial populations in the intestine, soil or water. Several other advantages and disadvantages are listed in the review by Lorenzen and LaPatra (2005). Initial studies in goldfish suggested, however, that pCMV-lacZ (a plasmid that contains the cytomegalovirus (CMV)-promoter-driven lacZ reporter gene) could induce long-term foreign protein expression (70 d) with both humoral and cell-mediated immunity without any signs of autoimmunity or integration (Kanellos et al., 1999). Nevertheless, this may be different in other species or when using a different construct.

DNA VACCINES, DECISION MAKING AND ETHICS

Decision making related to the introduction of novel technologies and their products, such as DNA vaccines, is based on an evaluation of the proposed benefits versus the possible risk that the use of this new product may represent. Risk evaluation generally includes an assessment of the risks involved, followed by risk management and risk communication. Risk assessment is primarily considered as a scientific task, in order to document the impacts of introducing a new product and consists of hazard identification, risk characterization and risk estimation. This provides the information upon which the competent authorities base their decision (risk management). Through risk communication the public and interested parties can be informed and included in the decision making process. In the following section, ethical aspects will be discussed that are of relevance when conducting each of these tasks, focusing on the implications of taking a precautionary approach in decision making.

What are Risks?

The term "risk" originates from the theory of probability and is commonly defined as outcome (possibility of hazard) multiplied by the probability or likelihood of the outcome to occur. The purpose of risk assessment is to identify these possible hazards and their associated probabilities. Often technological approaches to risk are employed where the frequencies of specific events are calculated through observations, statistics and/or modelling (Renn, 1998). Such approaches place a high degree of trust on the experts in the field. Consequently, technical concepts used in risk assessments presume that probabilities of events can be accurately predicted, and that uncertainties can be identified, managed and sufficiently reduced through further research. The numerical conditions of magnitude and probabilities assume equal weight for both components.

In this context quantitative methods, such as statistical measures and prediction error, are most commonly used to characterize uncertainty. It is however increasingly recognized that the use of new technologies and introduction of novel products, such as DNA vaccines, into complex systems, such as the aquatic ecosystem, may cause hazards that are difficult, if not impossible, to identify in advance. Lack of scientific understanding allows for interpretive flexibility and thus a diversity of opinions on the issue, which in turn can lead to disagreement among the scientists involved in the risk assessment (Krayer von Krauss et al., 2005; Kvakkestad et al., 2007). Furthermore, it is increasingly being recognized that in environmental risk assessments, conducting further research may reveal larger uncertainties rather than reduce those uncertainties identified, and that these uncertainties cannot be adequately characterized in statistical terms. Hence, there is a growing awareness that quantitative methods might not cover all aspects of uncertainty (Funtowicz and Ravetz, 1990; Wynne, 1992; Stirling 2001; Walker et al., 2003). Efforts have been made to understand uncertainty in terms of sources and types. For instance, different typologies of uncertainty have been developed with the purpose to contextualize the broader scientific uncertainties found in risk assessment (Wynne, 1992; Stirling, 2001;Walker et al., 2003):

- *Risk* represents a condition where the possible outcomes are identified and the relative likelihood of the outcomes is expressed in probabilities.
- *Uncertainty* refers to situations where we do not know or cannot estimate the probability of hazard, even though the hazards to consider are known. This may be due to the novelty of the activity, or to the variability or complexity involved.
- *Ignorance* represents situations where the hazards to assess remain unknown, i.e. completely unexpected hazards may emerge. This has historically been experienced with for instance BSE, dioxins, and pesticides (Harremoës et al., 2001).
- *Ambiguity* arises when there are different interpretations of the description of a system or a phenomenon, or that there are strong disagreements about definitions of terms. For instance, most of the scientific disputes within risk assessment and managements do not

refer to differences in methodology but to the questions of what all this means for human health, the environmental protection and management requirements, and which questions, dimensions and variables are of importance.

The various types of uncertainties arise from different sources (e.g. lack of knowledge, or from the natural variability of complex systems) and the distinction of the typologies is important as it may help to identify appropriate processes and mechanisms that can be used to broaden the scope of risk assessment.

As we have shown the introduction of DNA vaccines in aquaculture can result in a number of unintended impacts. Uncertainties still persist regarding: (i) unintended immunological responses such as autoimmunity, tolerance or cross protection, (ii) distribution, persistence and integration of the injected DNA vaccine, (iii) spreading of the DNA vaccine to the aquatic system (Myhr and Dalmo, 2005). In addition there are social and ethical uncertainties such as impacts on animal welfare, market acceptance and which regulatory framework to apply. More research and improved methodologies will provide sufficient knowledge to deal with some of these uncertainties. Still, due to the complexity of the aquatic system and novelty of DNA vaccines, decision makers will be faced with uncertainties and therefore they will need to adopt flexible management strategies accordingly.

The Regulatory Framework of DNA Vaccines

An appropriate set of regulatory requirements for DNA vaccines is still under development by various national and regional authorities. Since most national and regional Acts regulating the application of gene technology and genetically modified organisms (GMOs) were adopted before DNA vaccines were developed, there is at present disagreement on whether the application of DNA vaccines are properly governed by the existing regulations (Foss and Rogne, 2003). The most contested issue related to DNA vaccine regulation is whether an animal treated with DNA vaccines is to be regulated as a GMO. Various regulatory authorities solve this problem in different ways. In the U.S., the Food and Drug Administration (FDA) has asserted that genetic constructs distributed to animals fall under the legal definition of a drug substance, a practice that corresponds with the regulation in Europe where the European Agency for Evaluation of Medical Products (EMEA) authorize pharmaceuticals based on genetic engineering through a centralized procedure. The Norwegian Directorate for Nature Management (DN) has on the contrary concluded that a DNA vaccinated animal is to be considered as a GMO as long as the added DNA is present in the animal (Foss and Rogne, 2003).

Thus, uncertainties as to whether DNA plasmid constructs are stored and possibly integrate with the genome of the host organism become an issue of policy implications. This will further influence requirements for labelling and traceability, as well as public acceptance, thereby affecting the future of DNA vaccine development (Myhr and Dalmo, 2005).

Adopting a Precautionary Approach in Decision Making

The precautionary principle is a normative principle with historic roots that specifically relates to the acknowledgement of scientific uncertainties in decision making (Foster et al., 2000; Sandin, 2004). With regard to GMO regulations, the precautionary principle plays an important role in the Cartagena Protocol on biosafety, which intends to contribute to safe transfer, handling and use of living genetically modified organisms. The principle is also employed in regulations such as the EU directive 2001/18/EC on the deliberate release into the environment of genetically modified organisms and the Norwegian Gene Technology Act of 1993. A classical formulation of the principle is stated in the Rio Declaration on Environment and Development, Principle 15:

> "In order to protect the environment, the precautionary approach shall be widely applied by States according to their capabilities. Where there are threats of serious or irreversible damage lack of full scientific certainty shall not be used as a reason for postponing cost-effective measures to prevent environmental degradation" (Rio Declaration on Environment and Development, 1992)".

A wide range of alternative formulations, with diverging legal status, exists (Foster et al., 2000). These are often referred to as either weak or strong versions of the principle. The strong versions generally promote an active approach and is often based on ecocentric values, whereas the weak versions (like the one stated in the Rio Declaration) do not necessarily suggest any actions and reflect a more anthropocentric worldview. Even though not specifically stated in the version from the Rio Declaration, four central components are commonly associated with the principle: (i) taking preventive action in the face of uncertainty, (ii) shifting the burden of proof to the proponents of an activity, (iii) exploring a wide range of alternatives to possible harmful actions, and (iv) increasing public participation in decision making (Kriebel et al., 2001). The precautionary principle and its practical implications is a contested issue. At present there are discussions on how the precautionary principle should be implemented and how precautionary approaches are to be facilitated (Peterson, 2007; Renn, 2007; Stirling, 2007). Some argue that current regulatory procedures are already precautionary and that the precautionary principle is unscientific as it requires actions without scientific proof, while others worry that it will reduce the incentives for development and economic growth, as the developer of a product has to prove its safety (Morris, 2002). Another contested issue is whether the principle should only be applied in risk management, or if it is relevant and should be included as a guiding principle also in risk assessment (Peterson, 2007; Renn, 2007; Stirling, 2007). Stirling (2007) argues that the precautionary principle is of practical relevance as much to risk assessment as risk management, and supports his argument by pointing to the various choices scientists make when framing scientific risk assessments, choices that are shaped by the scientists' values and interests.

Ethical Considerations Related to Risk Assessment

FDA in the US and EMEA in EU have drafted guidelines for the veterinary use of DNA vaccines (EMEA, Committee for Veterinary Medicinal Products, 2000; FDA, 2007). The European guidelines mention several special aspects, which ought to be investigated in order to conduct a risk assessment of DNA vaccines intended for clinical testing or market release. This includes: (i) the possibility of pDNA integrating into the chromosome, (ii) concerns about possible adverse effects on the immune system, for example auto-immune reactions, (iii) risks posed by the additional use of genes encoding cytokines or co-stimulatory molecules or (iv) undesirable biological activity by the expressed antigen itself. The guidelines prepared by the FDA (US) does also point to some of these areas, and suggests that safety testing should include tests on vaccine immunogenicity, effects from cytokines and other immunomodulatory genes, autoimmunity, local reactogenicity and systemic toxicity and studies of biodistribution, persistence and integration. Thus, both guidelines emphasize on several of the areas that are referred to as associated with scientific uncertainties related to the introduction of DNA vaccines in aquaculture (Myhr and Dalmo, 2005). The crucial question regarding biosafety is how the knowledge responding to these areas is obtained. Stirling (2007) points to the fact that the outcome of a risk assessment depends on how the analysis is framed. Throughout the process of conducting scientific research, scientists have to make a number of choices for example regarding how to formulate the research question, which method to apply when collecting data, and how to interpret and communicate findings. Studies show that scientists' judgements of uncertainties and risks of introducing GMOs may depend on their academic background, funding and work context (Krayer von Krauss, 2004, 2005; Kvakkestad, 2007). Thus, the scientists' values and interests come into play.

According to Kriebel et al. (2001), a precautionary approach to research implies a broadening of the problem framing and the promotion of interdisciplinary research projects in order to develop more comprehensive models that better represents complex ecological systems. Furthermore, they emphasize the need to apply new statistical methods that include qualitative aspects of scientific uncertainty. Foster et al. (2000) underline that a precautionary influenced decision, must be followed up by evaluation, monitoring and risk associated research. Research that intends to investigate possible risks does however involve different approaches compared to research that aim to document a product's safety. Research that aims at creating new innovations is generally driven by a strong belief in the need and benefit of the product. Thus, the same research groups may not possess similar creativity when investigating the potential risks of the product. Risk associated research therefore needs to be conducted by researchers without economic or political interests. An obstacle for independent risk assessment is however the difficulty of obtaining access to information and material, since it is often claimed to be confidential business information.

Ethical Considerations Related to Risk Management

Risk-cost-benefit-analysis is a commonly used decision making tool when policy makers are deciding which management strategy to apply in order to solve a specific problem. It implies that alternative solutions to the problem are identified and that the risks, costs and benefits associated with each of these alternatives are characterized and quantified (commonly in economical terms). Then the costs and benefits of each alternative can be compared and the preferred alternative (i.e. the alternative representing maximum benefits) is chosen. Even though such cost effective measures are promoted in the definition of the precautionary principle as stated in the Rio Declaration, it may be argued that it is not in accordance with a precautionary approach and that it does not properly acknowledge the different types of scientific uncertainties involved (Aslaksen and Myhr, 2006). In situations characterized with uncertainty, ambiguity or ignorance, the possible risks, benefits and costs are, as we have already discussed, are difficult to define. This becomes even more evident when taking long term effects into account. This ultimately raise the question of who should be involved in the definition of unaccepted harm, how they should be expressed and what should be set as the normative baselines (Jensen et al., 2003; Aslaksen and Myhr, 2006; von Scomberg, 2006). An integrated assessment of these questions implies the handling of technical facts and social issues that almost always are incommensurable. This represents a challenge to the methods and analysis to be used. Multi-criteria methods have been suggested as a useful analytical tool for mapping social preferences, and has been developed as a software based technique for exploring the links between scientific/expert analysis and divergent social values and interest (see for example, Stirling, 1999; Munda, 2004).

Ethical Consideration Related to Risk Communication

Strong versions of the precautionary principle promote public participation and involvement of stakeholders in decision making processes, and recognize the need to include other forms of knowledge when dealing with problems associated with uncertainties, ambiguities or ignorance (Aslaksen and Myhr, 2006). By including stakeholders' review of scientific data and their opinion on relevant solutions and relevant criteria for evaluating these alternatives, the knowledge base upon which policy decisions are made may be improved (Pew Initiative on Food and Biotechnology, 2003). Furthermore, transparency and public involvement may help to ensure sound decision making processes concerning DNA vaccines, (for example through peoples conferences), represents valuable support to policymakers when they are left to make decisions in face of uncertainty. This requires that means of risk communication are carried out in an open and transparent manner. Practice shows, however, that this is often hindered by companies who wish to protect proprietary interests when developing new products. The data supporting the applications of DNA vaccine in order to protect the Californian Condor, which was allowed by the U.S. Center for Disease Control, as well as the IHNV DNA vaccine for salmon which was approved by the Canadian Food Inspection Agency in 2005, were never made available for the public. This lack of transparency is also reflected in the EU, where GM medicinal products lack the public openness central to other GMO applications. Such an approach will inevitably limit the public's opportunity to participate in the decision making process, which may reduce the public's credibility and confidence regarding safety aspects of DNA vaccines.

In general, the introduction of gene technology has caused significant scientific and public debates. Gene technology associated with medical applications is more accepted than GM plants, whereas transgenic animals receive most opposition (Anonymous, 2003). The public's conception of risk related to gene technology are not limited to technological and economic risks, but also concern factors such as social, cultural and religious values (Sagar et al., 2000; Wynne, 2001; Power, 2003). Hence, there may be several issues regarding how consumers will respond to gene technology in aquaculture that pull in different directions. Food safety and quality are perhaps the most obvious consumer concerns. Animal welfare and the naturalness of using gene technology are other factors important for the public. For instance, the notion that animals, apart from having utility or instrumental value, also have intrinsic values and thereby rights at ethical and legal levels is gaining ground. In a gene technology context, the violation of an animal's integrity may take place at different levels: the level of the individual animal (phenotypic integrity), the level of the species-specific nature (genomic integrity) and the level of animality (self-maintenance, self-organization, independence, animal-type integrity) (Verhoog, 2001). New products need market acceptance which is ultimately determined by consumers. Thus, public feelings and morals are concerns that must be addressed. These aspects have been discussed in more detail in Chapter 9 of this volume.

CONCLUSIONS

Further development of vaccination for fish is a necessity to secure future growth of the aquaculture industry, which will contribute to world food supplies. DNA vaccines have shown promising results as an effective strategy to combat viral diseases in fish. There are several advantages of DNA vaccines; they induce both a humoral and cell mediated immune response and have proven to be efficient against viral pathogens where alternative strategies for protection are still missing. It is however a need for further investigation of safety issues related to DNA vaccination. More basic research is needed to increase our understanding of the working mechanisms of DNA vaccines, how intended immune response is stimulated and the level of distribution, expression and integration of the injected pDNA in the host organism, and whether there are environmental impacts from unintended spread to the environment. Furthermore, decision making processes need to acknowledge the scientific uncertainties associated with the introduction of DNA vaccines in aquaculture. This can be facilitated by adopting a precautionary approach both in risk assessment and risk management and by including the public and relevant interest groups in decision making processes.

ABBREVIATIONS

AIDS   Acquired immuno deficiency syndrome

APC   Antigen presenting cells

CpG   Regions of DNA where a cytosine nucleotide occurs next to a guanine nucleotide and is linked by a phosphodiester bond and have immunomodulatory properties.

CTL   Cytotoxic T cell

DN   Norwegian Directorate for Natural Management
EMEA   European Agency for Evaluation of Medical Products
FDA   Federal Drug Administration (USA)
GMO   Genetically modified organisms
IHNV   Infectious Hematopoietic Necrosis Virus
IPNV   Infectious pancreatic necrosis virus
IgD   Immunoglobulin D
IgM   Immunoglobulin M
MHC   Major histocompatibility complex
NK   Natural killer
ODN   Oligodeoxynucleotides
pDNA   Circular plasmid DNA
TCR   T cell receptor
Th1   T helper 1
TLR 9   Toll-like receptor 9
VHSV   Viral haemorrhagic septicaemia virus

REFERENCES

Anonymous, H. (2003). Europeans and Biotechnology in 2002. European Commission. Eurobarometer 58.

Andersson, E. and Matsunaga, T. (1993). Complete cDNA sequence of a rainbow trout IgM gene and evolution of vertebrate IgM constant domains. Immunogenetics 38: 243-250.

Anderson, E.D., Mourich, D.V. and Leong, J.A., (1996). Gene expression in rainbow trout (*Oncorhynchus mykiss*) following intramuscular injection of DNA. Mol. Mar. Biol. Biotechnol. 5: 105-113.

Aslaksen, I. and Myhr, A.I. (2006). "The worth of a wildflower": Precautionary perspectives on the environmental risk of GMOs. Ecol. Econ. 60: 489-497.

Banchereau, J. and Steinman, R.M. (1998). Dendritic cells and the control of immunity. Nature 392: 245-252.

Babiuk, L.A,, van Drunen Littel-van den Hurk, S. and Babiuk S.L. (1999). Immunization of animals: from DNA to the dinner plate. Vet. Immunol. Immunop. 72: 189-202.

Bauer, S., Kirschning, C.J., Hacker, H., Redecke, V., Hausmann, S. and Akira, S. (2001). Human TLR9 confers responsiveness to bacterial DNA via species-specific CpG motif recognition. Proc. Natl. Acad. Sci., USA 98: 9237-9242.

Bot, A. (2000). DNA vaccination and the immune responsiveness of neonates. Int. Rev. Immunol. 19: 221-245.

Bouchie, A. (2003). DNA vaccines deployed for endangered condors. Nat. Biotechnol. 21: 11.

Boudinot, P., Blanco, M., de Kinkelin, P. and Benmansour, A. (1998). Combined DNA immunization with the glycoprotein gene of viral hemorrhagic septicemia virus and infectious hematopoietic necrosis virus induces double-specific protective immunity and nonspecific response in rainbow trout. Virology 249: 297-306.

Carrington, A.C. and Secombes, C.J. (2006). A review of CpGs and their relevance to aquaculture. Vet. Immunol. Immunop. 112: 87-101.

Carrington, A.C., Collet, B., Holland, J.W. and Secombes, C.J. (2004). CpG Oligodeoxynucleotides stimulate immune cell proliferation but not specific antibody production in rainbow trout (*Oncorhynchus mykiss*). Vet. Immunol. Immunop. 101: 211-222.

Casares, S., Inaba, K., Brumeanu, T.D., Steinman, R.M. and Bona, C.A. (1997). Antigen presentation by dendritic cells after immunization with DNA encoding a major histocompatibility complex class II-restricted viral epitope. J. Exp. Med. 186: 1481-1486.

Chattergoon, M.A., Robinson, T.M., Boyer, J.D. and Weiner, D.B. (1998). Specific immune induction following DNA-based immunization through *in vivo* transfection and activation of macrophages/antigen-presenting cells. J. Immunol. 160: 5707-5718.

Clem, L.W., Miller, M.W. and Bly, J.E. (1991). Evolution of lymphocyte subpopulations, their interactions and temperature sensitivities. In: Phylogenesis of Immune Functions. G.W. Warr and N. Cohen (eds), Boca Raton, CRC Press, Florida, USA, pp. 191-213.

Coombes, B.K. and Mahony, J.B. (2001). Dendritic cell discoveries provide new insight into the cellular immunobiology of DNA vaccines. Immunol. Lett. 78: 103-111.

Dafour, V. (2001). DNA vaccines: new applications for veterinary medicine. Online publication in Veterinary Sciences Tomorrow. (www.vetscite.org/publish/articles/000020/index .html)

Danilova, N., Hohman, V.S., Kim, E.H. and Steiner, L.A. (2000). Immunoglobulin variable-region diversity in the zebrafish. Immunogenetics 52: 81-91.

Danilova, N., Bussmann, J., Jekosch, K. and Steiner, L.A. (2005). The immunoglobulin heavy-chain locus in zebrafish: identification and expression of a previously unknown isotype, immunoglobulin Z. Nat. Immunol. 6: 295-302.

De Guerra, A. and Charlemagne, J. (1997). Genomic organization of the TcR beta-chain diversity (Dbeta) and joining (Jbeta) segments in the rainbow trout: presence of many repeated sequences. Mol. Immunol. 34: 653-662.

Dijkstra, J.M., Kiryu, I., Kollner, B., Yoshiura, Y. and Ototake, M. (2003). MHC class II Invariant chain homologues in rainbow trout (*Oncorhynchus mykiss*). Fish Shellfish Immunol. 15: 91-105.

Dijkstra, J.M., Somamoto, T., Moore, L., Hordvik, I., Ototake, M. and Fischer, U. (2006). Identification and characterization of a second CD4-like gene in teleost fish. Mol. Immunol. 43: 410-419.

Directive 2001/18/EC of the European Parliament and of the Council on the deliberate release into the environment of genetically modified organisms and repealing Council Directive 90/220/EEC.(2001). Official Journal of the European Union. L 106.

Edholm, E.S., Stafford, J.L., Quiniou, S.M., Waldbieser, G., Miller, N.W. and Bengten, E. (2007). Channel catfish, *Ictalurus punctatus*, CD4-like molecules. Dev. Comp. Immunol. 31: 172-187.

Fischer, C., Bouneau, L., Ozouf-Costaz, C., Crnogorac-Jurcevic, T., Weissenbach, J. and Bernot, A. (2002) Conservation of the T-cell receptor [alpha]/[delta] linkage in the teleost fish, *Tetraodon nigroviridis*. Genomics 79: 241-248.

Food and Drug Administration (FDA), Center for Biologics Evaluation and Research (2007). Guidance for Industry. Considerations for Plasmid DNA Vaccines for Infectious Disease Indications.

Foss, G.S. (2003). Regulation of DNA vaccines and gene therapy on animals. The Norwegian Biotechnology Advisory Board. Website: (http://www.bion.no/publikasjoner/regulation_of_DNA_vaccines.pdf) (accessed on 5[th]October 2007).

Foss, G.S. and Rogne, S. (2003). Gene medification or genetic modification? The devil is in the details. Nat. Biotechnol. 21: 1280-1281.

Foster, K.R., Vecchia, P. and Repacholi, M.H. (2000). Science and the precautionary principle. Science 288: 979-981.

Franch, R., Cardazzo, B., Antonello, J., Castagnaro, M., Patarnello, T. and Bargelloni, L. (2006). Full-length sequence and expression analysis of Toll-like receptor 9 in the gilthead seabream (*Sparus aurata* L.). Gene 378: 42-51.

Funtowicz, S.O. and Ravetz, J.R. (1990). Uncertainty and Quality in Science for Policy. Kluwer, Dordrecht, The Netherlands, pp. 7-16.

Garver, K., Conway, C., Elliott, D. and Kurath, G. (2005). Analysis of DNA-vaccinated fish reveals viral antigen in muscle, kidney and thymus, and transient histopathologic changes. Mar. Biotechnol. 7: 540-553.

Giri, M., Ugen, K.E. and Weiner, D.B. (2004). DNA vaccines against human immunodeficiency virus type 1 in the past decade. Clin. Microbiol. Rev. 17: 370-389.

Goebels, N., Michaelis, D., Wekerle, H. and Hohlfeld, R. (1992). Human myoblasts as antigen- presenting cells. J. Immunol. 149: 661-667.

Gregoriadis, G. (1998). Genetic vaccines: strategies for optimization. Pharm. Res. 15 : 661-670.

Griffiths, E. (1995). Assuring the safety and efficacy of DNA vaccines. Ann. New York Acad. Sci. 772: 164-169.

Grimholt, U., Hordvik, I., Fosse, V.M., Olsaker, I., Endresen, C. and Lie O. (1993). Molecular cloning of major histocompatibility complex class I cDNAs from Atlantic salmon (*Salmo salar*). Immunogenetics 37: 469-473.

Hansen, E., Fernandes, K., Goldspink, G., Butterworth, P., Umeda, P.K. and Chang, K.C. (1991). Strong expression of foreign genes following direct injection into fish muscle. FEBS Lett. 290: 73-76.

Hansen, J.D. and Strassburger, P. (2000). Description of an ectothermic TCR coreceptor, CD8 alpha, in rainbow trout. J. Immunol. 164: 3132-3139.

Hansen, J.D., Strassburger, P., Thorgaard, G.H., Young, W.P. and Du Pasquier, L. (1999). Expression, linkage, and polymorphism of MHC-related genes in rainbow trout, *Oncorhynchus mykiss*. J. Immunol. 163: 774-786.

Hansen, J.D., Landis, E.D. and Phillips, R.B. (2005). Discovery of a unique Ig heavy-chain isotype (IgT) in rainbow trout: Implications for a distinctive B cell developmental pathway in teleost fish. Proc. Natl. Acad. Sci., USA 102: 6919-6924.

Harremoës, P., Gee, D., KacGarvin, M., Stirling, A., Keys, J., Wynne, B. and Guedes Vaz, S. (2001). The Precautionary Principle in the 20th Century. Late lessons from early warnings. European Environmental Agency. Earthscan, London, UK.

Hashimoto, K., Nakanishi, T. and Kurosawa, Y. (1990). Isolation of carp genes encoding major histocompatibility complex antigens. Proc. Natl. Acad. Sci., USA 87: 6863-6867.

Hashimoto, K., Okamura, K., Yamaguchi, H., Ototake, M., Nakanishi, T. and Kurosaiva, Y. (1999). Conservation and diversification of MHC class I and its related molecules in vertebrates. Immunol. Rev. 167: 81-100.

Hemmi, H., Takeuchi, O., Kawai, T., Kaisho, T., Sato, S. and Sanjo, H. (2000). A Toll-like receptor recognizes bacterial DNA. Nature 408: 740-745.

Heppell, J. and Davis, H.L. (2000). Application of DNA vaccine technology to aquaculture. Adv. Drug Deliver. Rev. 43: 29-43.

Heppell, J., Lorenzen, N., Armstrong, N.K., Wu, T., Lorenzen, E., Einer-Jensen, K., Schorr, J. and Davis, H.L. (1998). Development of DNA vaccines for fish: vector design, intramuscular injection and antigen expression using viral haemorrhagic septicaemia virus genes as model. Fish Shellfish Immunol. 8: 271-286.

Hohlweg, U. and Doerfler, W. (2001). On the fate of plant or other foreign genes upon the uptake in food or after intramuscular injection in mice. Mol. Gene. Genom. 265: 225-233.

Hordvik, I., Voie, A.M., Glette, J., Male, R. and Endresen, C. (1992). Cloning and sequence analysis of two isotypic IgM heavy chain genes from Atlantic salmon, *Salmo salar* L. Eur. J. Immunol. 22: 2957-2962.

Hordvik, I., Jacob, A.L.J., Charlemagne, J. and Endresen, C. (1996). Cloning of T-cell antigen receptor beta chain cDNAs from Atlantic salmon (*Salmo salar*). Immunogenetics 45: 9-14.

Hordvik, I., Torvund, J., Moore, L. and Endresen, C. (2004). Structure and organization of the T cell receptor alpha chain genes in Atlantic salmon. Mol. Immunol. 41: 553-559.

Huggins, J., Pellegrin, T., Felgar, R.E., Wei, C., Brown, M., Zheng, B. (2007). CpG DNA activation and plasma-cell differentiation of CD27- naive human B cells. Blood 109: 1611-1619.

Ichino, M., Mor, G., Conover, J., Weiss, W.R., Takeno, M. and Ishii, K.J. (1999). Factors associated with the development of neonatal tolerance after the administration of a plasmid DNA vaccine. J. Immunol. 162: 3814-3818.

Jakob, T., Walker, P.S., Krieg, A.M., Udey, M.C. and Vogel, J.C. (1998). Activation of cutaneous dendritic cells by CpG-Containing oligodeoxynucleotides: A role for dendritic cells in the augmentation of Th1 responses by immunostimulatory DNA. J. Immunol. 161: 3042-3049.

Jensen, K.K., Gamborg, C., Hauge Madsen, K., Krayer von Krauss, M., Folker, A.P. and Sandøe, P. (2003). "Making the EU "Risk Window" Transparent: The normative foundations of the environmental risk assessment". Environ. Biol. Res. 2: 161-171.

Jørgensen, J.B., Johansen, A., Stenersen, B. and Sommer, A.I. (2001). CpG oligodeoxynucleotides and plasmid DNA stimulate Atlantic salmon (*Salmo salar* L.) leucocytes to produce supernatants with antiviral activity. Dev. Comp. Immunol. 25: 313-321.

Jørgensen, J.B., Zou, J., Johansen, A. and Secombes, C.J. (2001). Immunostimulatory CpGoligodeoxynucleotides stimulate expression of IL-1[beta] and interferon-like cytokines in rainbow trout macrophages via a chloroquine-sensitive mechanism. Fish Shellfish Immun. 11: 673-682.

Jørgensen, J.B., Johansen, L.H., Steiro, K. and Johansen, A. (2003). CpG DNA induces protective antiviral immune responses in Atlantic salmon (*Salmo salar* L.). J. Virol. 77: 11471-11479.

Kamper, S.M. and McKinney, C.E. (2002). Polymorphism and evolution in the constant region of the T-cell receptor beta chain in an advanced teleost fish. Immunogenetics. 53: 1047-1054.

Kanellos, T., Sylvester, I.D., Ambali, A.G., Howard, C.R. and Russell P.H. (1999). The safety and longevity of DNA vaccines for fish. Immunology. 96: 307-313.

Klinman, D.M., Yi, A.K., Beaucage, S.L., Conover, J. and Krieg, A.M. (1996). CpG motifs present in bacterial DNA rapidly induce lymphocytes to secrete interleukin 6, interleukin 12, and interferon gamma. Proc. Natl. Acad. Sci., USA 93: 2879-2883.

Koppang, E.O., Dannevig, B.H., Lie, O., Ronningen, K. and Press, C.M. (1999). Expression of Mhc class I and II mRNA in a macrophage-like cell line (SHK-1) derived from Atlantic salmon, *Salmo salar* L., head kidney. Fish Shellfish Immun. 9: 473-489.

Krayer von Krauss, K.M., Casman, E. and Small, M. (2004). Elicitation of expert judgments of uncertainty in the risk assessment of herbicide tolerant oilseed crops. J. Risk Anal. 24: 1515-1527.

Kriebel, D., Tickner, J., Epstein, P., Lemons, J., Levins, R., Loechler, E.L., Quinn, M., Rudel, R., Schettler, T. and Stoto. M. (2001). The precautionary principle in environmental science. Environ. Health Persp. 109: 871-876.

Krieg, A.M., Yi, A.K., Matson, S., Waldschmidt, T.J., Bishop, G.A., Teasdale, R. et al. (1995). CpG motifs in bacterial DNA trigger direct B-cell activation. Nature 374: 546-549.

Kurath, G., Garver, K.A., Corbeil, S., Elliott, D.G., Anderson, E.D. and LaPatra, S.E. (2006). Protective immunity and lack of histopathological damage two years after DNA vaccination against infectious hematopoietic necrosis virus in trout. Vaccine 24: 345-354.

Kvakkestad, V., Gillund, F. and Kjølberg, K.A. (2007). Scientists´ perspectives on the deliberate release of GM crops. Environ. Value 16: 79-104.

Latz, E., Schoenemeyer, A., Visintin, A., Fitzgerald, K.A., Monks, B.G., Knetter, C.F. et al. (2004). TLR9 signals after translocating from the ER to CpG DNA in the lysosome. Nature Immun. 5: 190-198.

Ledwith, B.J., Manam, S., Troilo, P.J., Barnum, A.B., Pauley, C.J. and Nichols, W.W. (2000). Plasmid DNA vaccines: assay for integration into host genomic DNA. Dev. Biol. Stand. 104: 33-43.

Lee, U.H., Pack, H.J., Do, J.W., Bang, J.D., Cho, H.R., Ko, B.K. et al. (2001). Flounder (*Paralichthys olivaceus*) cDNA encoding a secreted immunoglobulin M heavy chain. Fish Shellfish Immun. 11: 537-540.

Lee, C.H., Jeong, H.D., Chung, J.K., Lee, H.H. and Kim, K.H. (2003). CpG motif in synthetic ODN primes respiratory burst of olive flounder *Paralichthys olivaceus* phagocytes and enhances protection against *Edwardsiella tarda*. Dis. Aquat. Organ. 56: 43-48.

Leong, J.C. and Fryer, J.L. (1993). Viral vaccines for aquaculture. Annu. Rev. Fish. Dis. 3: 225-240.

Lillehaug, A., Lunder, T. and Poppe, T.T. (1992). Field testing of adjuvanted furunculosis vaccines in Atlantic salmon, *Salmo salar* L. J. Fish Dis. 15: 485-496.

Lorenzen, N. and LaPatra, S.E. (2005). DNA vaccines for aquacultured fish. Revue Scientifique et Technique (International Office of Epizootics). 24: 201-213.

Lorenzen, N., Lorenzen, E., Einer-Jensen, K. and LaPatra, S.E. (2002). Immunity induced shortly after DNA vaccination of rainbow trout against rhabdoviruses protects against heterologous virus but not against bacterial pathogens. Dev. Comp. Immunol. 26: 173-179.

Manam, S., Ledwith, B.J., Barnum, A.B., Troilo, P.J., Pauley, C.J., Harper, L.B. et al. (2000). Plasmid DNA vaccines: Tissue distribution and effects of DNA sequence, adjuvants and delivery method on integration into host DNA. Intervirology 43: 273-281.

Martin-Armas, M., Simon-Santamaria, Jaione, Pettersen, I., Moens, U., Smedsrød, B. and Sveinbjørnsson, B. (2006). Toll-like receptor 9 (TLR9) is present in murine liver sinusoidal endothelial cells (LSECs) and mediates the effect of CpG-oligonucleotides. J. Hepat. 44: 939-946.

Medzhitov, R. and Janeway, C.A. Jr. (1997). Innate immunity: impact on the adaptive immune response. Cur. Opin. Immunol. 9: 4-9.

Medzhitov, R. and Janeway, J.C. (2000). The Toll receptor family and microbial recognition. Trends Microbiol. 8: 452-456.

Meijer, A.H., Gabby Krens, S.F., Medina Rodriguez, IA, He, S., Bitter, W., Ewa Snaar-Jagalska, B. et al. (2004). Expression analysis of the Toll-like receptor and TIR domain adaptor families of zebrafish. Mol. Immunol. 40: 773-783.

Meng, Z., Shao, J. and Xiang, L. (2003). CpG oligodeoxynucleotides activate grass carp (*Ctenopharyngodon idellus*) macrophages. Dev. Comp. Immunol. 27: 313-321.

Midtlyng, P.J., Reitan, L.J. and Speilberg, L. (1996). Experimental studies on the efficacy and side-effects of intraperitoneal vaccination of Atlantic salmon (*Salmo salar* L.) against furunculosis. Fish Shellfish Immun. 6: 335-350.

Minor, P.D. (1995). Regulatory issues in the use of DNA vaccines. Annals New York Acad. Sci. 772: 170-177.

Miyadai, T., Ootani, M., Tahara, D., Aoki, M. and Saitoh, K. (2004). Monoclonal antibodies recognising serum immunoglobulins and surface immunoglobulin-positive cells of puffer fish, torafugu (*Takifugu rubripes*). Fish Shellfish Immun. 17: 211-222.

Moelling, K. (1997). DNA for genetic vaccination and therapy. Cytokines, Cel. Mol. Therap. 3: 127-136.

Moore, L.J., Somamoto, T., Lie, K.K., Dijkstra, J.M. and Hordvik I. (2005). Characterisation of salmon and trout CD8alpha and CD8beta. Mol. Immunol. 42: 1225-1234.

Mor, G. (1998). Plasmid DNA: A new era in vaccinology. Biochem. Pharmacol. 55: 1151-1153.

Mor, G. (2001). Plasmid DNA vaccines: immunology, tolerance, and autoimmunity. Mol. Biotechnol. 19: 245.

Morris, J. (2002). The relationships between risk analysis and the precautionary principle. Toxicology 181-182: 127-130.

Myhr, A.I. and Dalmo, R.A. (2005). Introduction of genetic engineering in aquaculture: ecological and ethical implications for science and governance. Aquaculture 250: 542-554.

Munda, G. (2004). Social Multi-Criteria Evaluation (SMCE): Methodological foundations and operational consequences. Eur. J. Oper. Res. 158: 662-677.

Nam, B.H., Hirono, I. and Aoki, T. (2003). The four TCR genes of teleost fish: the cDNA and genomic DNA analysis of Japanese flounder (*Paralichthys olivaceus*) TCR alpha-, beta-, gamma-, and delta-chains. J. Immunol. 170: 3081-3090.

Nichols, W.W., Ledwith, B.J., Manam, S. and Troilo, P.J. (1995). Potential DNA vaccine integration into host cell genome. Ann. N.Y. Acad. Sci. 772: 30-39.

Norwegian Ministry of Environment (1993). The Act relating to the production and use of genetically modified organisms (Gene Technology Act), Act No. 38 of 2 April 1993.

Novartis Animal Health Inc (2005). Novel Novartis vaccine to protect Canadian salmon farms from devastating viral disease. Press Release, 19th of July 2005.

Parker, S.E., Borellini, F., Wenk, M.L., Hobart, P., Hoffman, S.L., Hedstrom, R. et al. (1999). Plasmid DNA malaria vaccine: Tissue distribution and safety studies in mice and rabbits. Hum. Gene Ther. 10: 741-758.

Partula, S., de Guerra, A., Fellah, J.S. and Charlemagne, J. (1995). Structure and diversity of the T cell antigen receptor beta-chain in a teleost fish. J. Immunol. 155: 699-706.

Partula, S., de Guerra, A., Fellah, J.S. and Charlemagne J. (1996). Structure and diversity of the TCR alpha-chain in a teleost fish. J. Immunol. 157: 207-212.

Peterson, M. (2007). The precautionary principle should not be used as a basis for decision-making. EMBO J. 8: 305-308.

Pew Initiative on Food and Biotechnology. (2003). Future fish: issues in science and regulation of transgenic fish. Pew Initiative on Food and Biotechnology. Washington, USA.

Pisetsky, D.S., Reich, C., Crowley, S.D. and Halpern, M.D. (1995). Immunological properties of bacterial DNA. Ann. New York Acad. Sci. 772: 152-163.

Power, M. (2003). Lots of fish in the sea: Salmon Aquaculture, Genomics and Ethics, Electronic. Working Paper Series. W. Maurice Young Center for Applied Ethics. University of British Columbia.

Renn, O. (1998). Three decades of risk research: accomplishments and new challenges. J. Risk Res. 1: 44-71.

Renn, O. (2007). Precaution and analysis: two sides of the same coin? EMBO J. 8 : 303-304.

Restifo, N.P., Ying, H., Hwang, L. and Leitner, W.W. (2000). The promise of nucleic acid vaccines. Gene Therap. 7: 89-92.

Rio Declaration on Environment and Development (1992). Un.Doc/CpNF.151/5/Rev.1

Robertson, J.S. (1994). Safety considerations for nucleic acid vaccines. Vaccine 12: 1526-1528.

Rodrigues, P.N.S., Hermsen, T.T., Rombout, J.H.W.M. Egberts, E. and Stet, R.M. (1995). Detection of MHC class II transcripts in lymphoid tissues of the common carp (*Cyprinus carpio* L.). Dev. Comp. Immunol. 19: 483-496.

Sagar, A., Daemmrich, A. and Ashiya, M. (2000). The tragedy of the commoners: biotechnology and its public. Nat. Biotechnol. 18: 2-4.

Saha, N.R., Suetake, H. and Suzuki, Y. (2005). Analysis and characterization of the expression of the secretory and membrane forms of IgM heavy chains in the pufferfish, *Takifugu rubripes*. Mol. Immunol. 42: 113-124.

Salonius, K., Simard, N., Harland, R. and Ulmer J.B. (2007). The road to licensure of a DNA vaccine. Curr. Opin. Invest. Drugs. 8: 635-641.

Sandin, P. (2004). The precautionary principle and the concept of precaution. Environ. Value 13: 461-475.

Scapigliati, G., Romano, N., Abelli, L., Meloni, S., Ficca, A.G., Buonocore, F. et al. (2000). Immunopurification of T-cells from sea bass *Dicentrarchus labrax* (L.). Fish Shellfish Immun. 10: 329-341.

Shen, L., Stuge, T.B., Zhou, H., Khayat, M., Barker, K.S., Quiniou, S.M. et al. (2002). Channel catfish cytotoxic cells: a mini-review. Dev. Comp. Immunol. 26: 141-149.

Smith, H.A. and Klinman, D.M. (2001). The regulation of DNA vaccines. Cur. Opin. Biotechnol. 12: 299-303.

Somamoto, T., Yoshiura, Y., Nakanishi, T. and Ototake, M. (2005). Molecular cloning and characterization of two types of CD8alpha from ginbuna crucian carp, *Carassius auratus langsdorfii*. Dev. Comp. Immunol. 29: 693-702.

Sommerset, I., Krossøy, B., Biering, E. and Frost, P. (2005). Vaccination for fish in aquaculture. Expert Rev. Vaccines 4: 89-101.

Sparwasser, T., Koch, E.S., Vabulas, R.M., Heeg, K., Lipford, G.B., Ellwart, J.W. et al. (1998). Bacterial DNA and immunostimulatory CpG oligonucleotides trigger maturation and activation of murine dendritic cells. Euro. J. Immunol. 28: 2045-2054.

Stacey, K.J., Sweet, M.J. and Hume, D.A. (1996). Macrophages ingest and are activated by bacterial DNA. J. Immunol. 157: 2116-2122.

Stevenson, F.K. (2004). DNA vaccines and adjuvants. Immunol. Rev. 199: 5-8.

Stirling, A. (2001). On science and precaution on risk in the management of technological risk. An ESTO Project Report Prepared for the European Commission-JRC. Institute Prospective Technological Studies. Seville, Spain.

Stirling, A. (2007). Risk, precaution and science: towards a more constructive policy debate. EMBO J. 8: 309-312.

Stirling, A. and Mayer, S. (1999). Rethinking Risk: A Pilot Multi-criteria Mapping of Genetically Modified Crop in Agricultural Systems in the UK, Science and Technology Policy Research, University of Sussex, Brighton, UK.

Suetake, H., Araki, K. and Suzuki, Y.(2004). Cloning, expression, and characterization of fugu CD4, the first ectothermic animal CD4. Immunogenetics 56: 368-374.

The European Agency for the Evaluation of Medical Products, Committee for Veterinary Medicinal Products (EMEA). (2000). Note for Guidance: DNA vaccines Non- amplifiable in eukaryotic cells for veterinary use. Fisheries and Aquaculture Department Food and Agricultural Organization of the United Nations (FAO). (2007). The State of World Fisheries and Aquaculture 2006. Rome, Italy.

Tonheim, T.C., Leirvik, J., Løvoll, M., Myhr, A.I., Bøgwald, J. and Dalmo, R.A. (2007). Detection of supercoiled plasmid DNA and luciferase expression in Atlantic salmon (*Salmo salar* L.) 535 days after injection. Fish Shellfish Immun. 23: 867-876.

Underhill, D.M. and Ozinsky, A. (2002). Toll-like receptors: key mediators of microbe detection. Cur. Opin. Immunol. 14: 103-110.

Utke, K., Bergmann, S., Lorenzen, N., Kollner, B., Ototake, M. and Fischer, U. (2007). Cell-mediated cytotoxicity in rainbow trout, *Oncorhynchus mykiss*, infected with viral haemorrhagic septicaemia virus. Fish Shellfish Immun. 22: 182-196.

Verhoog, H. (2001). The intrinsic value of animals: its implementation in governmental regulations in the Netherlands and its implications for plants. In: Intrinsic Value and Integrity of Plants in the Context of GE. Proceedings from Ifgene workshop. D. Heaf and J. Wirz (eds). Dornach, Switzerland, pp 15-18.

Von Schomberg, R. (2006). The precautionary principle and its normative challenges. In: Implementing the Precautionary Principle: Perspectives and Prospects. E. Fisher, J. Jones and R. von Schomberg (eds). Cheltenham, UK, pp. 19-42.

Walker, W.E., Harremoeës, P., Rotmans, J., van der Sluijs, J.P., van Asselt, M.B.A., Janssen, P. and Kraye von Krauss, M.P. (2003). Defining uncertainty; a conceptual basis for uncertainty management in model based decision support. J. Integrated Assess. 4: 5-17.

Wang, Z., Troilo, P.J., Griffiths, II, T.G., Pacchione, S.J., Barnum, A.B., Harper, L.B. et al. (2004). Detection of integration of plasmid DNA into host genomic DNA following intramuscular injection and electroporation. Gene Therap. 11: 711-721.

Weiner, D.B. and Kennedy, R.C. (1999). Genetic vaccines. Sci. Am. 281: 50-57.

Wermenstam, N.E. and Pilstrom L. (2001). T-cell antigen receptors in Atlantic cod (*Gadus morhua* L.): structure, organisation and expression of TCR [alpha] and [beta] genes. Dev. Comp. Immunol. 25: 117-135.

Wilson, M.R., Zhou, H., Bengten, E., Clem, L.W., Stuge, T.B., Warr, G.W. et al. (1998). T-cell receptors in channel catfish: structure and expression of TCR [alpha] and [beta] genes. Mol. Immunol. 35: 545-557.

Whitton, J.L., Rodriguez, F., Zhang, J. and Hassett, D.E. (1999). DNA immunization: mechanistic studies. Vaccine 17: 1612-1619.

Wynne, B. (1992). Uncertainty and environmental learning: reconciving science and policy in the preventive paradigm. Global Environ. Chang. 2: 111-127.

Wynne, B. (2001). Creating public alienation: expert cultures of risk and ethics on GMOs. Science as Culture 10: 445-481.

Zhao, H., Hemmi, H., Akira, S., Cheng, S.H., Scheule, R.K. and Yew, N.S. (2004). Contribution of Toll-like receptor 9 signaling to the acute inflammatory response to nonviral vectors. Mol. Therap. 9: 241-248.

Zhou, H., Bengtén, E., Miller, N.W., Warr, G.W., Clem, L.W. and Wilson, M.R. (1997). T cell receptor sequences in the channel catfish. Dev. Comp. Immunol. 21: 238.

7

The Use of Probiotics in Aquaculture

B. Austin* and J.W. Brunt**

INTRODUCTION

Probiotics have entered everyday language largely as a result of their common use on various milk products. Here, the term probiotic refers to the bacteria, i.e. lactic acid bacteria namely *Bifidobacterium*, *Lactobacillus* and *Streptococcus* (Fuller, 1987; Smoragiewicz et al., 1993), used to produce milk products, notably yogurts (Shinohara et al., 2002) and cultured buttermilk (Rodas et al., 2002). The problem for the consumer is that there is usually negligible information to support the nature or number of the organisms in the product, and there could well be issues with quality control. Moreover, it is relevant to enquire whether or not storage could affect the microbial populations in these milk products? So what is a probiotic?

By definition, a probiotic is a cultured product or live microbial feed supplement, which beneficially affects the host by improving its intestinal (microbial) balance and should be capable of commercialization (Fuller, 1987). In 2001, the Food and Agriculture Organization (FAO) of the United Nations and the World Health Organization (WHO) refined this definition, and stated that probiotics are "live microorganisms which when administered in adequate amounts confer a health benefit on the host" (FAO/WHO Report, 2001). Still further new or modified definitions are being proposed to encompass the different aspects and action of

School of Life Sciences, Heriot-Watt University, Riccarton, Edinburgh, Scotland, EH14 4AS (U.K.)
*Corresponding author: E-mail: b.austin@hw.ac.uk
**Present address: Institute of Food Research, Norwich Research Park, Colney, Norwich NR4 7UA, England

probiotics, and are set to continue as our understanding and development of probiotics continues. Probiotics have become widely accepted in human and veterinary use, with products available in many countries for cattle (Khuntia and Chaudhary, 2002) and poultry, e.g. ducks (Fulton et al., 2002). Use of probiotics in aquaculture is comparatively new and has an unprecise origin, but it is clear that products have become widely used for finfish (Austin and Austin, 2007) and invertebrates (Jorquera et al., 2001; Macey and Coyne, 2005; Farzanfar, 2006; Preetha et al., 2007), particularly in South America (Chile and Ecuador) and Asia (especially China and India). However, the concept of a probiotic as used in aquaculture has been broadened to include a wide range of organisms not just lactic acid bacteria and their products, and the application is not always *via* feed (Gram et al., 1999; Salminen et al., 1999) but may include the waterborne route (Moriarty, 1999; Makridis et al., 2000). In this context, there could be confusion between a probiotic and a biological control agent administered by water (Maeda et al., 1997). Nevertheless, the literature supports the notion that microorganisms are beneficial to aquatic animals with use leading to:

- Enhanced appetite and growth (Macey and Coyne, 2005; El-Haroun et al., 2006; Kumar et al., 2006) even when use of the probiotic has stopped (Brunt and Austin, 2005). This aspect of a probiotic has added value for the user.
- A reduction in nonspecific and specific diseases leading to herd immunity (Irianto and Austin, 2002a, b).
- Reduction in the need for costly chemotherapy (Wang et al., 1999; Irianto and Austin, 2002a, b).

Although probiotics should not be harmful to the host, supportive data are not always published (Salminen et al., 1999). For some organisms, particularly those belonging to Gram-negative taxa, there is a realistic concern about the possible acquisition of virulence genes by horizontal gene transfer particularly when the cultures enter the aquatic environment.

PROBIOTICS CONSIDERED FOR USE IN AQUACULTURE

For all uses, except for aquaculture, probiotics centre on the Gram-positive lactic acid producing bacteria, i.e. the lactobacilli. However, a diverse range of microorganisms has been considered for aquatic animals (Table 7.1) (e.g. Riquelme et al., 1997, 2000; Araya et al., 1999; Ruiz-Ponte et al., 1999; Skjermo and Vadstein, 1999; Gatesoupe, 2000; Gomez-Gil et al., 2000; Makridis et al., 2000; Verschuere et al., 2000; Huys et al., 2001; El-Sersy et al., 2006), and include representatives of Gram-negative and

Table 7.1 Organisms evaluated as probiotics for use in aquaculture

Probiotic identity	Isolation/source	Used with	Application	Reference
Gram-positive				
Bacillus subtilis	Gastrointestinal bacterium isolated from *Cirrhinus mrigala*	Indian major carp, *Labeo rohita*	Premix with feed	Kumar et al. (2006)
Bacillus subtilis BT23	Isolated from shrimp culture ponds	Black tiger shrimp, *Penaeus monodon*	Mixed with water	Vaseeharan and Ramasamy (2003)
Bacillus S11	Gastrointestinal tract of *P. monodon* broodstock	Black tiger shrimp, *P. monodon*	Premix with feed	Rengipipat et al. (2003)
Bacillus subtilis (CECT 35)	-	Gilthead sea bream, *Sparus aurata* L.	Premix with feed	Salinas et al. (2005)
Bacillus P64	Hepatopancreas of adult shrimp *Penaeus vannamei*	Adult shrimp, *P. vannamei*	Added to water	Gullian et al. (2004)
Bacillus (BIOSTART)	Commercial product	Catfish, *Ictalurus punctatus*	Added to water	Queiroz and Boyd (1998)
Bacillus (DMS series)	Commercial product	Shrimp, *P. monodon*	Added to water	Moriarty (1998)
Mixed culture, mostly *Bacillus* spp.	Commercial product	*Brachionus plicatilis*	Mixed with water	Hirata et al. (1998)
Bacillus sp. 48	Common snook	*Centropomus undecimalis*	Added to water	Kennedy et al. (1998)
Bacillus sp. S11	*Penaeus monodon*	*P. monodon*	Premix with feed	Rengipat et al. (1998)
Carnobacterium sp. BA211	*Oncorhynchus mykiss* digestive tract	*O. mykiss*	Premix with feed	Irianto and Austin (2002a)
Bacillus subtilis	-	Tilapia, *Oreochromis niloticus*	Premix with feed	Gunther and Jimenez-Montealegre (2004)
Enterococcus faecium SF 68	Commercial product	*Anguilla anguilla*	Oral administration	Chang and Liu (2002)

Table 7.1 contd...

Probiotic	Source	Target species	Delivery method	Reference
Lactobacillus helveticus	Turbot larvae	*Scophthalmus maximus*	Indirectly *via* rotifers	Gatesoupe (1991)
Lactobacillus plantarum	Turbot larvae	*S. maximus*	Indirectly *via* rotifers	Gatesoupe (1991)
Lactobacillus plantarum (906)	Human faeces	Sea bream *Sparus aurata*	*Brachionus plicatilis* and/or *Artemia salina* and dry feed as vectors	Carnevali et al. (2004)
Lactobacillus sakei	Intestines of healthy salmonids	Rainbow trout, *O. mykiss*	Premix with dry feed	Balcazar et al. (2006)
Lactobacillus rhamnosus ATCC 53103	Culture collection	*O. mykiss*	Mixed with feed	Nikoskelainen et al. (2001)
Lactobacillus rhamnosus (ATCC 53103)	Culture collection	Tilapia, *O. niloticus*	Incorporated into commercial dry pellets	Pirirat et al. (2006)
Lactobacillus sp. DS-12	Flounder intestine	*Paralichthys olivaceus*	Premix with feed	Byun et al. (1997)
Lactobacillus sp.	Tilapia intestine	*Oreochromis niloticus*	Premix with feed	Suyanandana et al. (1998)
Carnobacterium divergens	Atlantic salmon intestine	*Gadus morhua*	Feed	Gildberg et al. (1997)
Lactobacillus lactis AR21	Rotifer mass culture	*Brachionus plicatilis*	Feed additive	Harzevili et al. (1998)
Carnobacterium inhibens K1	Atlantic salmon intestine	Salmonids	Premix with feed	Jöborn et al. (1997)
Micrococcus luteus	Rainbow trout intestine	Rainbow trout	Premix with feed	Irianto and Austin (2002a)
Micrococcus MCCB 104	From hatchery water	*Macrobrachium rosenbergii* larvae	Added to water	Jayaprakash et al. (2005)
Gram-negative				
Pseudomonas fluorescens	*Salmo trutta*	*Salmo salar*	Bath	Smith and Davey (1993)
Pseudomonas I-2	Estuarine water sample	*P. monodon*	Added to water	Chythanya et al. (2002)
Pseudomonas fluorescens AH2	*O. mykiss*	*O. mykiss*	Mixed with water	Gram et al. (2001)
Pseudomonas sp.	Skin, gills and intestine *O. mykiss*	*O. mykiss*	Via water	Spanggaard et al. (2001)
Pseudomonas sp. PS-102	Brackish water lagoon	*P. monodon* larvae	?	Vijayan et al. (2006)

Table 7.1 contd...

Pseudomonas sp. PM 11	Gut contents of farm-reared sub-adult tiger shrimp, *P. monodon*	*P. monodon*	Added to water	Alavandi et al. (2004)
Alteromonas CA2	-	*Crassostrea gigas*	Mixed with water	Douillet and Langdon (1994)
Vibrio proteolyticus	-	Juvenile turbot *Scophthalmus maximus*	Liquid mixture	Deschrijver and Ollevier (2000)
Aeromonas media A199	-	*C. gigas*	Mixed with water	Gibson et al. (1998)
Vibrio fluvialis	*O. mykiss* digestive tract	*O. mykiss*	Premix with feed	Irianto and Austin (2002 a, b)
Vibrio fluvialis PM 17	Gut contents of farm-reared sub-adult tiger shrimp. *P. monodon*	*P. monodon*	Added to water	Alavandi et al. (2004)
A. sobria	*O. mykiss* digestive tract	*O. mykiss*	Premix with feed	Brunt and Austin (2005)
Roseobacter sp. BS 107	Marine sample	Scallop larvae	Mixed with water	Ruiz-Ponte et al. (1999)
Roseobacter strain 27-4	Turbot hatchery	Turbot *Scophthalmus maximus* L.	Rotifers enriched with *Roseobacter* 27-4	Planas et al. (2006)
Synechocystis MCCB 114 and 115	Seawater	*Penaeus monodon* post-larvae	Mixed with feed	Preetha et al. (2007)
Bacteriophage				
Representative of (Myoviridae and Podoviridae)	Diseased ayu and pond water	*Plecoglossus altivelis*	Premix with feed	Park et al. (2000)
Yeasts				
Saccharomyces cerevisiae, S. exiguous, Phaffia rhodozyma	Commercial product	*Litopenaeus vannamei*	Premix with feed	Scholz et al. (1999)
S. cerevisiae	-	Nile tilapia *Oreochromis niloticus*	Premix with feed	Meurer et al. (2006)
Debaryomyces hansenii HF1 (DH)	Gut of rainbow trout	Sea bass *Dicentrarchus labrax*	Premix with basal diet	Tovar et al. (2002)

Table 7.1 contd...

Unicellular marine algae

Isochryisis galbana	-	Sea bass *Dicentrarchus labrax*	Added to water	Cahu et al. (1998)
Dunaliella extract (Algro Natural®)	Commercial product	Black tiger shrimp, *Penaeus monodon*	Premix with commercial shrimp feed	Supamattaya et al. (2005)
Tetraselmis suecica	Commercial product	Penaeids, *S. salar*	Feed	Austin et al. (1992)
Dunaliella tertiolecta	Commercial product	Gnotobiotic *Artemia*	Feed	Marques et al. (2006)

Gram-positive bacteria, as well as bacteriophages, unicellular algae, yeasts, and their products. Application of these probiotics has included use of:

- Live feeds, namely on *Artemia* and rotifers (Gatesoupe, 1991; Harzevili et al., 1998),
- Artificial diets (e.g. Robertson et al., 2000; Brunt and Austin, 2005; Kim and Austin, 2006),
- The waterborne route (Austin et al., 1995; Moriarty, 1999; Ringø and Birkbeck, 1999) including their use in biofilters (Gross et al., 2003).

Gram-positive Bacteria

Among the Gram-positive bacteria considered as probiotics in aquaculture, there has been some interest in *Arthrobacter* (Plante et al., 2007) and considerable success with endospore-formers of the genus *Bacillus* (Hong et al., 2005) as feed additives (Kumar et al., 2006) in addition to their role in inhibiting pathogens (Wang et al., 1999) and improving water quality (Wang et al., 1999). Use of live preparations of *Bacillus pumulis* led to enhanced survival of *Penaeus japonicus* larvae (El-Sersy et al., 2006). Isolates of *Bacillus subtilis* have been especially successful, and an application at 1.5×10^7 colony forming units (CFU)/g in feed for 2 wk led to the control of *Aeromonas hydrophila* infection in Indian major carp, *Labeo rohita* (Kumar et al., 2006). Separately, *B. subtilis* BT23 and *Bacillus* S11 controlled *Vibrio harveyi* infections in black tiger shrimp, *Penaeus monodon* (Vaseeharan and Ramasamy, 2003; Rengpipat et al., 2003), whereas *B. subtilis* was documented as a probiotic for use with gilthead sea bream (Salinas et al., 2005). *Bacillus* P64 was reported to be inhibitory and immunomodulatory to *Penaeus vannamei* (Gullian et al., 2004). Commercial preparations containing *Bacillus* have been used in catfish (Queiroz and Boyd, 1998) and shrimp (Moriarty, 1998). Moreover, mixed cultures with *Bacillus* improved the performance of the rotifer *Brachionus plicatilis* in water (Hirata et al., 1998). Furthermore, *Bacillus* 48 enhanced the survival of larvae, increased food absorption by raising protease levels and conferred better growth on common snook, *Centropomus undecimalis* (Kennedy et al., 1998). Yet, there was not any evidence of benefit during the 100-d administration in feed of *Bacillus* S11 as wet or lyophilized cells, or saline suspensions to penaeids (Rengpipat et al., 1998). Although probiotics are often associated with increased appetite (e.g. Irianto and Austin, 2002a), sometimes a retardation of growth has been documented (Gunther and Jimenez-Montealegre, 2004).

A commercial product containing *Enterococcus faecium* SF 68 reduced edwardsiellosis, caused by *Edwardsiella tarda*, in European eel, *Anguilla*

anguilla (Chang and Liu, 2002). Thus, after feeding for 58 d at a dose of 2×10^8 cells/g of feed, *Enterococcus faecium* improved the growth of sheat fish, *Silurus glanis*, by 11% (Bogut et al., 2000).

An increasing range of lactic acid bacteria, including *Lactobacillus delbrueckii*, *Lactobacillus plantarum* and *Lactobacillus helveticus*, have been used extensively in aquaculture (Gatesoupe, 1991; Gildberg and Mikkelson, 1998; Carnevali et al., 2004; Salinas et al., 2005) when there is the added bonus of enhanced growth performance (Gatesoupe, 1991). Additionally, there has been reference to *Lactococcus lactis* subsp. *lactis*, *Lactobacillus sakei*, and *Leuconostoc mesenteroides* (Balcazar et al., 2006; 2007a, b). As an example, *Lactobacillus rhamnosus* ATCC 53101 was administered orally at 10^9 cells/g of feed to rainbow trout for 51 d with the outcome that mortalities caused by *A. salmonicida* were reduced from ~53 to ~19% (Nikoskelainen et al., 2001). In particular, this dose led to better protection than 10^{12} cells/g. In a separate study, *Lc. mesenteroides* CLFP 196 and *Lb. plantarum* CLFP 238 were administered orally to rainbow trout at 10^7 CFU/g feed. Following challenge with *Lactococcus garvieae*, probiotic supplementation reduced fish mortality significantly from 78% in the control group to 46–54% in the probiotic treated groups (Vendrell et al., 2007). Furthermore, *Lb. rhamnosus* was successful in combating *Edwardsiella tarda* infection in tilapia, *Oreochromis niloticus* (Pirirat et al., 2006). *Lb. fructivorans* and *Lb. plantarum*, which were isolated from sea bream *(Sparus aurata)* gut and human faeces, respectively, were fed to sea bream using *Brachionus plicatilis* and/or *Artemia* as vectors (Carnevali et al., 2004). The results revealed an effect on gut colonization and in particular stimulation of HSP 70, which points to an improvement in stress tolerance (Rollo et al., 2006). Another study pointed to the usefulness of *Lactobacillus fructivorans* and *Lb. plantarum* leading to significantly decreased larval and fry sea bream mortality (Carnevali et al., 2004). Similarly, *Lactobacillus* DS-12 was beneficial to flounder (Byun et al., 1997). Another *Lactobacillus* culture was useful to tilapia (Suyunandana et al., 1998). Moreover in Atlantic cod *(Gadus morhua)* fry, *Carnobacterium divergens* reduced vibriosis caused by *Vibrio anguillarum* (Gildberg et al., 1997). *Lactococcus lactis* AR21 stimulated the growth of rotifers and inhibited *V. anguillarum* (Harzevili et al., 1998). *Carnobacterium inhibens* K1, which was isolated from the digestive tract of Atlantic salmon, *Salmo salar*, inhibited a range of bacterial fish pathogens *in vitro* (Jöborn et al., 1997; 1999). *In vivo* experiments demonstrated that the organism was active in intestinal mucus and faeces. Oral administration of *Carnobacterium inhibens* K1 enhanced appetite, reduced minor health problems such as fin and tail rot, and virtually eliminated mortalities in salmonids caused by *Aeromonas salmonicida*, *Vibrio ordalii* and *Yersinia ruckeri* (Robertson et al., 2000).

Micrococcus luteus reduced *A. salmonicida* infections in rainbow trout, *Oncorhynchus mykiss* (Irianto and Austin, 2002a). Furthermore, *Micrococcus* MCCB 104 inhibited *Aeromonas* sp., *Vibrio alginolyticus*, *V. cholerae*, *V. fluvialis*, *V. mediterranei*, *V. nereis*, *V. parahaemolyticus*, *V. proteolyticus* and *V. vulnificus*, and was inferred to be useful in freshwater prawn, *Macrobrachium rosenbergii*, larvae (Jayaprakash et al., 2005).

Streptomyces has been used as a probiotic in shrimp, *Penaeus monodon* (Das et al., 2006). Following a 25 d feeding regime at 10 g of cells/kg of feed, improved growth (both length and weight) and better water quality ensued (Das et al., 2006). *Weissella hellenica* DS-12 (Cai et al., 1998) was antagonistic to some bacterial fish pathogens, and was regarded to have potential as a probiotic (Byun et al., 1997).

Gram-negative Bacteria

In contrast to terrestrial agriculture and human medicine, a wide range of Gram-negative bacteria have been considered for use in aquaculture. *Pseudomonas fluorescens* inhibited *Saprolegnia* sp. and *A. salmonicida* infection (Smith and Davey, 1993; Bly et al., 1997), and by means of low molecular weight inhibitors *Pseudomonas* I-2 was harmful to shrimp pathogenic *Vibrio harveyi*, *V. fluvialis*, *V. parahaemolyticus*, *V. vulnificus* and *Photobacterium damselae* (Chythanya et al., 2002). Bathing rainbow trout for 6 d in *Pseudomonas fluorescens* AH2 reduced mortalities caused by *Vibrio anguillarum* from 47 to 32% (Gram et al., 1999) Again *Pseudomonas* featured in the control of vibriosis in rainbow trout albeit after application *via* water (Spanggaard et al., 2001). Yet, *P. fluorescens* AH2, which was an effective probiotic protecting rainbow trout against vibriosis, was not beneficial to Atlantic salmon against infection with *A. salmonicida* despite *in vitro* methods involving iron-depletion indicating inhibition (Gram et al., 2001). *Pseudomonas* PS-102 was reported as being antagonistic to pathogenic vibrios of relevance to penaeid culture (Vijayan et al., 2006). In addition, *Pseudomonas* PM11 was considered as a probiotic for use with tiger shrimp, *Penaeus monodon* (Alavandi et al., 2004).

Other Gram-negative bacteria have been reported to improve the culture of larval crab, Pacific oyster, *Crassostrea gigas*, and turbot (Nogami and Maeda, 1992; Douillet and Langdon, 1994; Gatesoupe, 1994) with *Vibrio proteolyticus* improving protein digestion in juvenile turbot after oral intubation (Deschrijver and Ollevier, 2000). *Vibrio midae* was described as a probiotic, which colonized the digestive tract of abalone, *Haliotis midae* (Macey and Coyne, 2006). An organism presumptively identified as *Alteromonas* CA2 increased survival of Pacific oyster when administered in water (Douillet and Langdon, 1994). Moreover, *Aeromonas media* A199 was effective at controlling *Vibrio tubiashii* infection in Pacific oyster larvae by means of bacteriocin-like inhibitory substances which were effective against several pathogenic bacteria in culture (Gibson et al., 1998; Gibson, 1999). In addition, *A. hydrophila* and *Vibrio fluvialis* were effective at controlling infections by *A. salmonicida* in rainbow trout (Irianto and Austin, 2002a). Also along with *Bacillus pumulis*, *Vibrio fluvialis* was reported to substantially enhance survival of *Penaeus japonicus* larvae (El-Sersy et al., 2006). *Vibrio fluvialis* PM17 was mentioned as a probiotic for use with tiger shrimp, *Penaeus monodon* (Alavandi et al., 2004).

Live cells of *Aeromonas sobria* GC2, which were incorporated into feed at 5×10^7 cells/g of feed and fed to rainbow trout for 14 d, conferred protection against challenge by *Lactococcus garvieae* and *Streptococcus iniae* (Brunt and Austin, 2005). Meanwhile, formalized and sonicated preparations of *A. sobria* GC2 and cell-free supernatants fared less well (Brunt and Austin, 2005). Furthermore, *Roseobacter* (BS 107), which in co-culture with *Vibrio anguillarum*, was inhibitory to *Vibrio*, with cell extracts enhancing the survival of larval scallop (Ruiz-Ponte et al., 1999). Another representative, *Roseobacter* 27-4, applied in rotifers, was effective at controlling *V. anguillarum* infection in larval turbot, *Scophthalmus maximus* (Planas et al., 2006).

Cyanobacteria, namely *Synechocystis* spp., have been evaluated for the control of *Vibrio harveyi* infection in *Penaeus monodon* postlarvae (Preetha et al., 2007). Feeding with *Synechocystis* was reported to lead to a substantial reduction or complete removal of vibrios from the intestine of shrimp, and a concomitant increase in survival following challenge with *V. harveyi* (Preetha et al., 2007).

Bacteriophage

Two bacteriophage cultures representing the families Myoviridae and Podoviridae were obtained from diseased ayu, *Plecoglossus altivelis*, and applied orally leading to a decline in numbers (in water and kidney) and protection against *Pseudomonas plecoglossicida* (Park et al., 2000).

Yeasts

Bacteria and yeasts led to increased weight and survival of catla, *Catla catla* (Mohanty et al., 1996). However, cells and ß-glucan of *Saccharomyces cerevisiae*, *S. exiguous* cells containing xeaxanthin (HPPR1) and *Phaffia rhodozyma* improved resistance of juvenile penaeids to vibriosis (Scholz et al., 1999). In particular, the diets containing *Phaffia rhodozyma* greatly improved larval survival. *S. cerevisiae* was highlighted by Meurer et al. (2006) for use as a probiotic for Nile tilapia, *Oreochromis niloticus*. Also, a polyamine (spermine and spermidine) producing yeast recovered from the digestive tract of fish, i.e. *Debaryomyces hansenii*, improved survival but reduced the growth of larval sea bass, *Dicentrarchus labrax* (L.), when administered orally (Tovar et al., 2002). The presence of the yeast, which adhered to the gut, led to enhanced amylase secretion and a stimulation of brush border membrane enzymes in the 27-d-old larvae (Tovar et al., 2002). Furthermore, *D. hansenii* and *Cryptococcus* sp. were cited as probiotics, which colonized the digestive tract of abalone, *Haliotis midae* (Macey and Coyne, 2006).

Unicellular Marine Algae

Various studies have shown enhanced growth, survival or improved health status of marine larvae cultured with microalgae (e.g., *Isochrysis galbana*, *Tetraselmis suecica*, *Phaedactylum tricornutum*, *Dunaliella salina* or *Dunaliella tertiolecta*) (e.g. Nass et al., 1992, Reitan et al., 1997, Cahu et al., 1998; Supamattaya et al., 2005; Marques et al., 2006). A heterotrophically grown, spray-dried unicellular marine alga, *Tetraselmis suecica*, was useful as a feed for penaeids and as a feed-additive for salmonids with data revealing a reduction in the level of bacterial diseases (Austin and Day, 1990; Austin et al., 1992). In a subsequent development, Marques et al. (2006) revealed that the microalga, *Dunaliella tertiolecta* enhanced protection of gnotobiotic *Artemia* against two pathogenic bacterial strains, i.e. *Vibrio campbellii* and *V. proteolyticus*.

The Need for Single or Multiple Cultures of Probiotics

Should probiotics be used as single or multiple cultures? This aspect has been comparatively ignored, to date, although some studies have addressed the use of combinations of probiotics, for example, *B. subtilis* with *Lactobacillus delbrueckii* (Salinas et al., 2005). Certainly, the relevance of combining probiotics with prebiotics needs to be addressed (Salminen et al., 1998).

Viable versus Inactivated and Subcellular Preparations

There is an assumption that a live preparation of probiotics remains viable throughout the period of use of the feedstuff. However in one study, data revealed that inactivated cells were just as effective as live preparations (Irianto and Austin, 2003). Also, subcellular components and extracellular products may be beneficial (Brunt and Austin, 2005). However, the converse is true in other cases when only viable cells were considered to be effective (Taoka et al., 2006), but this may reflect the taxonomic status of the organism with some taxa surviving better than others. The attraction about inactivated, subcellular and cell-free preparation is that there should be less safety concerns, i.e. the products should not be capable of gaining pathogenicity and therefore causing harm to the host.

SOURCE OF AND REASONS FOR SELECTING PROBIOTIC CULTURES

Generally, pututative probiotics are obtained from the aquatic environment, such as water or sand (e.g. Preetha et al., 2007) and from the digestive tract of aquatic animals (e.g. Irianto and Austin, 2002b), although culture collections are sometimes used (Hjelm et al., 2004). Ecuadorian shrimp larval aquaculture has achieved widespread success with the use of probiotics often derived from seawater or sand, and producing round yellow colonies of 3–5 mm in diameter on thiosulphate citrate bile salts sucrose agar, with a concomitant reduction in the use of and need for antimicrobial compounds (Garriques and Arevalo, 1995). The probiotics were equated with *Vibrio alginolyticus* often by means of the API 20E rapid identification system. Indeed, a more detailed taxonomy of 23 isolates confirmed the identity as *V. alginolyticus* but regarded some as comprising *Vibrio* spp. (Vandenberghe et al., 1999). One of the *V. alginolyticus* isolates inhibited *Vibrio ordalii*, *V. anguillarum*, *A. salmonicida* and *Y. ruckeri*, and protected Atlantic salmon after challenge with *A. salmonicida* and to a lesser extent *V. anguillarum* and *V. ordalii* (Austin et al., 1995).

Fish guts are often used as a source of probiotics, including *Carnobacterium* and an unidentified Gram-positive coccus, which were dosed at ~10^7 cells/g of feed for 14 d, for the control of *A. salmonicida* (Irianto and Austin, 2002a) with the data revealing a complete lack of harm after intramuscular or intraperitoneal injection and enhanced feeding activity within 1 d of initiating feeding. Again, there was complete success with controlling mortalities caused by *A. salmonicida* (Irianto and Austin, 2002a).

The means of choosing candidate probiotics has often reflected inhibitory activity against target pathogens *in vitro* (e.g. Hjelm et al., 2004; Bourouni et al., 2007). The inhibition due to such compounds is highly dependent on the experimental conditions, which are different *in vitro* and *in vivo*. However, it may be argued that inhibition *in vitro* is not a sufficient criterion to select candidate probiotics (Riquelme et al., 1997; Verschuere et al., 2000) nor is the absence of antagonistic activity sufficient to exclude potentially useful strains (Rico-Mora et al., 1998). Other criteria have been adopted by some workers including an effect of the potential probiotic on the mucus adhesion ability of the target pathogen (Vine et al., 2004a; Balcazar et al., 2007b). Indeed, a ranking index, which was based on the *in vitro* characteristics—including the doubling time and lag period of the cultures, was proposed by Vine et al. (2004a) to enable the meaningful selection of putative probiotics.

MODE OF ACTION OF PROBIOTICS

In many publications, the case for proposing probiotic activity is based on laboratory rather than field-work with data highlighting reductions in mortalities and increased resistance to disease (e.g. Gatesoupe, 1994; Moriarty, 1998; Skjermo and Vadstein, 1999; Chang and Liu, 2002). In comparison with terrestrial use, it is often assumed — sometimes with supporting evidence - that the mode of action reflected competitive exclusion by which the probiotic occupies niches (= colonizes) in the digestive tract (Jöborn et al., 1997; Macey and Coyne, 2006) and antagonizes any potential pathogen (Jöborn et al., 1997) by the production of inhibitory compounds or by competition for nutrients, space (= adhesion sites in the digestive tract) or oxygen (Fuller, 1987). There is some evidence that this may apply to fish insofar as the application of some probiotics reduced the ability of pathogenic bacteria, i.e. *V. alginolyticus* (Vine et al., 2004b) and *V. harveyi* (Chabrillon et al., 2005b), to attach to the skin and intestinal mucus. However, the longevity of the probiotics in the digestive tract may well reflect the age and health status of the host. For example, there is anecdotal evidence that the feeding of probiotics to juvenile fish or to older animals immediately after antibiotic treatment may lead to prolonged colonization by the probiotic. Competitive exclusion may apply in some situations with fish, but in addition, a range of other possibilities has been broached including:

- Improved digestibility of feed by the action of microbial [digestive (Tovar et al., 2002)] enzymes (Lin et al., 2004; Ziaei-Nejad et al., 2006) including amylases and proteases (Fuller and Turvy, 1971; Parker, 1974; Roach and Tannock, 1980; Fuller, 1992; Smoragiewicz et al., 1993; Sugita et al., 1996; Hoshina et al., 1997), thus removing potentially indigestible components in the diet (Chabrillon et al., 2005a). With this scenario, the enzymes from the probiotic cultures breakdown feed components in the digestive tract thereby improving nutrition. Enhanced protease activity was also recorded in abalone after the administration of probiotics (Macey and Coyne, 2005). Additionally, *S. cerevisiae* when administered as probiotics to Nile tilapia (*Oreochromis niloticus*) stimulated growth performance and feed efficiency (Lara-Flores et al., 2003). Perhaps, essential macro- or micro-nutrients may also be generated, e.g. biotin and vitamin B_{12} (Sugita et al., 1991, 1992) and fatty acids (Vine et al., 2006), which could improve the health of the recipient animal.

- Stimulation of the immune system, or immunomodulation, is considered an important mechanism to support probiotic activity (Hong et al., 2005). In particular, stimulation of innate immunity, including increased phagocytic and lysozyme activities (Gibson et al., 1997; Irianto and Austin, 2002b; Brunt and Austin, 2005; Salinas et al., 2005; Balcazar et al., 2006; Kim and Austin, 2006; Balcazar et al., 2007a, b), increased numbers of leucocytes (Brunt and Austin, 2005), migration of neutrophils and plasma bactericidal activity (Taoka et al., 2006), complement activity (Panigrahi et al., 2004), cytotoxicity (Salinas et al., 2005) and enhanced respiratory burst and leucocyte peroxidase activity (Salinas et al., 2006). As an example, feeding with Gram-positive and Gram-negative probiotics at 10^7 cells/g of feed led to an increase in the number of erythrocytes, macrophages, lymphocytes and enhanced lysozyme activity within 2 wk of feeding with probiotics (Brunt and Austin, 2005). It is likely that probiotics contain immunostimulatory molecules, with the mode of action reflecting the presence and administration of these compounds. Lipopolysaccharide from Gram-negative bacteria, vibrio vaccines, *Clostridium butyricum* spores, and glucan from yeast cell walls have been considered as immunostimulants for use in aquaculture (Sakai, 1998). These compounds have been found to stimulate the host's defence system by increasing phagocytosis, antibody production, increasing the chemiluminescent response and by superoxide anion production (Sakai, 1998). In the case of invertebrates, an association with immunity is sometimes more difficult to discern. Indeed after administering *Pseudomonas* PM11 and *V. fluvialis* PM17 to tiger shrimp, Alavandi et al. (2004) reported a decline in immunological traits, specifically haemocyte counts, phenol oxidase and antibacterial activity. Nevertheless, Macey and Coyne (2005) reported an increase in circulating and phagocytic haemocytes in abalone which had been administered with probiotics. These animals had increased survival against challenge with *V. anguillarum*. Proteomic and genomic approaches have recently been applied to studies of the Acute Phase Response (APR) in some teleosts (Gerwick et al., 2000) The APR has been defined as a rapid, orchestrated, physiologically induced response to tissue injury, infection, neoplasia, trauma and stress (Baumann and Gauldie, 1994; Jensen et al., 1997). This involves a large number of acute phase proteins (APP) and functions in a variety of defence-related activities, such as limiting the dispersal of infectious agents, repairing of tissue damage, the killing of microorganisms and restoration of a healthy state (Gerwick et al., 2000; Larsen et al., 2001; Gerwick et al., 2002). Several of the mammalian positive APP has been identified, but it is anticipated that more are still unknown because the host response is complex, and has received only marginal attention. Brunt et al. (2007) used a proteomics approach to observe plasma proteins whose levels change during exposure to a probiotic in comparison to sera from control fish. In all individual fish that were treated with GC2, three proteins were increased (Pt1, Pt2 and Pt3). These proteomic techniques show great promise in increasing our understanding of probiotic action (Brunt et al., 2007).

• A placebo effect? This is an interesting possibility, and needs to be considered seriously.

Until recently, the mode of action of probiotics used in aquaculture was not considered in detail. The length of feeding time and the long-term probiotic effect is also generally ignored by researchers. A purely descriptive approach was adopted in which the name, dose and target pathogen of the beneficial organism was reported to the exclusion of other important information. It is certainly possible that any probiotic could lead to multiple responses by the host. In addition, different probiotics could lead to greatly differing effects. For example, with *Carnobacterium inhibens* K, it was appreciated that the organism produced weak antimicrobial activity, and the cells were capable of remaining in the digestive tract during feeding regimes (Robertson et al., 2000).

In addition to Ecuador, probiotics are finding increasing use in Chinese aquaculture. For example, phototrophic bacteria of the genus *Photorhodobacterium* are used in grow out of ponds culturing *Penaeus chinensis* (Xu, pers. comm.).

FUTURE DEVELOPMENTS AND CONCLUSIONS

There is an overriding concern about the legal status of probiotics – are they feed additives or medicines? There is a blurred distinction between the meaning of a probiotic, an oral vaccine and an immunostimulant. If probiotics are commercialized and are used in a viable form then there is a realistic possibility than unscrupulous individuals could acquire and exploit the cultures. Notwithstanding there is ample evidence that microorganisms have an important contribution to make to aquaculture as potential probiotics. Nevertheless, it is essential that the probiotics need to be chosen prudently, with attention clearly focused on the effects of the microorganisms on the host, i.e. it is essential to ensure that the culture is harmless. This raises a concern regarding the possibility that some seemingly harmless organisms could gain virulence by horizontal gene transfer.

Future considerations should include the possibility of using probiotics in combination with other immunostimulants, vaccines, probiotics or as a mix containing several probiotic cultures. Moreover, it is apparent from the published literature that the precise mechanism observed is a multi-complex process, encompassing a variety of possible mechanisms of action, and the need for more research in this area is valid. Finally, if probiotics are to become a real alternative control strategy, meticulous assessments of probiotic functionality have to be pursued and the suitability of a given probiotic for different species and its environments needs to be fully explored. Finally, large scale field trials need to be performed and any concerns of pathogenicity need to be addressed before any consideration for commercialization is proposed.

ABBREVIATIONS

APP acute phase protein
APR acute phase response
CFU colony-forming unit
ECP extracellular product
HSP heat shock protein
LPS lipopolysaccharide

REFERENCES

Alavandi, S.V., Vijayan, K.K., Santiago, T.C., Poornima, M., Jithendran, K.P., Ali, S.A. and Rajan, J.J.S. (2004). Evaluation of *Pseudomonas* sp. PM11 and *Vibrio fluvialis* PM17 on immune indices of tiger shrimp, *Penaeus monodon*. Fish Shellfish Immunol. 17: 115-120.

Araya, R.A., Jorquera, M.A. and Riquelme, C.E. (1999). Association of bacteria to the life cycle of *Argopecten purpuratus*. Rev. Chilena Hist. Nat. 72: 261-271.

Austin, B. and Day, J.G. (1990). Inhibition of prawn pathogenic *Vibrio* spp. by a commercial spray-dried preparation of *Tetraselmis suecica*. Aquaculture 90: 389-392.

Austin, B. and Austin, D.A. (2007). Bacterial Fish Pathogens Disease of Farmed and Wild Fish, 4th Edition. Springer Praxis, Godalming, U.K.

Austin, B., Baudet, E. and Stobie, M.B.C. (1992). Inhibition of bacterial fish pathogens by *Tetraselmis suecica*. J. Fish Dis. 15: 55-61.

Austin, B., Stuckey, L.F., Robertson, P.A.W., Effendi, I. and Griffith, D.R.W. (1995). A probiotic strain of *Vibrio alginolyticus* effective in reducing diseases caused by *Aeromonas salmonicida*, *Vibrio anguillarum* and *Vibrio ordalii*. J. Fish Dis. 18: 93-96.

Balcazar, J.L., Vendrell, D., de Blas, I., Ruiz-Zarzuela, I., Girones, O. and Muzquiz, J.L. (2006). Immune modulation by probiotic strains: quantification of phagocytosis of *Aeromonas salmonicida* by leukocytes isolated from gut of rainbow trout *(Oncorhynchus mykiss)* using a radiolabelling assay. Comp. Immunol. Microbiol. Infect. Dis. 29: 335-343.

Balcazar, J.L., de Blas, I., Ruiz-Zarzuela, I., Vendrell, D., Calvo, A.C., Marquez, I., Girones, O. and Muzquiz, J.L. (2007a). Changes in intestinal microbiota and humoral immune response following probiotic administration in brown trout (*Salmo trutta*). Brit. J. Nutr. 97: 522-527.

Balcazar, J.L., Vendrell, D., de Blas, I., Ruiz-Zarzuela, I., Girones, O. and Muzquiz, J.L. (2007b). *In vitro* competitive adhesion and production of antagonistic compounds by lactic acid bacteria against fish pathogens. Vet. Microbiol. 122: 373-380.

Baumann, H. and Gauldie, J. (1994). The acute phase response. Immunol. Today 15: 74-80.

Bly, J.E., Quiniou, S.M.-A., Lawson, L.A. and Clem, L.W. (1997). Inhibition of *Saprolegnia* pathogenic for fish by *Pseudomonas fluorescens*. J. Fish Dis. 20: 35-40.

Bogut, I., Milakovic, Z., Brkic, S., Novoselic, D. and Bukvic, Z. (2000). Effects of *Enterococcus faecium* on the growth rate and content of intestinal microflora in sheat fish (*Silurus glanis*). Vet. Med. 45: 107-109.

Bourouni, O.C., El Bour, M., Mraouna, R., Abdennaceur, H. and Boudabous, A. (2007). Preliminary selection study of potential probiotic bacteria from aquaculture area in Tunisia. Annal. Microbiol. 57: 185-190.

Brunt, J. and Austin, B. (2005). Use of a probiotic to control lactococcosis and streptococcosis in rainbow trout, *Oncorhynchus mykiss* (Walbaum). J. Fish Dis. 28: 693-701.

Brunt, J., Hansen, R., Jamieson, D.J. and Austin, B. (2007). Proteomic analysis of rainbow trout (*Oncorhynchus mykiss*, Walbaum) serum after administration of probiotics in diets. Vet. Immunol. Immunopathol. In Press, Corrected Proof, Available online 2 October 2007.

Byun, J.W., Park, S.C., Benno, Y. and Oh, T.K. (1997). Probiotic effect of *Lactobacillus* sp. DS-12 in flounder (*Paralichthys olivaceus*). J. Gen. Appl. Microbiol. 43: 305-308.

Cahu, C., Zambonino-Infante, J., Peres, A., Quazuguel, P. and Le Gall, M. (1998). Algal addition in sea bass *Dicentrarchus labrax* larvae rearing: effect on digestive enzymes. Aquaculture 161: 479-489.

Cai, Y.M., Benno, Y., Nakase, T. and Oh, T.K. (1998). Specific probiotic characterization of *Weissella hellenica* DS-12 isolated from flounder intestine. J. Gen. Appl. Microbiol. 44: 311-316.

Carnevali, O., Zamponi, M.C., Sulpizio, R., Rollo, A., Nardi, M., Orpianesi, C., Silvi, S., Caggiano, M., Polzonetti, A.M. and Cresci, A. (2004). Administration of probiotic strain to improve sea bream wellness during development. Aquacult. Int. 12: 377-386.

Chabrillon, M., Rico, R.M., Balebona, M.C. and Morinigo, M.A. (2005a). Adhesion to sole, *Solea senegalensis* Kaup, mucus of microorganisms isolated from farmed fish, and their interaction with *Photobacterium damselae* subsp. *piscicida*. J. Fish Dis. 28: 229-237.

Chabrillon, M., Rico, R.M., Arijo, S., Diaz-Rosales, P., Balebona, M.C. and Morinigo, M.A. (2005b). Interactions of microorganisms isolated from gilthead seabream, *Sparus aurata* L., on *Vibrio harveyi*, a pathogen of farmed Senegalese sole, *Solea senegalensis* (Kaup). J. Fish Dis. 28: 531-537.

Chang, C-I. and Liu, W-Y. (2002). An evaluation of two probiotic bacterial strains, *Enterococcus faecium* SF68 and *Bacillus toyoi*, for reducing edwardsiellosis in cultured European eel, *Anguilla anguilla*, L. J. Fish Dis. 25: 311-315.

Chythanya, R., Karunasagar, I. and Karunasagar, I. (2002). Inhibition of shrimp pathogenic vibrios by a marine *Pseudomonas* I-2 strain. Aquaculture 208: 1-10.

Das, S., Lyla, P.S. and Khan, S.A. (2006). Application of *Streptomyces* as a probiotic in the laboratory culture of *Penaeus monodon* (Fabricius). Israeli J. Aquacult. – Bamidgeh 58: 198-204.

DeSchrijver, R. and Ollevier, F. (2000). Protein digestion in juvenile turbot (*Scophthalmus maximus*) and effects of dietary administration of *Vibrio proteolyticus*. Aquaculture 186: 107-116.

Douillet, P.A. and Langdon, C.J. (1994). Use of probiotic for the culture of larvae of the Pacific oyster (*Crassostrea gigas* Thurnberg). Aquaculture 119: 25-40.

El-Haroun, E.R., Goda, A.M.A.S. and Chowdhury, M.A.K. (2006). Effect of dietary probiotic Biogen (R) supplementation as a growth promoter on growth performance and feed utilization of Nile tilapia *Oreochromis niloticus* (L.). Aquacult. Res. 37: 1473-1480.

El-Sersy, N.A., Abdelrazek, F.A. and Taha, S.M. (2006). Evaluation of various probiotic bacteria for the survival of *Penaeus japonicus* larvae. Fresenius Environ. Bull. 15: 1506-1511.

FAO/WHO (2001). Health and nutritional properties of probiotics in food including powder milk with liver lactic acid bacteria. Food and Agriculture Organization and World Health Organization Joint report. 34 pp.

Farzanfar, A. (2006). The use of probiotics in shrimp aquaculture. FEMS Immunol. Med. Microbiol. 48: 149-158.

Fuller R. (1987). A review, probiotics in man and animals. J. Appl. Bacteriol. 66: 365-378.

Fuller, R. (1992). Probiotics. The Scientific Basis, Chapman and Hall, London, U.K.

Fuller, R. and Turvy, A. (1971). Bacteria associated with the intestinal wall of the fowl (*Gallus domesticus*). J. Appl. Bacteriol. 34: 617-622.

Fulton, R.M., Nersessian, B.N. and Reed, W.M. (2002). Prevention of *Salmonella enteritidis* infection in commercial ducklings by oral chicken egg-derived antibody alone or in combination with probiotics. Poult. Sci. 81: 34-40.

Garriques, D. and Arevalo, G. (1995). An evaluation of the production and use of a live bacterial isolate to manipulate the microbial flora in the commercial production of *Penaeus vannamei* postlarvae in Ecuador. In: Swimming Through Troubled Waters. Proceedings of the Special Session on Shrimp Farming. C.L. Browd and J.S. Hopkins (eds). World Aquaculture Society, Baton Rouge, LA, USA, pp. 53-59.

Gatesoupe, F-J. (1991). Siderophore production and probiotic effect of *Vibrio* sp. associated with turbot larvae, *Scophthalmus maximus*. Aquat. Living Res. 10: 239-246.

Gatesoupe, F.-J. (1994). Lactic acid bacteria increase the resistance of turbot larvae, *Scophthalmus maximus* against pathogenic *Vibrio*. Aquat. Living Res. 7: 277-282.

Gatesoupe, F.-J. (2000). The use of probiotics in aquaculture. Aquaculture 180: 147-165.

Gerwick, L., Regnolds, W. and Bayne, C.J. (2000). A pre-cerebellin-like protein is part of the acute phase response in rainbow trout, *Oncorhynchus mykiss*. Devel. Comp. Immunol. 24: 597-607.

Gerwick, L., Steinhauer, R., Lapatra, S., Sandell, T., Ortuno, J., Hajiseyedjavadi, N. and Bayne, C.J. (2002). The acute phase response of rainbow trout (*Oncorhynchus mykiss*) plasma proteins to viral, bacterial and fungal inflammatory agents. Fish Shellfish Immunol. 12: 229-242.

Gibson, L.F. (1999). Bacteriocin activity and probiotic activity of *Aeromonas media*. J. Appl. Microbiol. 85: S243-S248.

Gibson, G.R., Saavendra, J.M., MacFarlane, S. and MacFarlane, G.T. (1997). Probiotics and intestinal infections, In: Probiotics 2, Application and Practical Aspects. R. Fuller (ed). Chapman and Hall, London, U.K., pp. 10-31.

Gibson, L.F., Woodworth, J. and George, A.M. (1998). Probiotic activity of *Aeromonas media* when challenged with *Vibrio tubiashii*. Aquaculture 169: 111-120.

Gildberg, A. and Mikkelsen, H. (1998). Effects of supplementing the feed to Atlantic cod (*Gadus morhua*) fry with lactic acid bacteria and immunostimulating peptides during a challenge trial with *Vibrio anguillarum*. Aquaculture 167: 103-113.

Gildberg, A., Mikkelsen, H., Sandaker, E. and Ringø, E. (1997). Probiotic effect of lactic acid bacteria in the feed on growth and survival of fry of Atlantic cod (*Gadus morhua*). Hydrobiologia 352: 279-285.

Gomez-Gil, B., Roque, A. and Turnbull, J.F. (2000). The use and selection of probiotic bacteria for use in the culture of larval aquatic organisms. Aquaculture 191: 259-270.

Gram, L., Melchiorsen, J., Spanggaard, B., Huber, I. and Nielsen, T.F. (1999). Inhibition of *Vibrio anguillarum* by *Pseudomonas fluorescens* AH2, a possible probiotic treatment of fish. Appl. Environ. Microbiol. 65: 969-973.

Gram, L., Lovold, T., Nielsen, J., Melchiorsen, J. and Spanggaard, B. (2001). *In vitro* antagonism of the probiont *Pseudomonas fluorescens* strain AH2 against *Aeromonas salmonicida* does not confer protection of salmon against furunculosis. Aquaculture 199: 1-11.

Gross, A., Nemirovsky, A., Zilberg, D., Khaimov, A., Brenner, A., Snir, E., Ronen, Z. and Nejidat, A. (2003). Soil nitrifying enrichments as biofilter starters in intensive recirculating saline water aquaculture. Aquaculture 223: 51-62.

Gullian, M., Thompson, F. and Rodriguez, J. (2004). Selection of probiotic bacteria and study of their immunostimulatory effect in *Penaeus vannamei*. Aquaculture 233: 1-14.

Gunther, J. and Jimenez-Montealegre, R. (2004). Effect of the probiotic *Bacillus subtilis* on the growth and food utilization of tilapia (*Oreochromis niloticus*) and prawn (*Macrobrachium rosenbergii*) under laboratory conditions. Rev. Biol. Trop. 52: 937-943.

Harzevili, A.R.S., Van Duffel, H., Dhert, P., Swings, J. and Sorgeloos, P. (1998). Use of a potential probiotic *Lactococcus lactis* AR21 strain for the enhancement of growth in the rotifer *Brachionus plicatilis* (Muller). Aquacult. Res. 29: 411-417.

Hirata, H., Murata, O., Yamada, S., Ishitani, H. and Wachi, M. (1998). Probiotic culture of the rotifer *Branchionus plicatilis*. Hydrobiologia 387/388: 495-498.

Hjelm, M., Bergh, O., Riaza, A., Nielsen, J., Melchiorsen, J., Jensen, S., Duncan, H., Ahrens, P., Birkbeck, H. and Gram, L. (2004). Selection and identification of autochthonous potential probiotic bacteria from turbot larvae (*Scophthalmus maximus*) rearing units. System. Appl. Microbiol. 27: 360-371.

Hong, H.A., Duc, L.H. and Cutting, S.M. (2005,). The use of bacterial spore formers as probiotics. FEMS Microbiol. Rev. 29: 813-835.

Hoshino, T., Ishizaki, K., Sakamoto, T., Kumeta, H., Yumoto, I., Matsuyama, H. and Ohgiya, S. (1997). Isolation of a *Pseudomonas* species from fish intestine that produces a protease active at low temperature. Lett. Appl. Microbiol. 25: 70-72.

Huys, L., Dhert, P., Robles, R., Ollevier, F., Sorgeloos, P. and Swings, J. (2001). Search for beneficial bacterial strains for turbot (*Scophthalmus maximus* L.) larviculture. Aquaculture 193: 25-37.

Irianto, A. and Austin, B. (2002a). Use of probiotics to control furunculosis in rainbow trout, *Oncorhynchus mykiss* (Walbaum). J. Fish Dis. 25: 1-10.

Irianto, A. and Austin, B. (2002b). Probiotics in aquaculture. J. Fish Dis. 25: 633-642.

Irianto, A. and Austin, B. (2003). Use of dead probiotic cells to control furunculosis in rainbow trout, *Oncorhynchus mykiss* (Walbaum). J. Fish Dis. 26: 59-62.

Jayaprakash, N.S., Pai, S.S., Anas, A., Preetha, R., Philip, R. and Singh, I.S.B. (2005). A marine bacterium, *Micrococcus* MCCB 104, antagonistic to vibrios in prawn larval rearing systems. Dis. Aquat. Org. 68: 39-45.

Jensen, L.E., Hiney, M.P., Shields, D.C., Uhlar, C.M., Lindsay, A.J. and Whitehead, A.S. (1997). Acute phase proteins in salmonids: evolutionary analyses and acute phase response. J. Immunol. 158: 384-392.

Jöborn, A., Olsson, J.C., Westerdahl, A., Conway, P.L. and Kjelleberg, S. (1997). Colonisation in the fish intestinal tract and production of inhibitory substances in intestinal mucus and faecal extracts by *Carnobacterium* sp. K1. J. Fish Dis. 20: 383-392.

Jöborn, A., Dorsch, M., Christer, O.J., Westerdahl, A. and Kjelleberg, S. (1999). *Carnobacterium inhibens* sp. nov., isolated from the intestine of Atlantic salmon (*Salmo salar*). Int. J. System. Bacteriol. 49: 1891-1898.

Jorquera, M.A., Silva, F.R. and Riquelme, C.E. (2001). Bacteria in the culture of the scallop *Argopecten purpuratus* (Lamarck, 1819). Aquacult. Int. 9: 285-303.

Kennedy, S.B., Tucker, J.W., Neidic, C.L., Vermeer, G.K., Cooper, V.R., Jarrell, J.L. and Sennett, D.G. (1998). Bacterial management strategies for stock enhancement of warm water marine fish: a case study with common snook (*Centropomus undecimalis*). Bull. Mar. Sci. 62: 573-588.

Khuntia, A. and Chaudhary, L.C. (2002). Performance of male crossbred calves as influenced by substitution of grain by wheat bran and the addition of lactic acid bacteria to diet. Asian-Aust. J. Anim. Sci. 15: 188-194.

Kim, D.-H. and Austin, B. (2006). Innate immune responses in rainbow trout (*Oncorhynchus mykiss*, Walbaum) induced by probiotics. Fish Shellfish Immunol. 21: 513-524.

Kumar, R., Mukherjee, S.C., Prasad, K.P. and Pal, A.K. (2006). Evaluation of *Bacillus subtilis* as a probiotic to Indian major carp *Labeo rohita* (Ham.). Aquacult. Res. 37: 1215-1221.

Lara-Flores, M., Miguel, A. Olvera-Novoa, Beatríz, E., Guzmán-Méndez, B.E. and López-Madrid, W. (2003). Use of the bacteria *Streptococcus faecium* and *Lactobacillus acidophilus*, and the yeast *Saccharomyces cerevisiae* as growth promoters in Nile tilapia (*Oreochromis niloticus*). Aquaculture 216: 193-201.

Larsen, M.H., Larsen, J.L. and Olren, J.E. (2001). Chemotaxin of *Vibrio anguillarum* to fish mucus, role of the origin of the fish mucus, the fish species and the serogroup of the pathogen. FEMS Microbiol. Ecol. 38: 77-80.

Lin, H.Z., Guo, Z.X., Yang, Y.Y., Zheng, W.H. and Li, Z.J.J. (2004). Effect of dietary probiotics on apparent digestibility coefficients of nutrients of white shrimp *Litopenaeus vannamei* Boone. Aquacult. Res. 35: 1441-1447.

Macey, B.M. and Coyne, V.E. (2005). Improved growth rate and disease resistance in fanned *Haliotis midae* through probiotic treatment. Aquaculture 245: 249-261.

Macey, B.M. and Coyne, V.E. (2006). Colonization of the gastrointestinal tract of the farmed South African abalone *Haliotis midae* by the probionts *Vibrio midae* SY9, *Cryptococcus* sp. SS1, and *Debaryomyces hansenii* AY1. Mar. Biotechnol. 8: 246-259.

Maeda, M., Nogami, K., Kanematsu, M. and Hirayama, K. (1997). The concept of biological control methods in aquaculture. Hydrobiologia 358: 285-290.

Makridis, P., Jon Fjellheim, A., Skjermo, J. and Vadstein, O. (2000). Colonization of the gut in first feeding turbot by bacterial strains added to the water or bioencapsulation in rotifers. Aquacult. Int. 8: 267-380.

Marques, A., Thanh, T.H., Sorgeloos, P. and Bossier, P. (2006). Use of microalgae and bacteria to enhance protection of gnotobiotic *Artemia* against different pathogens. Aquaculture 258: 116-126.

Meuer, F., Hayashi, C., da Costa, M.M., Mauerwek, V.L. and Freccia, A. (2006). *Saccharomyces cerevisiae* as probiotic for Nile tilapia during the sexual reversion phase under a sanitary challenge. Rev. Brasil. Zootech. 35: 1881-1886.

Mohanty, S.N., Swain, S.K. and Tripathi, S.D. (1996). Rearing of catla (*Catla catla* Ham.) spawn on formulated diets. J. Aquacult. Tropics 11: 253-258.

Moriarty, D.J.W. (1998). Control of luminous *Vibrio* species in penaeid aquaculture ponds. Aquaculture 164: 351-358.

Moriarty, D.J.W. (1999). Diseases control in shrimp aquaculture with probiotic bacteria. In: Microbial Biosystems: New Frontiers, Proceedings of the 8th International Symposium on Microbial Ecology. C.R. Bell, M. Brylinsky and P. Johnson-Green (eds). Atlantic Canada Society for Microbial Ecology, Halifax, Canada, pp. 237-243.

Nass, K., Naess, T. and Harboe, T. (1992). Enhanced first feeding of halibut larvae *Hippoglossus hippoglossus*. L. in green water. Aquaculture 105: 143-156.

Nikoskelainen, S., Ouwehand, A., Salminen, S. and Bylund, G. (2001). Protection of rainbow trout (*Oncorhynchus mykiss*) from furunculosis by *Lactobacillus rhamnosus*. Aquaculture 198: 229-236.

Nogami, K. and Maeda, M. (1992). Bacteria as biocontrol agents for rearing larvae of the crab *Portunus tribeculatus*. Can. J. Fish. Aquat. Sci. 49: 2373-2376.

Panigrahi, A., Kiron, V., Kobayashi, T., Puangkaew, J., Satoh, S. and Sugita, H. (2004). Immune responses in rainbow trout *Oncorhynchus mykiss* induced by a potential probiotic bacteria *Lactobacillus rhamnosus* JCM 1136. Vet. Immunol. Immunopathol. 102: 379-388.

Park, S.C., Shimamura, I., Fukunaga, M., Mori, K. and Nakai, T. (2000). Isolation of bacteriophages specific to a fish pathogen, *Pseudomonas plecoglossicida*, as a candidate for disease control. Appl. Environ. Microbiol. 66: 1416-1422.

Parker, R.B. (1974). Probiotics, the other half of the antibiotic story. Anim. Nutr. Health 29: 4-8.

Pirarat, N., Kobayashi, T., Katagiri, T., Maita, M. and Endo, M. (2006). Protective effects and mechanisms of a probiotic bacterium *Lactobacillus rhamnosus* against experimental *Edwardsiella tarda* infection in tilapia (*Oreochromis niloticus*). Vet. Immunol. Immunopathol. 113: 339-347.

Planas, M., Perez-Lorenzo, M., Hjelm, M., Gram, L., Fiksdal, I.U., Bergh, O. and Pintado, J. (2006). Probiotic effect *in vivo* of *Roseobacter* strain 27-4 against *Vibrio (Listonella) anguillarum* infections in turbot (*Scophthalmus maximus* L.) larvae. Aquaculture 225: 323-333.

Plante, S., Pernet, F., Hache, R., Ritchie, R., Ji, B.J. and McIntosh, D. (2007). Ontogenetic variations in lipid class and fatty acid composition of haddock larvae *Melanogrammus aeglefinus* in relation to changes in diet and microbial environment. Aquaculture 263: 107-121.

Preetha, R., Jayaprakash, N.S. and Singh, I.S.B. (2007). *Synechocystis* MCCB 114 and 115 as putative probionts for *Penaeus monodon* post-larvae. Dis. Aquat. Organ. 74: 243-247.

Queiroz, J.F. and Boyd, C.E. (1998). Effects of bacterial inoculum in channel catfish ponds. J. World Aquacult. Soc. 29: 67-73.

Reitan, K., Rainuzzo, J., Oie., G. and Olsen, Y. (1997). A review of the nutritional effects of algae in marine fish larvae. Aquaculture 155: 207-221.

Rengpipat, S., Phianphak, W., Piyatiratitivorakul, S. and Menasveta, P. (1998). Effects of a probiotic bacterium in black tiger shrimp *Penaeus monodon* survival and growth. Aquaculture 167: 301-313.

Rengpipat, S., Tunyanun, A., Fast, A.W., Piyatiratitivorakul, S. and Menasveta, P. (2003). Enhanced growth and resistance to *Vibrio* challenge in pond-reared black tiger shrimp *Penaeus monodon* fed a *Bacillus* probiotic. Dis. Aquat. Org. 55: 169-173.

Rico-Mora, R., Voltolina, D. and Villaescusa-Celaya, J.A. (1998). Biological control of *Vibrio alginolyticus* in *Skeletonema costatum* (Bacillariophyceae) cultures. Aquacult. Eng. 19: 1-6.

Ringø, E. and Birkbeck, T.H. (1999). Intestinal microflora of fish larvae and fry. Aquacult. Res. 30: 73-93.

Riquelme, C., Araya, R., Vergara, N., Rojas, A., Guaita, M. and Candia, M. (1997). Potential probiotic strains in culture of the Chilean scallop *Argopecten purpuratus* (Lamarck, 1819). Aquaculture 154: 17-26.

Riquelme, C., Araya, R. and Escribano, R. (2000). Selective incorporation of bacteria by *Argopecten purpuratus* larvae: implications for the use of probiotics in culturing systems of the Chilean scallop. Aquaculture 181: 25-36.

Roach, S. and Tannock, G.W. (1980). Indigenous bacteria that influence the number of *Salmonella typhimurium* in the spleen of intravenously challenged mice. Can. J. Microbiol. 26: 408-411.

Robertson, P.A.W., O'Dowd, C., Burrells, C., Williams, P. and Austin, B. (2000). Use of *Carnobacterium* sp. as a probiotic for Atlantic salmon (*Salmo salar* L.) and rainbow trout (*Oncorhynchus mykiss* Walbaum). Aquaculture 185: 235-243.

Rodas, B.A., Angulo, J.O., de la Cruz, J. and Garcia, H.S. (2002). Preparation of probiotic buttermilk with *Lactobacillus reuteri*. Milchwissen. Milk Sci. Int. 57: 26-28.

Rollo, A., Sulpizio, R., Nardi, M., Silvi, S., Orpianesi, C., Caggiano, M., Cresci, A. and Carnevali, O. (2006). Live microbial feed supplement in aquaculture for improvement of stress tolerance. Fish Physiol. Biochem. 32: 167-177.

Ruiz-Ponte, C., Samain, J.F., Sanchez, J.L. and Nicolas, J.L. (1999). The benefit of a *Roseobacter* species on the survival of scallop larvae. Mar. Biotech. 1: 52-59.

Sakai, M. (1998). Current research status of fish immunostimulants. Aquaculture 172: 63-92.

Salinas, I., Cuesta, A., Esteban, M.A. and Meseguer, J. (2005). Dietary administration of *Lactobacillus delbrueckii* and *Bacillus subtilis*, single or combined, on gilthead seabream cellular innate immune responses. Fish Shellfish Immunol. 19: 67-77.

Salinas, I., Diaz-Rosales, P., Cuesta, A., Meseguer, J., Chabrillon, M., Morinigo, M.A. and Esteban, M.A. (2006). Effect of heat-inactivated fish and non-fish derived probiotics on the innate immune parameters of a teleost fish (*Sparus aurata* L.). Vet. Immunol. Immunopathol. 111: 279-286.

Salminen, S., Ouwehand, A.C. and Isolauri, E. (1998). Clinical applications of probiotic bacteria. Int. Dairy J. 8: 563-572.

Salminen, S., Ouwehand, A., Benno, Y. and Lee ,Y.K. (1999). Probiotics: how should they be defined? Trends Fd. Sci. Technol. 10: 107-110.

Scholz, U., Garcia Diaz, G., Ricque, D., Cruz Suarez, L.E., Vargas Albores, F. and Latchford, J. (1999). Enhancement of vibriosis resistance in juvenile *Penaeus vannamei* by supplementation of diets with different yeast products. Aquaculture 176: 271-283.

Shinohara, M., Matsumoto, K., Ushiyama, Y., Wakiguchi, H., Akasawa, A. and Saito, H. (2002). Effect of dietary probiotic lactobacillus-fermented milk and yogurt on the development of atopic diseases in early infancy. J. Allergy Clin. Immunol. 109: 191.

Skjermo, J. and Vadstein, O. (1999). Techniques for microbial control in the intensive rearing of marine larvae. Aquaculture 177: 333-343.

Smith, P. and Davey, S. (1993). Evidence for the competitive exclusion of *Aeromonas salmonicida* from fish with stress-inducible furunculosis by a fluorescent pseudomonad. J. Fish Dis. 16: 521-524.

Smoragiewicz, W., Bielecka, M., Babuchowski, A., Boutard, A. and Dubeau, H. (1993). Les probiotiques. Can. J. Microbiol. 39: 1089-1095.

Spanggaard, B., Huber, I., Nielsen, J., Sick, E.B., Pipper, C.B., Martinussen, T., Slierendrecht, W.J. and Gram, L. (2001). The probiotic potential against vibriosis of the indigenous microflora of rainbow trout. Environ. Microbiol. 3: 755-765.

Sugita, H., Miyajima, C. and Deguchi, H. (1991). The vitamin B_{12}-producing ability of the intestinal microflora of freshwater fish. Aquaculture 92: 267-276.

Sugita, H., Takahashi, J. and Deguchi, H. (1992). Production and consumption of biotin by the intestinal microflora of cultured freshwater fishes. Biosci. Biotechnol. Biochem. 56: 1678-1679.

Sugita, H., Kawasaki, J., Kumazawa, J. and Deguchi, Y. (1996). Production of amylase by the intestinal bacteria of Japanese coastal animals. Lett. Appl. Microbiol. 23: 174-178.

Supamattaya, K., Kiriratnikom, S., Boonyaratpalin, M. and Borowitzka, L. (2005). Effect of a *Dunaliella* extract on growth performance, health condition, immune response and disease resistance in black tiger shrimp (*Penaeus monodon*). Aquaculture 248: 207-216.

Suyanandana, P., Budhaka, P., Sassanarakkit, S., Saman, P., Disayaboot. P., Cai, Y. and Benno, Y. (1998). New probiotic lactobacilli and enterococci from fish intestine and their effect on fish production. In: Proceedings of International Conference on Asian Network on Microbial Researches, 23-25 February 1998. Yogyakarta. Indonesia.

Taoka, Y., Maeda, H., Jo, J.Y., Kim, S.M., Park, S.I., Yoshikawa, T. and Sakata, T. (2006). Use of live and dead probiotic cells in tilapia *Oreochromis niloticus*. Fish. Sci. 72: 755-766.

Tovar, D., Zambonino, J., Cahu, C., Gatesoupe, F.J., Vazquez-Juarez, R. and Lesel, R. (2002). Effect of yeast incorporation in compound diet on digestive enzyme activity in sea bass (*Dicentrarchus labrax*) larvae. Aquaculture 204: 113-123.

Vandenberghe, J., Verdonck, L., Robles-Arozarena, R., Rivera, G., Bolland, A., Balladares, M., Gomez-Gil, B., Calderon, J., Sorgeloos, P. and Swings, J. (1999). Vibrios associated with *Litopenaeus vannamei* larvae, postlarvae, broodstock, and hatchery probionts. Appl. Environ. Microbiol. 65: 2592-2597.

Vaseeharan, B. and Ramasamy, P. (2003). Control of pathogenic *Vibrio* spp. by *Bacillus subtilis* BT23, a possible probiotic treatment for black tiger shrimp *Penaeus monodon*. Lett. Appl. Microbiol. 36: 83-87.

Vendrell, D., Balcázar, J.L, Ruiz-Zarzuela, I., de Blas, I., Gironés, O. and Múzquiz, J.L. (2007). Safety and efficacy of an inactivated vaccine against *Lactococcus garvieae* in rainbow trout (*Oncorhynchus mykiss*). Prev. Vet. Med. 80: 222-229.

Verschuere, L., Rombaut, G., Sorgeloos, P. and Verstraete, W. (2000). Probiotic bacteria as biological control agents in aquaculture. Microbiol. Mol. Biol. Rev. 64: 655-671.

Vijayan, K.K., Singh, I.S.B., Jayaprakash, N.S., Alavandi, S.V., Pai, S.S., Preetha, R., Rajan, J.J.S. and Santiago, T.C. (2006). A brackishwater isolate of *Pseudomonas* PS-102, a potential antagonistic bacterium against pathogenic vibrios in penaeid and non-penaeid rearing systems. Aquaculture 251: 192-200.

Vine, N.G., Leukes, W.D. and Kaiser, H. (2004a). *In vitro* growth characteristics of five candidate aquaculture probiotics and two fish pathogens grown in fish intestinal mucus. FEMS Microbiol. Lett. 231: 145-152.

Vine, N.G., Leukes, W.D., Kaiser, H., Daya, S., Baxter, J. and Hecht, T. (2004b). Competition for attachment of aquaculture candidate probiotic and pathogenic bacteria on fish intestinal mucus. J. Fish Dis. 27: 319-326.

Vine, N.G., Leukes, W.D. and Kaiser, H. (2006). Probiotics in marine larviculture. FEMS Microbiol. Rev. 30: 404-427.

Wang, X.-H., Ji, W.-S. and Xu, H.-S. (1999). Application of Probiotic in Aquaculture. Aiken Murray Corp. (Internet).

Ziaei-Nejad, S., Rezaei, M.H., Takami, G.A., Lovett, D.L., Mirvaghefi, A.R. and Shakouri, M. (2006). The effect of *Bacillus* spp. bacteria used as probiotics on digestive enzyme activity, survival and growth in the Indian white shrimp *Fenneropenaeus indicus*. Aquaculture 252: 516-524.

8

Lignocellulose Biotechnology: Issues of Bioconversion and Utilization in Freshwater Aquaculture

S.K. Barik[1] and S. Ayyappan[2]*

INTRODUCTION

Utilization of lignocellulosic wastes by the aquatic animals is considered to be negligible in aquaculture, because much of the carbon within plant tissue (lignocellulose) is not readily digested and assimilated (Mann, 1988; Olah et al., 2006). The bulk of plant tissue eventually enters the detrital pool, where microorganisms (bacteria and fungi) are involved in its breakdown and mineralization. Plant litter is typically colonized by microorganisms before mineralization or used by higher trophic levels (Suberkropp, 1997). Consequently traditional tropical aquaculture systems have often been described as detritus-based, where decaying plant matter is considered an important energy source (Boyd and Silapajarn, 2006). Since detritus-based food chain is so important in freshwater aquaculture, the breakdown of the above mentioned substrates and factors that affect the decay in these habitats need attention.

[1]*The Patent Office, Intellectual Property Building, CP-2, Sector-V, Salt Lake City, Kolkata 700 091, West Bengal, India*

[2]*Indian Council of Agricultural Research, Krishi Anusandhan Bhawan-II, Pusa, New Delhi 110012, India*

Tel.: 011-25846738; Fax: 011-25841955; E-mail: ayyapans@yahoo.co.uk

*Corresponding author: Tel.: 033-23671987; Fax: 033-23671988; E-mail: skb_ipindia@yahoo.co.in

Organic manuring forms the mainstay of fertilization in Asian freshwater aquaculture. An important pathway of energy transfer from macrophytes to animals is microbial decomposition into dissolved organic matter forming aggregates that are assimilated efficiently by fish. Organic substrates in aquatic systems are known to support fish production as they provide sites for epiphytic microbial production, consequently eaten by fish food organisms and fish. Microbial film that develops on the surface of the submerged substrate comprises complex communities of autotrophic and heterotrophic organisms such as bacteria, protozoa, fungi and algae embedded in extracellular polysaccharide matrix secreted by bacteria. Though the basic objective of fertilization has been to support plankton growth through the photosynthetic food chain, the importance of decomposition in sustaining the detritus food chain is well established in fish pond system. Hence, the microorganisms and especially heterotrophic bacteria play an important role in aquatic detrital food chain. They could be considered as a single trophic group utilizing the primary production that has entered the detritus pool.

The productivity of cultured animals in aquatic systems largely depends upon the efficiency with which heterotrophic and autotrophic production components are converted into biomass at different trophic levels in the food chain. Oxygen and organic substrate availability limit the heterotrophic food chain, whereas light and nutrient availability limit the autotrophic food chain. Nutrient and organic substrate requirements can be met by pond inputs and oxygen availability could be augmented as a byproduct of autotrophic production. The above account brings out the importance of enrichment of detritus levels of aquatic body to increase the productivity, through recycling of lignocellulosic organic matter (Fig. 8.1). But, in order to achieve proper pond environmental management on a sustainable basis, it is necessary to ensure optimal application of organic manure with a balanced C : N ratio. The major problems could be high accumulation of recalcitrant residues that might interfere in the normal sediment-water interaction with a negative impact on the system, because the sediment layers in the aquatic ecosystem virtually regulate the productivity of the overlying waters, for they can be nutrient sources or sinks depending on the pH, redox potential (Eh), organic carbon content, bacterial and other biotic activities. Hence, for incorporation in the aquatic system, the lignocellulosics need prior processing, so as to reduce the recalcitrancy to the maximum possible extent and to render themselves manurially more efficient in terms of low C:N ratio. The decomposition patterns of different organic manures like water hyacinth, crop residues, etc. and detritus levels in fish ponds were studied (Olah et al., 1990; Ayyappan et al., 1992).

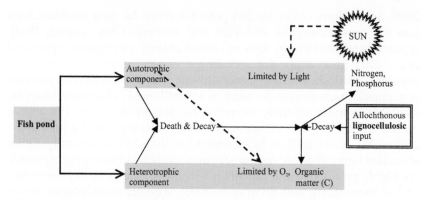

Fig. 8.1 A scheme showing possible incorporation of lignocellulosic input to further enrich the nutrient status in fish culture system.

In this chapter, the recent progress made on bioconversion and utilization of lignocellulosic residues in freshwater aquaculture ponds and the underlying mechanisms have been reviewed.

LIGNOCELLULOSIC WASTES AS RESOURCES FOR AQUACULTURE

The approximate global lignocellulose availability for recycling is a staggering 1550 Tg annually with a significant contribution from the Asian continent (Table 8.1) without considering their common use as animal feed, compost, soil conditioner and burnt as fuel. While efforts are on to produce bio-energy in terms of bio-fuel, enzymes and other high-value products (xylitol, furfural, etc.) from those wastes (Ray et al., 2006,

Table 8.1 Approximate global availability of lignocellulosic biomass (Tg) (Kim and Dale, 2004)

	Africa	Asia	Europe	North America	Central America	Oceania	South Africa	Sub-total
Corn stover	-	33.90	28.61	133.66	-	0.24	7.20	203.61
Barley straw	-	1.97	44.24	9.85	0.16	1.93	0.29	58.44
Oat straw	-	0.27	6.83	2.80	0.03	0.47	0.21	10.61
Rice straw	20.93	667.59	3.92	10.95	2.77	1.68	23.51	731.35
Wheat straw	5.34	145.20	132.59	50.05	2.79	8.57	9.80	354.34
Sorghum straw	-	-	0.35	6.97	1.16	0.32	1.52	10.32
Bagasse	11.73	74.88	0.01	4.62	19.23	6.49	63.77	180.73
Sub-total	38.00	923.81	216.55	218.90	26.14	19.70	106.30	**1549.40**

2008), aquaculture offers another potential scope for their recycling back into the ecosystem (Van der Valk and Attiwill, 1984; Simons, 1994). Aquaculture practices in Asia are based mainly on organic inputs and hence offer great scope for recycling a variety of animal wastes like cowdung, cattleshed refuse, lignocellulosic wastes such as crop residues and aquatic macrophytes. As much as 450 million MT of agro-residues are available in India annually for recycling (Ray et al., 2006). The animal excreta available for recycling is abundant, working out to a tune of 7.5 million MT N, 4.0 million MT P and 5.3 million MT K, annually. Other abundant lignocellulosic residues/resources such as macrophytes/weeds are highly productive and can attain a biomass yield of 10 dry tonne ha^{-1} yr^{-1} or more (Garg and Bhatnagar, 2005). Aquatic macrophytes are an important component of biotic community in tropical waters, with weed cover comprising mainly *Eichhornia* (water hyacinth), ranging up to 100% of the pond surface at times and grow even to the extent that they hinder water management activities, as the intense growth of the same makes the water nutrient-deficient, by absorbing nitrogen and phosphorus from water and sediment (Ayyappan et al., 1986; Boyd and Silapajarn, 2006). Water ferns like *Salvinia cucullata* and *Salvinia molesta* grow vigorously during the rainy season in tropical and sub-tropical zones including India. These are recognized as the world's most intransigent weeds, mainly because of their propensity to grow very rapidly at the expense of other weeds, even water hyacinth. It is difficult to control or destroy the weeds through chemical or biological agents owing to high costs and environmental backlash. Hence, periodical harvesting and utilization through different production practices like aquaculture is apparently the best strategy for keeping the weeds under control. About 65-80% of dry matter in cereal straws remains in the form of utilizable energy, i.e., as cellulose, hemicellulose and cell contents like soluble proteins, sugars and other substances, while the rest non-utilizable dry matter is constituted of lignin, tannins, ash, etc. (Howard, 2008).

Composition of Lignocelluloses

With respect to carbon, it is estimated that the photosynthetic process produces 1.5×10^{11} tonnes of dry plant material annually, of which about 50% is cellulose (Lynd et al., 2002). Lignocellulosic residues are mainly composed of cellulose, hemicellulose and lignin. Straw and other lignocellulosic plant byproducts are widely recognized as valuable sources of cellulose but their exploitation has been limited and they pose an increasing concern for their disposal mainly for reasons like the protective coating of lignin which along with hemicellulose, encrusts the bundles or microfibrils of cellulose chains and the highly crystalline nature of cellulose hinders the access to molecules of water (Durrant,

2005). Comprehensive description on the composition of lignocellulose is beyond the scope of this chapter and the subject is reviewed elsewhere (Howard et al., 2003; Durrant, 2005; Howard, 2008).

LIGNOCELLULOLYTIC MICROORGANISMS AND THEIR ENZYMES

Approximately 200 Tg of CO_2 are fixed on earth every year and the equivalent amount of organic material has to be degraded, out of which 70% is done by microorganisms (Gottschalk, 1988). The plant polysaccharide is used as an energy source by numerous microorganisms, including fungi and bacteria occupying a variety of habitats. The cellulose and hemicellulose degradation is feasible by aerobes and anaerobes whereas lignin is degraded primarily by aerobes (Durrant, 2005).

Some bacteria, several actinomycetes and fungi are able to degrade lignocellulose into fermentable sugars. The ability of many streptomycetes to degrade lignocellulose has been well established. Though microorganisms can degrade cellulose and hemicellulose, their ability to degrade and metabolize lignin is restricted to groups like white-rot fungi and some bacteria (Kaal et al., 1993). The lignin degradation by white rot fungi is strictly a secondary metabolism. However, some groups of fungi are known with certainty to degrade lignin. Extracellular fungal ligninase is expressed widely amongst the white-rot fungi and now preferential lignin degraders also have been identified. The degradation of polymeric lignin needs oxygen and the ligninolytic fungi need a second readily oxidizable substrate such as glucose or cellulose in order to attack lignin (Durrant, 2005).

The degradation of cellulose, hemicellulose and lignin is brought about by the action of extracellular hydrolytic enzymes, *viz.*, cellulase, xylanase and oxidative enzymes, *viz.*, ligninase, Mn-peroxidase and laccase (Phutela et al., 1996; Durrant, 2005). Complex substrates such as rice and wheat straw induce a complete set of enzymes required for degradation of cellulose and hemicellulose components (Paul and Varma, 1993). However, pure cellulose such as cotton and avicel has been reported to be the best inducer for producing a well balanced cellulase system with high yields of enzymes. *Trichoderma reesi* has an efficient cellulolytic system (Sriram and Ray, 2005). The elaboration on composition, characterization, classification and mode of action of these enzymes are not warranted in this chapter. However, readers are suggested to refer to the following reviews for a general account on fungal and bacterial lignocellulolytic enzymes (Howard et al., 2003; Howard, 2008). The relevant discussion on enzymes pertaining to the aquatic environment is dealt with later in this chapter.

PROCESSING OF LIGNOCELLULOSIC WASTES

Physical and Chemical Methods

The number of glucosidic bonds available for enzymatic action depends to a large extent on the degree of swelling of cellulose. For this reason, mechanical pretreatment to reduce particle size, chemical pretreatment to remove non-cellulosic components and physical pretreatments like steam treatment, milling and ultrasonic treatments are effective in causing native cellulose to swell (Ray et al., 2006). Different chemical procedures like treatments with alkali, ammonia, hot acid, alkaline hydrogen peroxide, para-acetic acid and acidified sodium chlorite to remove lignin have been suggested. Physical methods are energy-intensive on a large scale whereas chemical treatment has shown much promise but are uneconomical. In the last few years, there has been an increasing emphasis on identifying the processes by which lignocellulosic material are degraded naturally, because biologically based technologies have scope for greater substrate specificity, lower energy requirement and lower pollution generation (Howard et al., 2003).

Bioprocessing of Lignocellulosics

The double stage 'Karnal Process' (Gupta, 1988) aims at improvement of the N-status of urea-treated lignocellulosic substrates by using the fungus, *Coprinus fimetarius*, which converts exogenous N-compounds into fungal N, with a minimum loss of organic matter (Walli et al., 1993). The urea treatment at the first stage helps to loosen the lignocellulosic bonds, and in the second stage the ammonia compounds are captured/incorporated by fungus for conversion into a more stable form of N.

Solid state fermentation (SSF) process of lignocelluloses is described as a polyfactorial event in which the fungal species, the physical structure of the substrate and physiological parameters of the fermentation have an important role in controlling the degradation. Kondo (1996) suggested SSF by white rot fungi for upgradation of agrowastes, as SSF needs less energy input than liquid fermentation. For possible application in aquaculture, three lignocellulosic wastes, *viz.*, rice straw, *Eichhornia* and *Salvinia* were processed biologically. They were collected, fragmented, sun-dried and further processed by a two-step 'Karnal Process' involving an initial urea treatment (4%) and ensilage (keeping in cemented tanks under polythene cover for 30 d at 40% moisture). Ensiled products were subjected to SSF carried out for 10 d using fungus *C. fimetarius* in specially constructed enclosures with 65% moisture, 0.1% CaO and 1% single super phosphate (Barik et al., 1997; Barik, 2001). The final products were sun-dried and utilized for aquaculture. It was found that bioprocessing of these lignocellulosic substrates with *C. fimetarius* was efficient and resulted in loss of dry matter and more importantly decrease in C/N ratio through increased fungal biomass and incorporation of nitrogen by the substrates.

DECOMPOSITION OF LIGNOCELLULOSICS IN AQUATIC ENVIRONMENT

Detrital Food Chain in Aquatic Environment

Vascular plant litter is mainly composed of high molecular weight polymeric structural compounds, i.e., cellulose, hemicellulose, lignin and smaller fraction of soluble matter. Enzymatic hydrolysis of these organic materials in the aquatic system is the first event in the mineralization process and is a leading force in biological transformations (Sabil et al., 1993). Bacteria and fungi in the aquatic environment secrete extracellular enzymes (Tanaka, 1991) onto the litter cells to cause enzymatic depolymerization producing carbohydrates and less abundant peptides and amino acids which again could be assimilated by the decomposers. An average of 32% of the bacterial biomass in marsh water and sediment was predicted to be produced at the expense of lignocellulosic carbon (Moran et al., 1988). Also, some inorganic ions are leached from the plant litter into the pool of dissolved nutrients available to the decomposers (Harrison, 1989).

'Detritus' in a broad sense comprises a range of whole dead materials to dissolved molecules and particulate matter (Olah et al., 2006). The physical detrital processing includes changes like fragmentation of large particles, flocculation and sedimentation whereas biological detrital processing includes microbial decay, mineralization, shredding and grinding by animals and digestion in animal guts (Odum, 1984). The microbial exudates like muco-polysaccharides may contribute to the food value of detritus; as a result, the detritivores can speed up the microbial decay by fragmenting large pieces of detritus, thus enhancing the surface area available for attack by microbial enzymes (Harrison, 1977). Epibiota is also considered to be an integral part of the detritus complex, which necessarily includes micro- and macro-algae along with invertebrates and heterotrophic microbes at times (Godshalk and Wetzel, 1978). Several models (a typical model is presented in Fig. 8.2) for the decay and re-mineralization of detritus have been proposed, that emphasize on the chemical composition of the litter in determining the loss of biomass from leaching and microbial decay and the build-up of refractory or humic material. Part of the litter may never be decomposed or mineralized if the production of aromatic compounds leads slowly to the formation of

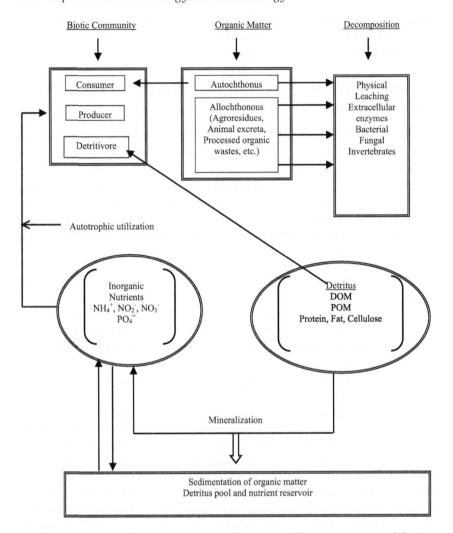

Fig. 8.2 A simplified scheme showing interaction among different components of detritus food chain.

humus from the aging detritus (Melillo et al., 1984). An accurate measurement of invertebrate secondary production is necessary to understand energy flow through heterotrophic or detrital food chain, which is also the main energy transfer pathway. Several stream studies have reported high correlations between the amount of detritus and invertebrates (Duggins et al. 1989).

Microbial Consideration and Detrital Processing in Aquatic Environment

Organic detritus is one of the most suitable nutrient resources for aquatic invertebrate detritivores (Suberkropp, 1992) and it fulfils an important role in the trophic processes of aquatic ecosystems. In nature, large amounts of macrophytes die, decompose and enter the detritus food chain. The macrophyte vegetation is recycled into the ecosystem on death and decay, which contribute to the production process, but is yet to be quantified. Macrophyte decay has been studied with particular reference to sustenance of the detritus food chain. Aquatic microorganisms that mediate the degradation of plant litter have been recognized as an important link between primary and secondary production in detrital foodwebs in the aquatic environment (Suberkropp, 1997). Bacterial biomass produced at the expense of lignocellulosic detritus is likely to be an important nutrient source to food webs in aquatic ecosystems like fish ponds.

Cellulose, xylan as well as lignin are typical structural components of macrophyte litter and are hydrolyzed by microbially derived extracellular enzymes and hence the microorganisms exerting these enzymes play an important role in the decomposition process. The binding of extracellular endoglucanase, CMCase (carboxy methyl cellulase) produced by marine shipworm bacterium to insoluble cellulose substrates has been investigated and it was found that up to 70% of CMCase activity was bound to cellulosic substrates and less than 10% were bound to non-cellulosic substrates (Imam et al., 1993). Lignocellulolytic bacteria and fungi play an important role in the degradation of detritus and fungi are also known to increase the palatability and digestibility of plant detritus (Butler and Suberkropp, 1986). There occurs succession of microorganisms on litter and chemical change of litter during decomposition in aquatic habitats. Studies on microbial colonization of plant detritus in aquatic environment indicated that fungi are primary colonizers whereas bacteria are secondary colonizers (Rublee and Roman, 1982), which was also contradicted (Zemek et al., 1985). While some workers were of the opinion that fungi have a relatively higher contribution than bacteria (Mason, 1976), others indicated bacteria to be more important than fungi (Tanaka, 1991, 1993). Among the various functional groups of bacteria, cellulolytic bacteria play an important role in the decomposition of macrophyte litter, because cellulose is a major component of such litter (Tanaka, 1993; Lynd et al., 2002).

Considerable evidence has established that plant litter is colonized by microorganisms. There are reports on the biomass of bacteria or fungi and the activity of cellulolytic enzymes during decomposition of litter in the aquatic environment. Fungi have been shown to be an important functional assemblage that is involved in breakdown and use of leaf litter in streams (Sinsabaugh and Findlay, 1995). Fungi are instrumental in softening and fragmenting leaves as a result of their filamentous growth form, that allow penetration of leaf surfaces, and their ability to produce extracellular enzymes that degrade structural polysaccharides in plant tissues (Jenkins and Suberkropp, 1995). Fungi have been demonstrated to be the microbial group responsible for initial breakdown of leaves in streams and for modifying the leaf detritus into a more suitable food resource for detritivores (Suberkropp, 1995). A survey of fungal succession on decomposition of *Bauhinia purpuria* and animal waste in two aquatic bodies showed that species richness of aquatic hyphomycetes was higher in case of the former (27 spp.) than the latter (14 spp.) (Au et al., 1992). The role of lower fungi (Labyrinthulomycetes) in nutrient cycling and decay of mangrove leaf litter has been studied by Bremer (1995).

Bacteria mediated significantly higher rates of decay than fungi during microbial decay of water hyacinth, which was in contrast with terrestrial ecosystems where fungi are the major decomposers of lignocellulosic litter (Kirk et al., 1977). The dominance of bacteria during the initial period of macrophyte decomposition has earlier been reported in aquatic environments. Benner et al. (1986) opined that bacteria could mineralize lignin at high rates in the aquatic environment. Bacteria are also components of the decomposer assemblage of litter in streams, but little is known about their contribution in the breakdown process. Among the various functional groups of bacteria, cellulolytic bacteria aid in the process of decomposition of litter components, and the cellulolytic and amylolytic bacterial groups are important as they utilize carbohydrates in preference to proteinaceous substrates (Tanaka, 1993). As N and P content have a direct influence on the decay of litter, the occurrence of N_2-fixing, phosphate-solubilizing and phosphatase-producing bacteria associated with decomposing lignocellulosic substrates are also of significance, and again they form a part of sediment-associated bacterial assemblages which mediate the decomposition of plant litter (Moran et al., 1989) and generally the allochthonous plant materials are settled at the sediment during decomposition in fish pond systems. The net retention and transformation of phosphorus by the microbial community associated with leaf detritus was studied by Elwood et al. (1988). Fungal and bacterial production during breakdown of leaf in streams was studied in litter bags by Weyers and Suberkropp (1996).

Decomposition Loss Rates of Lignocellulosic Substrates in Aquatic Environment

The environmental and experimental conditions affecting the decay rates are length of the decay; physico-chemical conditions like pH, temperature, oxygen, nutrient level (Hietz, 1992); condition of decaying material (fresh/dried) and types of containers (*in-situ* litter bags/laboratory studies in closed chamber). The loss rate of water hyacinth was significantly higher during summer than winter (White and Trapani, 1982). The decay rates of sea grass fall within the range of 0.03-10% per day for different plant parts (Wetzel, 1984). The highest loss rate of 10-11% per day was observed in *Halophila decipiens*, which has a high ratio of easily degradable photosynthetic tissue to resistant non-photosynthetic tissue. The lowest decay rate was 0.07% per day for dead leaves of *Thalassia testudinum* (Rice and Windom, 1982), which were dried, ground and incubated in laboratory vessels.

The microbial degradation of plant residues rich in cellulose is usually limited by the low nitrogen content of these products. Nitrogen content of the litter should reflect the decay rate and it was suggested that the decay rates for aquatic macrophytes in general are positively correlated with nitrogen content. The rate of decay is determined by the chemical composition of the litter, e.g., N, P and lignin content. Jiang et al. (1988) studied the decomposition rates of dead plant materials and changes in their nutritive components. During the decay process, bacteria also contribute some nitrogen to the aging detritus, both through their cells and extracellular products (Harrison, 1989). Microorganisms (cellulolytes) were able to assimilate N and P from dissolved organic pool when grown on lignocellulosic material, which is largely deficient in these elements (Melchiorri-Santolini et al., 1991).

Decomposition of freshwater macrophyte, *Juncus effusus* (Kuehn and Suberkropp, 1998) and marsh plant, *Spartina alterniflora* (White and Trapani, 1982) was studied in litter bags. The use of litter bags will allow finely fragmented particulate detritus to escape and hence rates of mass loss reported from litter bags may overestimate the transformation of particles to dissolved material and another major proposition of the litter bag is facilitation of the entrance and grazing action of micro-invertebrates (Polunin, 1982). Hence, the decay rates from litter bags really represent a combination of escape, decay and detritivory and must be well considered. Barik et al. (1998, 2000) studied the decomposition of rice straw, *Eichhornia* and *Salvinia* in an *in-situ* litter bag kept submerged at the bottom of a freshwater fish pond. The mean daily dry matter loss rates were highest in *Eichhornia* followed by rice straw and *Salvinia* in both unprocessed and processed substrates. The differences in chemical constitution (C/N and C/P ratios) of the litters of the different substrates might have also contributed to the variations in the decay rates (Kok and Van der Velde, 1994). The N and P contents of the substrates were found to be increasing gradually after an initial decrease during the experiment with steady decline of C. The initial decrease might be due to leaching of these nutrients from the substrates (Van der Valk and Attiwill, 1984). The accumulation of cellular matter of microorganisms, which proliferated steadily with time during decay, appears to be also one of the reasons for N and P increase (Tanaka, 1991). Further, many bacteria in aquatic system are responsible for P mineralization; the mineralized P is readily taken up by them and other microorganisms for their cellular metabolism and sometimes they also store P as polyphosphate bodies in their cells (Barik and Purushothaman, 1998). Hence, the lack of any appreciable loss of N and P from substrates during the decay suggests that both nutrients were immobilized into microbial biomass and the gradual decrease in C/N and C/P ratios of litter was the result of C loss through microbial mineralization (Barik et al., 2000).

There are only a few reports on the characteristics of the lignocellulolytic enzymes associated with litter/detritus and the relationships between their change and dynamics and succession of the decomposer microorganisms. Also the aspect of relationship between enzyme activities and the disappearance of specific components of the litter has received little attention. Factors like C, N, pH and temperature have been reported to affect the production of extracellular cellulase in aquatic hyphomycetes, *Lunulospora curvula* and *Flagellospora penicilloides* (Chandrasekhar and Kaveriappa, 1991).

Microbial community whose action is partially concerted and partly separated, spatially and temporally affects the decay of non-woody lignocellulosic litters. The substrates provide sites for epiphytic microbial production consequently eaten by fish food organisms and fish (Ramesh et al., 1999). Therefore, primary microbial colonization determines the final biomass of attached microorganisms which a substrate can harbour. Microbial biofilm that develops on the submerged surface of a substrate, comprises complex communities of autotrophic and hetrotrophic organisms such as bacteria, protozoa, fungi and algae.

LIGNOCELLULOSICS IN AQUACULTURE AS MANURE AND FEED

Recycling of organic wastes has assumed great significance in aquaculture from the point of view of resource utilization as also enrichment of detritus food chain (Kumar et al., 1991). In this regard, traditional fisheries employing organic substrates are worthy of consideration. Using a system of suspending tree twigs of *Acadja*, fishermen have been harvesting fish ranging from 4-20 t ha^{-1} yr^{-1} in West Africa and potential of *Acadja* as a good source of nourishment to fish has been recognized (Hem and Avit, 1994). Studies with the use of sugarcane bagasse have clearly shown over a 50% increase in the growth of common carp, rohu and tilapia, which was attributed to bacterial colonization (Umesh et al., 1999). Periphyton developing on these substrates is believed to be contributing to increased fish yield. Some of the aquatic weeds have been reported to be eaten by carps, at least in the rotting condition. Composted aquatic macrophytes have been considered to have potential as organic fertilizer and feed in fish culture systems (Edwards et al., 1985). The common aquatic weed, *Eichhornia crassipes* (Mart.) Solms. is being recycled into fish culture systems in different forms (Pullin, 1986). Water hyacinth, sudan grass, mugort and water peanut were used in a fresh and fermented state in the rearing of silver carp, grass carp, common carp, crucian carp and big head carp and a high fish yield (3600-7600 kg ha^{-1}) was recorded (Shan and Wu, 1988). The efficiency of recycling of cattle waste and water hyacinth in the rearing ponds of Indian major carps [rohu (*Labeo rohita*), catla (*Catla catla*) and mrigal (*Cirrhinus mrigala*)] was studied without any other input (Mishra et al., 1988). Water hyacinth was fermentatively treated with cowdung and urea and used as feed and manure in carp culture with observations of high organic processing potentials and primary production sustenance capacities (Olah et al., 1990). Earthworm compost of water hyacinth, rice straw and banana trunks were tried as manure in hatchery-cum-nursery ponds of common carp (Sudiarto et al., 1990). The propagation of zooplankton using cowdung and rice straw mixture was tried by Ovie and Fali (1990). An important observation was made when Cooper et al. (1997) recorded inhibition of growth of some fish pathogenic aquatic saprolegniaceous fungi by application of lignocellulosic barley straw.

Melchiorri- Santolini et al. (1985) studied the feasibility of producing proteins for aquaculture by using wheat straw and feeding zooplankton with colonized cellulose. Trial of mixing *Salvinia* powder with conventional fish feed was made by Mohanty and Swamy (1986). Feed efficiency of 4.1 and 4.05 and protein efficiency ratio of 1.89 and 1.91, respectively, in small and large fingerlings of *Ctenopharyngodon idella* was observed, when they were offered with aquatic weed *Ceratophyllum* (Hajra, 1987). Feeding of *Tilapia mossambica* with *Eichhornia crassipes* was tried (Dey and Sarmah, 1982) and a 70% consumption rate was observed. Aquatic weed (*Ceratophyllum demersum*)-based pellets were formulated and used as feed for tilapia (Chiayvareesajja et al., 1990) and no significant differences were observed for feed conversion ratio. Tantikitti et al. (1988) worked on the economics of tilapia pen culture using fresh aquatic weed *Ceratophyllum demersum* as feed. Powdered leaf of a weed, *Clitoria ternatea* was tested for its efficiency as a protein source in feed of *Cyprinus carpio* and it was observed that for the efficient feed conversion, there was scope for incorporation of the weed at a maximum of 30% (Raj, 1989). Performance of three artificial diets comprising water hyacinth with other ingredients in different ratios was studied in *Cyprinus carpio* (Rahman and Mustafa, 1989). A study was made to evaluate the suitability of water hyacinth leaf meal as a partial substitute for dietary fishmeal protein for fry of Indian major carps (Hassan et al., 1990). Growth of Nile Tilapia (*Oreochromis niloticus*) fed with *Ceratophyllum demersum* (hornwort), water hyacinth and chest-nut (*Eleocharis ochrostachys*) pellet was studied (Klinnavee et al., 1990). Sumagaysay (1991) studied the effect of replacement of organic matter in the feed with rice straw compost on milk fish (*Chanos chanos*) growth and production and suggested that compost was not a satisfactory feed substitute. The leaves and swollen petioles of *Eichhornia* were assessed as a source of protein in the feed of *Clarias batrachus* and incorporation of water hyacinth as a protein supplement in the feed of common carp along with other ingredients has been suggested (Rath and Datta, 1991). Formulation, water stability and effect of storage on the quality of artificial feeds with incorporation of *Eichhornia*, *Pistia* and *Salvinia* powder were evaluated for carp culture (Murthy and Devaraj, 1992). *Salvinia cucullata* could be utilized by *Labeo rohita* fingerlings in pelleted form when incorporated into a conventional diet (Ray and Das, 1992). Hasan and Roy (1994) did not observe any major histopathological changes of fish fed with different levels of water hyacinth and concluded that diets containing 50% water hyacinth with molasses are superior to other diets in terms of feed cost and economic return. The digestibility pattern of six aquatic macrophytes *viz.*, *Lemna polyrhiza*, *Eichhornia crassipes*, *Pistia stratiotes*, *Salvinia cucullata*, *Hydrilla verticillata* and *Nymphoides cristatum* in *Labeo rohita* fingerlings were elucidated (Ray and Das, 1994). *Oreochromis niloticus* showed good growth when diets containing 18-20% groundnut oil cake and 14-15% water hyacinth meals were provided, however increase in dietary water hyacinth led to decline in growth (Ofojekwu et al., 1994). The fingerlings of *Oreochromis niloticus* were evaluated for growth using a feed, which was a 45% of water hyacinth replacement of groundnut cake (Keke et al., 1994). Crayfish were offered a variety of foods ranging from formulated diet to agricultural wastes (Brown et al., 1995).

FISH CULTURE WITH BIOPROCESSED LIGNOCELLULOSIC WASTES AS MANURIAL INPUTS: THE CIFA (CENTRAL INSTITUTE OF FRESHWATER AQUACULTURE, INDIA) EXPERIENCE

An investigation was made to evaluate and compare the relative efficiency of bio-processed lignocellulosic wastes (ensilage followed by SSF using the fungus *C. fimetarius*) on fish production when added singly by releasing nutrients into water through mineralization. The bioprocessed rice straw, *Eichhornia* and *Salvinia* were applied singly as manures (@ 250 kg N-ha^{-1} y^{-1} iso-nitrogenous basis in split doses) in separate fish ponds with stocking density of 5000 fingerlings (rohu:catla:mrigal = 4:3:3) per hectare. Cow dung treatment was kept as a control (Barik, 2001).

Physico-chemical Parameters

The total alkalinity level of all the four treatments was found to be higher during the initial months, after which it decreased gradually and almost, maintained steadily throughout. The mean NH$_4$-N was highest (0.325 mg l^{-1}) in water hyacinth treatment that could possibly be due to the highest content of total nitrogen in water hyacinth in comparison to other manures. The seasonal peak for PO$_4$-P was observed in summer in all treatments. The mean biochemical oxygen demand (12.49 mg l^{-1}) and chemical oxygen demand (39.00 mg l^{-1}) were highest in cowdung treated ponds followed by *Eichhornia*, rice straw and *Salvinia* treatment which could be due to the comparatively higher abundance of degradable organic matter content of cowdung.

Variations in Plankton

High net plankton counts (904 nos l^{-1}) were observed with treatment of cowdung and the net plankton counts in all the treatments were similar as observed in the freshwater fish ponds (Jinghong et al., 1991). Comparative abundance of less number of planktonic divisions in all the ponds might be due to practice of organic manuring, as also recorded by Simons (1994) without provision for artificial feed or fertilizer. In all the ponds the phytoplankton peaks were not very prominent, probably because of higher grazing pressure.

Variations in Microbiological Parameters

The microbial populations in general in sediment media of different ponds were always observed to be more than that of water column, which could largely be due to the organic richness and higher surface area of sediment particles (Barik et al., 2001). The amylolytic bacteria were more abundant during periods of higher sediment organic carbon, which indicates the regulation of carbon content of pond sediment. The mean cellulolytic bacterial populations in water (1.2×10^2 ml^{-1}) and sediment (17.41×10^{-5} g^{-1} dry weight) were found to be highest in the treatment with cowdung, confirming the relative abundance of cellulolytic bacteria in cowdung in comparison to the processed rice straw, water hyacinth and *Salvinia*. N_2-fixing bacteria represent a substantial autochthonous input of fixed nitrogen to the sediment of ponds contributing to the overall productivity of the systems without application of nitrogenous fertilizer. The factors responsible for abundance of phosphatase-producing bacteria in sediment media could be the organic manuring practices and the excreta being generated from fish. The observations suggest that the microorganisms play an important role in the metabolism of phosphorus (Barik and Purushothaman, 1999). Higher abundance of yeasts and molds in bottom sediments could be apparently due to higher content of organic matter in sapropelic sediments. The fungal populations of the ponds in the present study could be of both autochthonous (natural flora) or allochthonous (from terrestrial source through the fungal processed lignocellulosic manurial input) origin.

Variations in Productivity Status of Ponds

The gross primary production increased gradually from the beginning till the end (Table 8.2). While the gross productivity was moderate, the community respiration level was high resulting in low or sometimes zero net productivity values. The high community respiration also points to heterotrophy of the systems. It is possible that not all the allochthonous organic matter added to the systems undergoes complete decomposition and instead, sustains the heterotrophic food chain. High respiration rates at the primary trophic level indicate higher microbial activity, which can increase the production efficiencies with higher organic enrichment (Ayyappan et al., 1990) indicating the potentials of lignocellulosic inputs as manures in fish ponds. The fish production levels (1.16-2.10 t ha^{-1} yr^{-1}) achieved in this study with different treatments was in line with

Table 8.2 Ranges in productivity parameters of ponds manured with bio-processed lignocellulosic wastes (Barik, 2001)

Pond treatment	Gross primary productivity (g C/m³/d)	Net primary productivity (g C/m³/d)	Respiration (g C/m³/d)
Cow dung	0.67-4.73	0.07-1.56	0.60-3.17
Eichhornia	0.13-3.60	0.03-0.96	0.75-3.15
Salvinia	0.48-4.29	0.0-1.20	0.45-3.09
Rice straw	0.52-4.24	0.0-1.50	0.45-3.30

the production levels (1157 kg ha^{-1} 6 months^{-1}) worked out by Ayyappan et al. (1990). The conversion efficiency with regard to total carbon input (i.e., carbon input through manures and natural gross primary production) ranged from 0.95-1.89%. While an efficiency of 1.2% is considered as an index of good production, LeCren and Lowe-McConnell (1980) concluded that the efficiency generally varies between 0.1 and 1.6% in lakes and reservoirs. The conversion efficiencies (primary production to fish production) in different treatments (1.64-2.93%) are also in the ranges as achieved by Olah et al. (1987) in different treatments.

Intestinal Microflora and Enzymatic Activity of Fish

The microflora in guts of different fishes was found to gradually increase during this study (Table 8.3). In the present study a substantial part of gut microflora might have been contributed from the manures, which inhabit high density of microbial population. The high amylase activity in guts of different fishes (rohu, catla and mrigal) in the present study suggests that starch is readily and efficiently hydrolyzed and probably plays an important role in energy metabolism. Low cellulase and xylanase activity were detected in the guts of all fish species during the study, which might be due to direct ingestion of very low quantities of cellulosic or hemicellulosic (xylan) material. As there was no correlation between cellulase activity and microbial counts in gut during the study, it could be predicted that the enzyme was not solely derived from the ingested bacteria but also contributed from endogenous origin as proposed by Xue et al. (1999).

High activity of alkaline phosphatase (APase) in guts of fishes might be attributed to the localization of phosphatase-producing bacterial populations in the guts of fishes, which may be autochthonous or allochthonous. However all the APase activity might not be due to the bacterial extracellular secretion, but also from other phytoplankton which are fed upon by the fish and the inherent activity by membrane linked specific structures (in microvilli) containing enzymes (Kuz'-mina and Gelman, 1997). In this study, the gut enzymatic activities were enhanced over time in almost all cases, which is in line with observations of Kumar and Chakrabarti (1998) who reported increase in gut enzyme activity of mrigal with the growth of larvae.

From the pond study on manurial efficiency of processed lignocellulosic wastes, it was evident that the fish production was higher in water hyacinth treatment due to its quicker decay rate, as observed during litter bag experiment. The application of processed lignocellulosic wastes as manures resulted in the functioning of an efficient and intense detritus food chain in the ponds. The distribution of various groups of

Table 8.3　Ranges of gut microbial population and gut enzymatic profiles of different fish species in ponds manured with bio-processed lignocellulosic wastes (Barik, 2001)

Microbial population

Pond treatment	Fish species	Heterotrophic bacteria (nos. $\times 10^5$/g)	Phosphatase producing bacteria (nos. $\times 10^5$/g)	Amylolytic bacteria (nos. $\times 10^5$/g)
	Rohu	20-42	3-29	5-30
Cow dung	Catla	48-56	29-56	20-59
	Mrigal	42-74	16-32	8-38
	Rohu	76-92	22-35	28-35
Eichhornia	Catla	52-74	21-33	18-40
	Mrigal	78-108	17-36	21-42
	Rohu	6-24	6-26	10-30
Salvinia	Catla	18-43	8-26	12-28
	Mrigal	34-59	15-36	15-40
	Rohu	8-21	5-18	9-22
Rice straw	Catla	18-36	5-21	10-27
	Mrigal	46-58	15-28	24-36

Enzymatic activity

		Amylase (g mol maltose/ g/min)	CMCase (g mol glucose/ g/min)	Xylanase (g mol xylose/ g/min)	APase (m g p-NP/ ml/h)
Cow dung	Rohu	69.5-169.5	0-2.9	0-0.80	21.6-56.2
	Catla	49.0-83.2	2.67-3.20	0.30-0.90	24.5-57.2
	Mrigal	10.9-69.7	1.15-3.40	0.42-1.30	21.9-59.0
Eichhornia	Rohu	20.5-64.8	1.84-4.28	0-0.52	18.3-47.5
	Catla	19.6-46.2	1.67-3.69	0-0.69	17.2-32.2
	Mrigal	11.1-63.1	3.09-5.70	0.42-2.20	23.4-62.5
Salvinia	Rohu	12.3-36.5	0.90-3.40	0-0.60	15.3-40.3
	Catla	16.5-39.8	1.10-2.80	0-0.50	13.2-39.5
	Mrigal	18.9-40.2	0.98-2.90	0-1.70	18.3-45.2
Rice straw	Rohu	8.3-26.5	0.50-2.80	0-0.90	13.2-36.8
	Catla	11.2-29.2	1.40-2.60	0-0.80	11.6-32.1
	Mrigal	12.5-30.6	1.20-2.70	0.20-1.30	19.5-51.1

microheterotrophic populations in water and sediment media of pond was responsible for a dynamic nutrient cycling through decomposition and various enzymatic actions.

The salient observations of the above study are:

- The percentage of zooplankton population was always found to be higher than that of phytoplankton. Chlorophyceae was the most dominant group among phytoplankton in all the treatments followed by Myxophyceae and Bacillariophyceae. Copepoda was invariably the most dominant group among zooplankton in all the treatments.

- Microbial forms play a major role during the degradation of lignocellulosic manures in the aquatic environment, which is further supplemented by the fact that the microbial population is directly correlated with the loss rates of all the substrates during decomposition in the aquatic system.

- Further, the presence of high intensity of bacterial biomass and related enzymatic activities in pond sediment reiterates the establishment of an efficient detritus-based food chain in the aquatic system provided with processed lignocellulosic wastes.

- The results showed that bacteria are predominant mediators of degradation of lignocellulosic detritus in systems in which such detritus is a substantial indirect source of carbon and energy for aquatic animals. The predominance of bacteria in the degradation of detritus in this study suggests that the transfer of carbon and energy from detritus to animals occurs primarily *via* grazing of bacterial biomass. Hence, promoting bacterial/microbial biofilm/ colonization on surface of lignocellulosic substrates may be a viable low-cost strategy to boost the fish production.

CONCLUSION AND FUTURE PERSPECTIVE STUDY

Use of bioprocessed lignocellulosics is a considerably new approach in aquaculture. The importance of this approach lies in lignocellulosics being locally available in great abundance and the need for organic enrichment of aquaculture system as also environmental upkeepment. The above account can serve as a model which clearly brings out the advantage of application of processed lignocellulosic wastes in aquaculture as the former comes into equilibrium with the surrounding aquatic system more quickly. The reduction in oxygen uptake by bio-processed lignocellulosics indicates the advantage and need for prior processing of the same for recycling through manurial input in fish ponds to enrich the detritus pool without much oxygen stress. This could also provide means for combining low value agro-wastes with animal manures for aquaculture practice for enriching the detritus food chain for growth of detrivores and fish food organisms through leached nutrients during microbial processing. The gradual decrease in mold population with increase in micro-heterotrophic population indicates the significance of the latter during decay of lignocellulosics in freshwater aquaculture systems. The predominance of bacteria during decay of lignocellulosics suggests that the transfer of carbon and energy from the detritus to animals occurs primarily *via* grazing of bacterial biomass.

Considering the chemical composition of the substrates along with bacterial community and water quality of the system in which those were applied, water hyacinth appears to be superior to *Salvinia* and rice straw as manure. The use of processed lignocellulosics as manure in fish showed the practices to be at par with other traditional organic manuring practices and to be economically viable and practicable for backyard culture practices. Comparatively quickly biodegradable *Eichhornia* favoured better growth of fish through bacterial colonization (biofilm formation) than rice straw and *Salvinia*. The results of the study indicated vast scope for increasing fish production through promotion of microbial biofilm production followed by degradation in backyard ponds employing cheap and easily available lignocellulosic substrates as a low cost strategy. Considering the cost of artificial feeds, the present approach of promoting fish food within a pond ecosystem through application of regionally available bio-converted waste lignocellulosic biomass is seen as a better cost effective measure basically for small and marginal farmers having backyard culture facilities.

ABBREVIATIONS

CMC   Carboxymethyl cellulose
DOM   Dissolved organic matter
POM   Particulate organic matter
p-NP p-Nitrophenol
SSF   Solid state fermentation

REFERENCES

Au, D.W.T., Hodgkins, I.J. and Vrijmoed, L.L.P. (1992). Fungi and cellulolytic activity associated with decomposition of *Bauhinia purpurea* leaf litter in a polluted and unpolluted Hong Kong waterway. Can. J. Bot. 70: 1071-1079.

Ayyappan, S., Olah, J., Raghavan, S.L., Sinha, V.R.P. and Purushothaman, C.S. (1986). Macrophyte decomposition in two tropical lakes. Arch. Hydrobiol. 106: 219-231.

Ayyappan, S., Pandey, B.K. and Tripathi, S.D. (1990). Nutritive quality of microbial communities as feed components in experimental carp culture. Indian J. Exp. Biol. 28: 977-980.

Ayyappan, S., Tripathi, S.D., Vasheer, V.S., Das, M. and Bhandari, S. (1992). Decomposition patterns of manurial inputs in aquaculture and their model-simulating substrates. Int. J. Ecol. Environ. Sci. 18: 101-109.

Barik, S.K. (2001). Evaluation of processed lignocellulosic wastes as nutrient inputs in fish culture with particular reference to microbial quality. Ph.D. Thesis, Utkal University, India.

Barik, S.K. and Purushothaman, C.S. (1998). Phosphatase activity of two strains of bacteria on orthophosphate enrichment. In: Frontiers in Applied Environmental Microbiology. A. Mohandas and I.S.B. Singh (eds). SES, CUSAT, Cochin, India, pp. 165-170.

Barik, S.K. and Purushothaman, C.S. (1999). Occurrence and distribution of alkaline phosphatase producing bacteria in freshwater fishpond ecosystems. Indian J. Fish. 46: 273-280.

Barik, S.K., Mishra, S., Ayyappan, S. and Singh, K. (1997). Saprophytic ability of some selected lignocellulolytic fungal strains. Proc. Natl. Symp. Microbial Tech. Environmental Management and Resource Recovery, New Delhi, India.

Barik, S.K., Mishra, S. and Ayyappan, S. (1998). Bacterial populations associated with lignocellulosic decomposition in fish pond systems. Proc. 39th Ann. Conf. Association of Microbiologists of India, Mangalore, India, pp. 107-108.

Barik, S.K., Mishra, S. and Ayyappan, S. (2000). Decomposition patterns of unprocessed and processed lignocellulosics in a freshwater fish pond. Aquatic Ecol. 34: 185-204.

Barik, S.K., Purushothaman, C.S. and Mohanty, A.N. (2001). Phosphatase activity with reference to bacteria and phosphorus in tropical freshwater aquaculture pond systems. Aquacult. Res. 32: 819-832.

Benner, R., Moran, M.A. and Hodson, R.E. (1986). Biogeochemical cycling of lignocellulosic carbon in marine and fresh water ecosystems: relative contributions of procaryotes and eucaryotes. Limnol. Oceanogr. 31: 89-100.

Boyd, C.E. and Silapajarn, O. (2006). Influence of microorganisms/microbial products on water and sediment quality in aquaculture ponds. In: Microbial Biotechnology in Agriculture and Aquaculture. Volume 2. R.C. Ray (ed). Science Publishers Inc., New Hampshire, USA, pp. 261-286.

Bremer, G.B. (1995). Lower marine fungi (Labyrinthulomycetes) and the decay of mangrove leaf litter. In: Proc Asia Pacific Symp. Mangrove Ecosystems. Y.S.T. Wong and N.F.Y. Tam (eds). 295: 89-95.

Brown, P.B., Wilson, K.A., Wetzel, J.E. and Hoene, B. (1995). Increased densities result in reduced weight gain of crayfish *Orconectes virilis*. J. World Aquacult. Soc. 26: 165-171.

Butler, S.K. and Suberkropp, K. (1986). Aquatic hyphomycetes on oak leaves: comparison of growth, degradation and palatability. Mycologia. 78: 922-928.

Chandrashekhar, K.R. and Kaveriappa, K.M. (1991). Production of extracellular cellulase by *Lunulospora curvula* and *Flagellospora penicillioides*. Folia Microbiol. 36: 249-255.

Chiayvareesajja, S., Wongwit, C. and Tansakul, R. (1990). Cage culture of tilapia (*Oreochromis niloticus*) using aquatic weed-based pellets. In: Proc of the 2nd Asian Fisheries Forum. R. Hirano and I. Hanyu (eds). Tokyo, Japan, pp. 287-290.

Cooper, J.A., Pillinger, J.M. and Ridge, I. (1997). Barley straw inhibits growth of some aquatic saprolegniaceous fungi. Aquacult. 156: 157-163.

Dey, S.C. and Sarmah, S. (1982). Prospect of the water hyacinth (*Eichhornia crassipes*) feed to cultivable fishes — A preliminary study with *Tilapia mossambica* Peters. Matsya. 8: 40-44.

Duggins, D.O., Simenstand, C.A. and Estes, J.A. (1989). Magnification of secondary production by kelp detritus in coastal marine ecosystems. Sci. Was. 245: 170-173.

Durrant, L.R. (2005). Biotechnology applied for the utilization of lignocelluloses biomass. In: Microbial Biotechnology in Agriculture and Aquaculture. Volume 1. R.C. Ray (ed). Science Publishers Inc., New Hampshire, USA, pp. 405- 432.

Edwards, P., Kamal, M. and Wee, K.L. (1985). Incorporation of composted and dried water hyacinth in pelleted feed for the tilapia *Oreochromis niloticus* (Peters). Aquacult. Fish. Managed. 1: 233-248.

Elwood, J.W., Mulholl, P.J. and Newbold, J.D. (1988). Microbial activity and phosphorus uptake on decomposing leaf detritus in a heterotrophic stream. Verh. Internat. Verein. Limnol. 23: 1198-1208.

Garg, S.K. and Bhatnagar, A. (2005). Use of microbial biofertlizers for sustainable aquaculture/fish culture. In: Microbial Biotechnology in Agriculture and Aquaculture. Volume 1. R.C. Ray (ed). Science Publishers Inc., New Hampshire, USA, pp. 261-298.

Godshalk, G.L. and Wetzel, R.G. (1978). Decomposition in the littoral zone of lakes. In: Freshwater Wetlands: Ecological Processes and Management Potential. (eds). R.E. Good, D.F. Whigham and R.L. Simpson (eds). Academic Press, New York, pp. 131-143.

Gottschalk, G. (1988). Biochemistry and genetics of cellulose degradation. In: Proc. FEMS Symp. J.P. Aubert, P. Beguin and J. Millet (eds). Academic Press, London, pp. 1-3.

Gupta, B.N. (1988). Development and concept of Karnal Process for fungal treatment of cereal straws. In: Proc Int. Workshop Fibrous Crop Residues as Animal Feed. K. Singh and J.B. Schiere (eds). Bangalore, India, pp. 46-56.

Hajra, A. (1987). Biochemical investigations on the protein-calorie availability in grass carp (*Ctenopharyngodon idella* Val.) from an aquatic weed (*Ceratophyllum demersum* Linn.) in the tropics. Aquacult. 61: 113-120.

Harrison, P.G. (1977). Decomposition of macrophyte detritus in sea water: effects of grazing by amphipods. Oikos 28: 165-169.

Harrison, P.G. (1989). Detrital processing in seagrass systems: A review of factors affecting decay rates, remineralization and detritivory. Aquat. Bot. 23: 263-288.

Hasan, M.R. and Roy, P.K. (1994). Evaluation of water hyacinth meal as a dietary protein source for Indian major carp (*Labeo rohita*) fingerlings. Aquacult. 124: 63.

Hasan, M.R., Moniruzzan, M. and Omar Farooque, A.M. (1990). Evaluation of leucaena and water hyacinth leaf meal as dietary protein sources of the fry of Indian major carp *Labeo rohita* Hamilton. In: Proc. 2nd Asian Fish. Forum. R. Hirano and I. Hanyu (eds). Tokyo, Japan, pp. 275-278.

Hem, S. and Avit, J.L.B. (1994). First results on 'Acadja-enclosure' as an extensive aquaculture system (West Africa). Bull. Mar. Sci. 55: 1040-1051.

Hietz, P. (1992). Decomposition and nutrient dynamics of reed (*Phragmites australis* (Cav.) Trin. Ex Steud.) litter in Lake Neusiedl, Austria. Aquat. Bot. 43: 211-230.

Howard, R.L. (2008). Lignocellulose biotechnology: bioconversion and cultivation of edible mushrooms. In: Microbial Biotechnology in Horticulture. Volume 3. R.C. Ray and O.P. Ward (eds). Science Publishers Inc., New Hampshire, USA, pp. 181-230.

Howard, R.L., Abotsi, E., Jansen, V.E.R.L. and Howard, S. (2003). Lignocellulose biotechnology: issues of bioconversion and enzyme production. African J. Biotech. 2: 602-619.

Imam, S.H., Greene, R.V. and Griffin, H.L. (1993). Binding of extracellular carboxymethylcellulase activity from the marine shipworm bacterium to insoluble cellulosic substrates. Appl. Environ. Microbiol. 59: 1259-1263.

Jenkins, C.C. and Suberkropp, K. (1995). The influence of water chemistry on the enzymatic degradation of leaves in streams. Freshwat. Biol. 33: 245-253.

Jiang, F., Wang, W., Zhao, M. and Zhong, C. (1988). The decomposition rates of dead plant materials of two species of *Spartina* and the changes in their nutritive components. Acta Oceanol. pp. 119-125.

Jinghong, L., Ruixiang, C., Yangu, D., Mao, L. and Yijn, H. (1991). Ecological studies of zooplankton in Dongshan Bay. J. Oceanogr. 10: 205-212.

Kaal, E.E.J., de Jong, E. and Field, J.A. (1993). Stimulation of ligninolytic peroxidase activity by nitrogen nutrients in the white rot fungus *Bjerkandera* sp. strain BOS55. Appl. Environ. Microbiol. 59: 4031-4036.

Keke, I.R., Ofojekwu, C.P., Ufodike, E.B. and Asala, G.N. (1994). The effects of partial substitution of groundnut cake by water hyacinth (*Eichhornia crassipes*) on growth and food utilisation in the Nile tilapia, *Oreochromis niloticus niloticus* (L.). Acta Hydrobiol. Cracow. 36: 235-244.

Kim, S. and Dale Bruce E. (2004). Global potential bioethanol production from wasted crops and crop residues. Biomass and Bioenergy 26: 361-375.

Kirk, T.K., Connors, W.J. and Zeikus, J.G. (1977). Advances in understanding the microbial degradation of lignin. Recent Adv. Phytochem. 11: 369-394.

Klinnavee, S., Tansakul, R. and Promkuntong, W. (1990). Growth of Nile tilapia (*Oreochromis niloticus*) fed with aquatic plant mixtures. In: Proc 2nd Asian Fish Forum. R. Hirano and I. Hanyu (eds). Tokyo, Japan, pp. 283-286.

Kok, C.J. and Van der Velde, G. (1994). Decomposition and microinvertebrate colonisation of aquatic and terrestrial leaf material in alkaline and acid still water. Freshwat. Biol. 31: 65-75.

Kondo, R. (1996). Investigation on mechanism of biological delignification by solid substrate fermentation. J. Sci. Ind. Res. 55: 394-399.

Kuehn, K. and Suberkropp, K. (1998). Decomposition of standing leaf litter of the freshwater emergent macrophyte *Juncus effusus*. Freshwat. Biol. 40: 217-227.

Kumar, K., Ayyappan, S., Murjani, G. and Bhandari, S. (1991). Utilization of mashed water hyacinth as feed in carp rearing. Proc. Nat. Symp. Freshwat. Aqua. Bhubaneswar, India, pp. 89-91.

Kumar, S. and Chakrabarti, R. (1998). Ontogenic development of amylase activity in three species of Indian Major Carps, *Catla catla*, *Labeo rohita* and *Cirrhinus mrigala*, in relation to natural diet. Asian Fish. Sci. 10: 259-263.

Kuz'-mina, V.V. and Gelman, A.G. (1997). Membrane-linked digestion in fish. Rev. Fish. Sci. 5: 99-129.

Lynd, L.R., Weimer, P.J., van Jyl, W.H. and Pretorius, I.S. (2002). Microbial cellulose utilization: fundamentals and biotechnology. Microbiol. Mol. Biol. Rev. 66: 506-577.

LeCren, E.D. and Lowe McConell, R.H. (1980). The Functioning of Freshwater Aquasystems. IBP Hand Book No. 22, Cambridge University Press, UK.

Mann, K.H. (1988). Production and use of detritus in various freshwater, estuarine and coastal marine ecosystems. Limnol. Oceanogr. 33: 910-930.

Mason, C.F. (1976). Relative importance of fungi and bacteria in the decomposition of *Phragmites* leaves. Hydrobiol. 51: 65-69.

Melchiorri-Santolini, U., Contesini, M., Margaritis, B., Pinolini, M.L., Tozzi, S. and Zanetta, A. (1991). Particulate organic matter as agent in transport of inorganic nutrients to the food chain. In: Director's Report on the Scientific Activity of the Institute for the Year 1988. R.De. Bernardi (ed). Instituto Italiano di Idrobiologia, Italy, pp. 25-32.

Melchiorri-Santolini, U., Della-Sala-Merigo, C., Contesini, M. and Malara, G. (1985). Nitrogen compound transformation during cellulose decomposition processes. In: Director's Report on the Scientific Activity of the Institute for the Year 1984. R.De. Bernardi (ed). Instituto Italiano di Idrobiologia, Italy, pp. 120-121.

Melillo, J.M., Naiman, R.J., Arber, J.D. and Linkins, A.E. (1984). Factors controlling mass loss and nitrogen dynamics of plant litter decaying in northern streams. Bull. Mar. Sci. 35: 341-356.

Mishra, B.K., Sahu, A.K. and Pani, K.C. (1988). Recycling of the aquatic weed, water hyacinth, and animal wastes in the rearing of Indian major carps. Aquacult. 68: 59-64.

Mohanty, S.N. and Swamy, D.N. (1986). Enriched conventional feed for Indian major carps. In: Proc First Asian Fish Forum. J.L. Maclean, L.B. Dizon and L.V. Hosillos (eds). Manila, Philippines, pp. 597-598.

Moran, M.A., Benner, R., Hodson, R.E. and Legovic, T. (1988). Carbon flow from lignocelluloses: A simulation analysis of a detritus-based ecosystem. Ecology 69: 1525-1536.

Moran, M.A., Benner, R. and Hodson, R.E. (1989). Kinetics of microbial degradation of vascular plant material in two wetland ecosystems. Oecologia 79: 158-167.

Murthy, H.S. and Devaraj, K.V. (1992). Formulation, stability and effect of storage on the quality of three artificial feeds used in carp culture. Fish. Technol. Soc. Fish. Technol., Kochi 29: 107-110.

Odum, W.E. (1984). Dual-gradient concept of detritus transport and processing in estuaries. Bull. Mar. Sci. 35: 510-521.

Ofojekwu, P.C., Keke, I.R., Asala, G.N. and Anosike, J.C. (1994). Evaluation of water hyacinth (*Eichhornia crassipes*) and groundnut cake as dietary components in feeds for *Oreochromis niloticus niloticus* (L.). Acta Hydrobiol. Cracow. 36: 227-233.

Olah, J., Sinha, V.R.P., Ayyappan, S., Purushothaman, C.S. and Radheyshyam, S. (1987). Detritus associated respiration during macrophyte decomposition. Arch. Hydrobiol. 111: 309-315.

Olah, J., Ayyappan, S. and Purushothaman, C.S. (1990). Processing and utilisation of fermented water hyacinth, *Eichhornia crassipes* (Mart.) Solms. in carp culture. Aquacult. Hung. 6: 77-95.

Olah, J., Poliouakos, M. and Boyd, C.E. (2006). Linking ecotechnology and biotechnology in aquaculture. In: Microbial Biotechnology in Agriculture and Aquaculture. Volume II. R.C. Ray (ed). Science Publishers Inc., New Hampshire, USA, pp. 287-318.

Ovie, S.I. and Fali, A.I. (1990.) Preliminary investigation into the use of cow dung and rice straw mixture in the propagation of zooplankton. Ann. Rep. Natl. Inst. Freshwat. Fish. Res. Nigeria, pp. 140-149.

Paul, J. and Varma, A.K. (1993). Hydrolytic enzyme(s) production in *Micrococcus roseus* growing on different cellulosic substrates. Lett. Appl. Microbiol. 16: 167-169.

Phutela, R.P., Bhadauria, A., Sodhi, H.S. and Kapoor, S. (1996). Screening of Chinese mushroom (*Volvariella* spp.) strains for cellulases and xylanases production. Indian J. Microbiol. 36: 125-128.

Polunin, N.V.C. (1982). Processes contributing to the decay of reed (*Phragmites australis*) litter in freshwater. Arch. Hydrobiol. 94: 182-209.

Pullin, R.S.V. (1986). Aquaculture development in Nepal — pointers for success. NAGA, 9: 9-10.

Rahman, N. and Mustafa, S. (1989). Effects of artificial diet on growth and protein content in the carp *Cyprinus carpio*. J. Ecobiol. 1: 215-222.

Raj, S.P. (1989). Evaluation of *Clitoria* leaf as a protein supplement in the feed of *Cyprinus carpio* var. *communis*. J. Ecobiol. 1: 195-202.

Ramesh, M.R., Shankar, K.M., Mohan, C.V. and Varghese, T.J. (1999). Comparison of three plant substrates for enhancing carp growth through bacterial biofilm. Aquacult. Eng. 19: 119-131.

Rath, S.S. and Dutta, H. (1991). Use of the water hyacinth, *Eichhornia crassipes* as an ingredient in the feed of *Clarias batrachus*. Natl. Symp. on New Horizons in Freshwater Aquaculture, Bhubaneswar, India, pp. 98-99.

Ray, A.K. and Das, I. (1992). Utilisation of diets containing composted aquatic weed (*Salvinia cucullata*) by the Indian Major Carp, Rohu (*Labeo rohita* Ham.) fingerlings. Biores. Technol. 40: 67-72.

Ray, A.K. and Das, I. (1994). Apparent digestibility of some aquatic macrophytes in rohu, *Labeo rohita* (Ham.), fingerlings. J. Aquacult. Trop. 9: 335-342.

Ray, R.C., Sahoo, A.K., Asano, K. and Tomita, F. (2006). Microbial processing of agricultural residues for production of food, feed and food-additives. In: Microbial Biotechnology in Agriculture and Aquaculture. Volume 2. R.C. Ray (ed). Science Publishers Inc., New Hampshire, USA, pp. 511-552.

Ray, R.C., Shetty, K. and Ward, O.P. (2008). Solid-state fermentation and value-added utilization of horticultural processing wastes. In: Microbial Biotechnology in Horticulture. Volume 3. R.C. Ray and O.P. Ward (eds). Science Publishers Inc., New Hampshire, USA, pp. 231-272.

Rice, D.L. and Windom, H.L. (1982). Trace metal transfer associated with the decomposition of detritus derived form estuarine macrophytes. Bot. Mar. 25: 213-223.

Rublee, P.A. and Roman, M. (1982). Decomposition of turtle grass (*Thalassia testudinum* Konig) in flowing seawater tanks and litter bags; compositional changes and comparison with natural particulate matter. J. Exp. Mar. Biol. Ecol. 58: 47-58.

Sabil, N., Tagliapietra, D. and Coletti Previero, M.A. (1993). Insoluble biodegradative potential of the Venice Lagoon. Environ. Technol. 14: 1089-1095.

Shan, J. and Wu, S. (1988). Study on the effects of four types of fresh and fermented green manure on fish production. Aquacult Int. Cong. and Expo, Vancouver, British Columbia, Canada.

Simons, J. (1994). Field ecology of freshwater microalgae in pools and ditches, with special attention to eutrophication. Netherlands J. Aquat. Ecol. 28: 25-33.

Sinsabaugh, R.L. and Findlay, S. (1995). Microbial production, enzyme activity, and carbon turnover in surface sediments of the Hudson River estuary. Microb. Ecol. 30: 127-141.

Sriram, S. and Ray, R.C. (2005). *Trichoderma:* Systematics, molecular taxonomy and agricultural and industrial applications. In: Microbial Biotechnology in Agriculture and Aquaculture. Volume I. R.C. Ray (ed). Science Publishers Inc., New Hampshire, USA, pp. 335-376.

Suberkropp, K. (1992). Aquatic hyphomycetes communities. In: The Fungal Community. G.C. Carrol and D.T. Wicklow (eds). Marcel Dekker, Inc., New York, USA, pp. 729-747.

Suberkropp, K. (1995). The influence of nutrients on fungal growth, productivity and sporulation during leaf breakdown in streams. Can. J. Bot. 73: S1361-S1369.

Suberkropp, K. (1997). Annual production of leaf-decaying fungi in a woodland stream. Freshwat. Biol. 38: 169-178.

Sudiarto, B., Zakaria, B.R. and Costa-Pierce, B.A. (1990). Use of earthworm composts and introduction of new systems and techniques to improve production in hatcheries and nurseries for common carp (*Cyprinus carpio*). In: Reservoir Fisheries and Aquaculture Development for Resettlement in Indonesia. B.A. Costa-Pierce and O. Soemarwoto (eds). 23: 272-284.

Sumagaysay, N.S. (1991). Utilisation of feed and rice straw compost for milk fish, *Chanos chanos*, production in brackish water ponds. J. Appl. Ichthyol. 7: 230-237.

Tanaka, Y. (1991). Microbial decomposition of reed (*Phragmites communis*) leaves in a saline lake. Hydrobiol. 220: 119-129.

Tanaka, Y. (1993). Aerobic cellulolytic bacterial flora associated with decomposing *Phragmites* leaf litter in a seawater lake. Hydrobiol. 263: 145-154.

Tantikitti, C., Rittibhonbhun, N., Chaiyakum, K. and Tankasul, R. (1988). Economics of tilapia pen culture using various feeds in Thale Noi, Songkhla Lake, Thailand. In: The 2nd Int. Symp. Tilapia Aquacult. R.S.V. Pullin, T. Bhukaswan, K. Tonguthai and J.L. Macleam (eds). Bangkok, Thailand, pp. 569-574.

Umesh, N.R., Shankar, K.M. and Mohan, C.V. (1999). Enhancing growth of common carp, rohu and Mozambique tilapia through plant substrate: the role of bacterial biofilm. Aquacult. Internat. 7: 251-260.

Van der Valk, A.G. and Attiwill, P.M. (1984). Decomposition of leaf and root litter of *Avicennia marina* at Westernport Bay, Victoria, Australia. Aquat. Bot. 18: 205-221.

Walli, T.K., Rai, S.N., Singh, G.P. and Gupta, B.N. (1993). Nitrogen status of biologically treated straw and its N utilization by animals. In: Feeding of Ruminant on Fibrous Crop Residues. Proc. Internat. Symp. under Indo-Dutch Project on Bioconversion of Crop Residues, Karnal, India, pp. 237-247.

Wetzel, R.G. (1984). Detrital dissolved and particulate organic carbon functions in aquatic ecosystems. Bull. Mar. Sci. 35: 503-509.

Weyers, H.S. and Suberkropp, K. (1996). Fungal and bacterial production during the breakdown of yellow poplar leaves in 2 streams. J. N. Am. Benthol. Soc. 15: 408-420.

White, D.A. and Trapani, J.M. (1982). Factors influencing disappearance of *Spartina alterniflora* from litterbags. Ecology 63: 242-245.

Xue, X.M., Anderson, A.J., Richardson, N.A., Xue, G.P. and Mather, P.B. (1999). Characterisation of cellulase activity in the digestive system of the redclaw crayfish (*Cherax quadricarinatus*). Aquacult. 180: 373-386.

Zemek, J., Marvanova, L., Kuniak, L. and Kadlecikova, B. (1985). Hydrolytic enzymes in aquatic hyphomycetes. Folia. Microbiol. 30: 363-372.

9

Genetic Engineering in Aquaculture: Ecological and Ethical Implications

Ioannis S. Arvanitoyannis* and Aikaterini Kassaveti

INTRODUCTION

Genetically modified (GM) fish (or transgenic fish) offer the possibility of improving both the production and characteristics of conventional fish strains currently exploited in aquaculture (Fu et al., 2005; Chapter 1 of this volume). Aquatic animals attract more research attention than terrestrial livestock for two primary reasons: First, fish lay eggs in large quantities and those eggs are more easily manipulated, making it easier to insert novel DNA. Finfish and shellfish can be GM through gene transfer, chromosome set manipulation, interspecific hybridization, and other methods (Aerni, 2004). The main species that have been GM are: Atlantic salmon, coho salmon, common carp, tilapia and channel catfish (Anonymous, 2003). Many biotechnologies used in aquaculture are developed to increase production, reduce costs of production, manage disease outbreaks, raise the value of currently cultured organisms, or result in the culture of new species (U.S. Congress, Office of Technology Assessment, 1995).

Development of transgenic fish has reached the stage where several species are ready to be marketed in different countries. This is thought to be the case for GM tilapia in Cuba, GM salmon in USA and Canada, and GM carp in the People's Republic of China. However, Cuba appears

School of Agricultural Sciences, Department of Agriculture Ichthyology & Aquatic Environment, University of Thessaly, Fytokou str., 38446, Neo Ionia Magnesia, Volos Hellas (Greece)
Tel.: +30 24210 93104; Fax: +30 24210 93137
*Corresponding author: E-mail: parmenion@uth.gr

reluctant to go ahead with production at present, and in the USA the approval process at the Food and Drug Administration (FDA) has been unexpectedly delayed (Kaiser, 2005). In terms of EU position, the European Commission has stated that GM fish have potential to cause irreversible damage to fish stocks and to the marine environment in the event of escape. Up to date, the Commission has not received any notification with respect to experimental releases of GM fish or any commercial production license applications (Millar and Tomkins, 2007).

No new technology is risk-free but the benefits may vastly outweigh the risks (Fletcher et al., 2002). On the other hand, one should bear in mind that the existence of this new industry (transgenic fish) raises many ethical questions. Among these are questions about environmental effects, the safety and nutritional value of the product and effects on the traditional fishery (Needham and Lehman, 1991). Released transgenic fish stocks are thought to pose a risk not only to conspecifics, but also to other species through niche expansion (Kapuscinski and Hallerman, 1990, 1991) and even speciation. It is generally accepted that inadvertent release of farmed fish into natural ecosystems should be avoided, especially if the fish has been GM (Alestrom, 1996). The particular environmental concerns regarding transgenic fish include issues like live competition of transgenic stocks with wild populations, introgression of the transgene into wild gene pools and heightened predation of transgenics on prey populations (Aerni, 2004).

The concern for an escape of transgenic or genetically modified organism (GMO) is understandably much greater (de la Fuente et al., 1996). That is why most countries have strict regulations concerning a thorough risk assessment before releasing these animals into the environment or introducing them into the market (Kaiser, 2003). Risk assessment can be addressed using the Net Fitness Approach (NFA) whereby critical fitness parameters are estimated on transgenic fish relative to wild type and incorporated into a model to predict risk. The net fitness parameters include viability to sexual maturity, age at sexual maturity, mating success, fecundity, fertility, and longevity. In this way, it is possible for regulators to develop a set of unambiguous tests to assess risk (Muir and Hostetler, 2001).

GENETICALLY ENGINEERED (GE) FISH OR TRANSGENIC FISH

Genetically engineered (also called transgenic) fish are those that carry and transmit one or more copies of a recombinant DNA sequence (i.e. a DNA sequence produced in a laboratory using *in vitro* techniques)

(Van Eenennaam, 2005). According to the International Council for Exploration of the Seas (ICES), the definition of GMOs includes both transgenics and fish with modified chromosome sets, but not fish which have been improved through classical breeding (ICES, 1996). The first recorded instances of production of transgenics in aquatic species are those of Maclean and Talwar (1984) in rainbow trout and Zhu et al. (1985) in goldfish.

Many kinds of fish have been engineered, some for disease resistance but most for faster growth (Stokstad, 2002). A number of species are targeted for gene transfer experiments and can be divided into two main groups: fish used in aquaculture, and model fish used in basic research. Among the major food fish species are carp (*Cyprinus* sp.), tilapia (*Oreochromis* sp.), salmon (*Salmo* sp., *Oncorhynchus* sp.) and channel catfish (*Ictalurus punctatus*), while the zebrafish (*Danio rerio*), medaka (*Oryzias latipes*) and goldfish (*Carassius auratus*) are used in basic research (Alestrom, 1996).

TECHNIQUES TO PRODUCE TRANSGENIC FISH

The traditional and most popular method of producing transgenic fish is still microinjection despite the recent advancements where electroporation, pseudotyped retroviral vectors, lipofection and particle bombardment were used (Muir and Hostetler, 2001). A schematic representation of the production process of GM fish for aquaculture is shown in Figure 9.1.

In microinjection, where a construct is injected into the cytoplasm of embryos at the one or two cell stage using ultra fine glass needles (Winkler and Schartl, 1997). Although cytoplasmic microinjection is a successful technique, it is a tedious and time-consuming procedure and unsuitable for mass gene transfer (Chen and Powers, 1990). Electroporation is an alternative that alleviates some of these problems and makes gene transfer more efficient (Muir and Hostetler, 2001). Electroporation involves the introduction of the expression vector into fertilized eggs or host cells by application of pulses of electricity to induce transient pores in the membrane of host cells (FAO/WHO, 2003).

Another example of new and possibly more effective ways for gene transfer is the use of pantropic retroviral vectors. The pantropic retroviral vectors were derived from the hepatitis B virus and the vesicular stomatitis virus, a pathogen similar to hoof and mouth disease which infects mammals, insects and possibly plants (Dunham, 2003). The inclusion of DNA in lipophylic particles (lipofection) which fuse with the

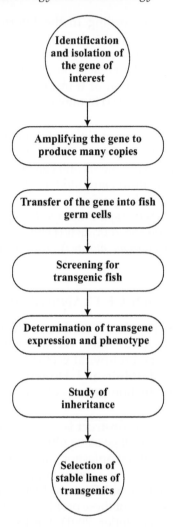

Fig. 9.1 Production of GM fish for aquaculture (adapted from Levy et al., 2000; Beardmore and Porter, 2003).

cell membrane efficiently transfects bacteria and eukaryotic cells. It is suitable for use with fish sperm and dechorionated embryos (Volckaert and Ollevier, 1998). Another less known technique is particle bombardment. This method involves the coating of the expression vector on to gold particles, and the introduction in host cells by bombardment with the particles (FAO/WHO, 2003).

CURRENT STATUS

Demand for seafood is rapidly rising, and it might double by 2040, according to the FAO (Stokstad, 2002). A key element to enhanced production of cultured species is the development of genetically superior broodstocks that are tailored both to their culture conditions and to the market (Hew and Fletcher, 1997).

To date, no transgenic animals have been approved for use as human food, although the Aqua Bounty transgenic salmon has been under regulatory review for more than five years (Van Eenennaam and Olin, 2006). A limited number of transgenic animals have been approved for rendering into animal feed components (FDA, 2006). The only transgenic fish that is commercially available today is not designed to be eaten (Muir, 2004). In 2003 a GE zebra danio that produces a red fluorescent protein became commercially available in most US pet shops. The zebra danio is a small aquarium species that has never survived outside captivity in the United States. The FDA determined not to formally regulate GloFish (Van Eenennaam, 2005).

Agencies responsible for overseeing environmental release of aquatic GMOs include the U.S. Fish and Wildlife Service (FWS), National Marine Fisheries Service (NMFS), and various state agencies overseeing aquatic resources (U.S. Congress, Office of Technology Assessment, 1995). Agencies such as the Organization for Economic Cooperation and Development (OECD), the International Council for Exploration of the Seas (ICES) and the Food and Agriculture Organization of the United Nations (FAO) have investigated policy issues raised by the release of aquatic GMOs (Kapuscinski and Hallerman, 1994).

ADVANTAGES AND DISADVANTAGES FROM THE USE OF GMOs IN AQUACULTURE

Both risks and benefits of genetic engineering are seen to apply mostly to future generations, as opposed to the average person or the self (Frewer and Shepherd, 1995).

Advantages

- Development of transgenic fish able to detect pollutants or mutagens in the aquatic environment. Some researchers in Singapore are looking at using transgenic zebrafish for this purpose (Pew Initiative on Food and Biotechnology, 2003).
- Faster-growing fish could make fish farming more productive, increasing yields while reducing the amount of feed needed, which

in turn could reduce waste (Borgatti and Buck, 2004). Studies have revealed enhancement of growth particularly in salmonids to an average of 3-5 times the size of non-transgenic controls (Devlin et al., 1994).

- Shellfish and finfish, genetically modified to improve disease resistance, would allow farmers to cut down further on antibiotics, insecticides or fungicides (Muir, 2004).

- Identification and removal of various proteins that make shellfish intolerable for some consumers and production of a line of shrimp or crab which individuals could consume (Pew Initiative on Food and Biotechnology, 2003).

- Increased cold resistance in fish could lead to the ability to grow seafood in previously inhospitable environments, allowing aquaculture to expand into previously unsuitable areas (Borgatti and Buck, 2004).

- The availability of the green fluorescent protein (GFP) as a reporter gene has enabled the use of transgenic organisms as continual qualitative biosensors for water contamination. Similar experiments have been carried out previously in yeast cells transformed with RAD54 fused to GFP; the cells turning green in response to DNA damage (Afanassiev et al., 2000).

Disadvantages

- The GM fish may interbreed in the wild with native fish leading to an undesirable ecological impact. This can range from slightly increased genetic load to extinction of local populations (Knibb, 1997; Maclean and Laight, 2000).

- GM fish with higher disease resistance and better use of nutrients could outcompete wild relatives and change predator–prey relationships, and they could therefore occupy new ecological niches where wild species would usually not survive (Muir, 2004).

- The GM fish, although do not interbreed in the wild, could be viewed as an undesirable new alien species analogous to an introduced species (Maclean and Laight, 2000).

- Dependence on external agencies for seed fish. If transgenic fish become widely grown because they are much more efficient, and if special broodstock are required to produce fry for on-growing to adults, which, cannot be used as broodstock, a dependency is created (Beardmore and Porter, 2003).

- GM fish and shellfish may contain new substances or high levels of substances that could be detrimental to human health such as the production of new proteins, or possibly even toxins, which can potentially cause allergic reactions (http://www.seaweb.org/ resources/aquaculturecenter/documents/Aquaculture. GMOD.pdf).

- Since genes may now be patented and therefore, enjoy commercial value, the opportunities for dispute about equitable treatment of stakeholders in cases where ownership of genes and strains is contested, are numerous (Beardmore and Porter, 2003).

ECOLOGICAL CONCERNS

Environmental concerns arising from emerging animal biotechnologies are largely speculative at this time because few products have been commercialized. Industrial developers of agricultural biotechnology might argue that more efficient production of animal-based feeds could reduce the resources necessary to produce food and, thereby, reduce the environmental burden of animal production (Cowan and Becker, 2006). The primary environmental concerns about releases of transgenic fish include competition with wild populations, movement of the transgene into the wild gene pool, and ecological disruptions due to changes in prey and other niche requirements in the transgenic variety versus the wild populations (FDA, 2006). The central theme in environmental concerns is the question whether escaped transgenic fish will transfer their special genes to wild and natural populations and whether these genetically modified organisms (GMOs) will be selected for (Volckaert and Ollevier, 1998).

Environment sustainability and biodiversity are the two major issues at stake with the introduction of GM products (Cockburn, 2002; Arvanitoyannis, 2006). Greece (86%), followed after a considerable gap by Denmark and Austria (67% each), France (66%) and Finland (65%) are most likely to support environmental protection organizations campaigning against biotechnology (Eurobarometer, 2000). Europeans are conscious of the impact of human actions and technology on nature (Moses, 1999). Yet, there is ambivalence about exploiting nature in the interests of progress (Table 9.1) (Gaskell et al., 2003). One of the causes of European opposition to GMOs is that GMOs are perceived above all as hardly useful, non-natural and risky (Tables 9.1, 9.2) (Eurobarometer, 2000; Hallman et al., 2002; Lassen et al., 2002).

Myhr and Dalmo (2005) endorsed the fact that the most prominent challenge to implementing genetic engineering in aquaculture in a sustainability context is how to ensure environmental protection and at the same time achieve economic benefits. If transgenic fish enter an

Table 9.1 Surveys related to environmental concerns of public about biotechnology

Country/Year	Respondents characteristics	Questions				References
Greece (2002)	Total number of respondents: 503 (273 males and 230 females) Age: 20–74 Other characteristics: 38% had a university education	*I believe that GM food production is in harmony with nature*	(Strongly) agree: 9.3%	Neither agree nor disagree: 22.8%	(Strongly) disagree: 68.0%	Arvanitoyannis and Krystallis, 2005
EU (1999)	Total number of respondents: 16000	*Even if GM food had benefits, it is fundamentally unnatural?*	Agree: 71.0%	Neither agree nor disagree: 12.0%	Disagree: 10.0% / Don't know: 7.0%	Eurobarometer, 2000
		GM food threatens the natural order of things?	Agree: 67.0%	Neither agree nor disagree: 13.0%	Disagree: 11.0% / Don't know: 8.0%	
EU (2002)	Total number of respondents: 15900	*Nature is fragile and easily damaged by human actions*	Tend to agree: 88.0%	Tend to disagree: 7.0%	Don't know: 5.0%	Gaskell et al., 2003
		Would buy GM if more environmentally friendly?	Tend to agree: 38.0%	Tend to disagree: 58.0%		
		Modern technology has upset the balance of nature	Tend to agree: 73.0%	Tend to disagree: 17.0%	Don't know: 10.0%	
US (2001)	Total number of respondents: 1233	*Do you agree or disagree with the idea of GM fish to improve efficiency of production?*	Total agree: 30%	Total disagree: 65%	Not sure: 6%	Pew Initiative on Food and Biotechnology, 2001
US (2002)	Total number of respondents: 1214	*The environmental risks or the environmental benefits of using biotechnology to GM plants, animals, fish or trees?*	Risks: 38%	Benefits: 38%	Risks and benefits are about the same: 21% / Not sure: 3%	Pew Initiative on Food and Biotechnology, 2002

Table 9.2   Surveys related to ethical concerns of public about biotechnology

Country/Year	Respondents characteristics	Questions				References
Greece (2002)	Total number of respondents: 503 (273 males and 230 females) Age: 20-74 Other characteristics: 38% had a university education	I believe that there exists GM fish	(Strongly) agree: 29.8%	Neither agree nor disagree: 41.9%	(Strongly) disagree: 28.3%	Arvanitoyannis and Krystallis, 2005
		I believe that I have consumed GM food despite my will	(Strongly) agree: 55.0%	Neither agree nor disagree: 29.8%	(Strongly) disagree: 15.3%	
		I would willingly purchase GM food	(Strongly) agree: 14.6%	Neither agree nor disagree: 29.5%	(Strongly) disagree: 55.9%	
		I believe that the use of GM in food is necessary	(Strongly) agree: 52.5%	Neither agree nor disagree: 23.9%	(Strongly) disagree: 23.6%	
		My general opinion about GM food is positive	(Strongly) agree: 28.8%	Neither agree nor disagree: 27.0%	(Strongly) disagree: 44.1%	
		I would offer GM food to my children without any particular concern	(Strongly) agree: 26.6%	Neither agree nor disagree: 37.5%	(Strongly) disagree: 35.9%	
		I believe that GM food production is not against bio-ethics	(Strongly) agree: 37.5%	Neither agree nor disagree: 16.3%	(Strongly) disagree: 46.2%	
		I believe that the GM components of food should be indicated on the label	(Strongly) agree: 61.4%	Neither agree nor disagree: 32.1%	(Strongly) disagree: 6.5%	
		I have noticed GM food components written on the label of some products	(Strongly) agree: 12.5%	Neither agree nor disagree: 28.8%	(Strongly) disagree: 58.7%	
EU (1996)	-	GM laboratory animals are useful (for medical applications)	(Strongly) agree: 27.0%		Agree: 31.0%	Pardo et al., 2002
		GM laboratory animals are risky (for medical applications)	(Strongly) agree: 22.0%		Agree: 33.0%	

Table 9.2 contd...

Study	Respondents	Statement					Reference
EU (1999)	Total number of respondents: 16000	GM laboratory animals are morally acceptable (for medical applications)	(Strongly) agree: 13.0%			Agree: 26.0%	Eurobarometer, 2000
		GM food is useful (for agricultural applications)	(Strongly) agree: 20.0%			Agree: 35.0%	
		GM food is risky (for agricultural applications)	(Strongly) agree: 25.0%			Agree: 35.0%	
		GM food is morally acceptable (for agricultural applications)	(Strongly) agree: 17.0%			Agree: 13.0%	
		GM food poses no danger for future generations	Agree: 13.0%	Neither agree nor disagree: 17.0%	Disagree: 52.0%	Don't know: 18.0%	
US (2001)	Total number of respondents: 1203	Do you approve or disapprove of creating hybrid animals using genetic modification?	Total approve: 28.0%	Total disapprove: 68.0%		Not sure: 5.0%	Hallman et al., 2002
		The idea of GM food causes me great concern	Total agree: 56.0%	Total disagree: 41.0%		Not sure: 2.0%	
		The risks involved in GM food are acceptable	Total agree: 40.0%	Total disagree: 52.0%		Not sure: 7.0%	
US (2004)	Total number of respondents: 1000	GM foods are basically safe	Agree: 30.0%			Disagree: 27%	Pew Initiative on Food and Biotechnology, 2004
US (2001-2005)	Telephone interviews: 1000 Age: 18 and older	During the past few months, have you taken any action or done anything because of any concerns you may have about foods produced using biotechnology? (March 2005)	Yes: 4%	No: 95%		Don't know/refused: 1%	IFIC, 2005

Table 9.2 contd...

	Question					Source
		Altered purchase behavior: 62%	Activism: 21%	Other: 14%	Don't know/Refused: 8%	
US (2002) Telephone survey Total number of respondents: 250 Age: 18 and older	*What action have you taken? (March 2005)*	Altered purchase behavior: 62%	Activism: 21%	Other: 14%	Don't know/Refused: 8%	Chern, 2002
	How important are ethical or religious concerns when you decide whether or not to consume GM foods?	Important: 36.3%	Neither: 15.2%	Unimportant: 46.9%	Don't know: 1.6%	
	How risky would you say GM foods are in terms of risk to human health?	Risky: 48.9%	Neither: 16.0%	Safe: 20.7%	Don't know: 14.5%	
Ireland (1999) Total number of respondents: 2592 Age: 15-74	*Are you concerned about GM food?*	Very concerned: 39%	Somewhat concerned: 43%	Unconcerned: 18%		Morris and Adley, 2001
	Does concern about the safety of GM food influence what you buy?	Always. I try to ensure that nothing I but is GM: 27%	Most of the time, but sometimes it is impossible to find out if something is GM: 33%	Some of the time. If I have time I will read the label but there are often times when it is too difficult: 27%	Never. Concern about GM foods doesn't affect my choice of products at all: 13%	

ecosystem that contains the same species, the genetics of that population will change if they interbreed. The population will acquire a new gene or set of genes that could alter the fitness of that population (Galli, 2002). Wild-type genes could be replaced by the introduction of new genes from the cultured organisms, resulting in a loss of natural genetic variation (Regal, 1994). Muir and Howard (2001) stated that transgenes could spread in populations despite high juvenile viability costs if transgenes also have sufficiently high positive effects on other fitness components. Furthermore, transgene effects on age at sexual maturity should have the greatest impact on transgene frequency, followed by juvenile viability, mating advantage, female fecundity, and male fertility, with changes in adult viability, resulting in the least impact.

The local extinction of a conspecific population could also have a disruptive effect on other species in a community such as releasing competing species from resource competition or prey species from predation; additionally, the survival of predatory species that depend on the eliminated species could be threatened. Thus, both extinction and invasion hazards could have a cascading impact on other species in the community with an unpredictable level of hazard (Muir and Howard, 2002). A study indicated that growth-enhanced transgenic and non-transgenic salmon were cohabitated and competed for different levels of food, transgenic salmon did not affect the growth of non-transgenic cohorts when food availability was high (daily feed ration equivalent to 7.5% of total fish biomass). When food availability was low (0.75% of total fish biomass), all groups containing transgenic salmon experienced population crashes or complete extinctions, whereas groups containing only non-transgenic salmon had good survival, and their population biomass continued to increase (Devlin et al., 2004).

Jönsson and co-workers (1998) investigated the agonistic behavior of transgenic and non-transgenic GH (growth hormone) juvenile rainbow trout (*Oncorhynchus mykiss*). The results revealed that GH did not appear to affect its fighting ability. However, GH affected aggression indirectly by increasing the swimming activity, and/or by inducing defense of a larger territory, thereby increasing the encounter rate between opponents. Since increased aggression can incur energetic and mortality costs, there may be selection against high GH levels in natural populations. In an experiment carried out by Wu et al. (2003) an "all-fish" construct *CAgcGH*, grass carp GH fused with common carp β-actin promoter, has been generated and transferred into Yellow River carp (*Cyprinus carpio*, a local strain in Yellow River) fertilized eggs. Under middle-scale trial, *CAgcGH*-transgenics revealed higher growth rate and food conversion efficiency than the controls, which was consistent with laboratory findings. Transgenic channel catfish (*Ictalurus punctatus*) containing salmonid growth hormone genes can grow 33% faster than normal channel catfish under aquaculture conditions with supplemental feeding. There was no significant difference in growth performance between transgenic and non-transgenic channel catfish in ponds without supplemental feeding; thus indicating equal foraging abilities, and the inability of transgenic catfish to exhibit their growth potential with limited feed (Chitmanat, 1996).

The transgenic Atlantic salmon carrying a Chinook salmon GH gene controlled by a cold-activated promoter from a third species, the ocean pout (Van Eenennaam and Olin, 2006) put on weight up to six times as fast as traditional hatchery salmon (Stokstad, 2002). The F4 generation of human growth hormone (hGH) transgenic red common carp *Cyprinus carpio* had significantly higher growth rates than the non-transgenic controls. Protein and energy intakes were significantly higher in the transgenic carp than in the controls fed the 20% protein diet, but were not different between the two strains fed diets with 30 and 40% protein (Fu et al., 1998). Devlin and co-workers (1999) studied the hypothesis that increased GH levels in GH-transgenic coho salmon (*Oncorhynchus kisutch*) increase competitive ability through higher feeding motivation. The results indicated that the transgenic fish consumed 2.9 times more pellets that the non-transgenic, indicating high feeding motivation of the transgenic fish throughout the feeding trials. Depending on how transgenic and wild individuals differ in other fitness-related characters, escaped GH transgenic fish may compete successfully with native fish in the wild.

Furthermore, by mating with wild fish, escaped GM fish could spread the transgene among the wild population, which could cause conflicting effects on mating success, viability in natural habitats and other fitness factors required for the species to survive (Muir, 2004). Recent researches suggest that the transgene could be purged within a few generations or could spread through the natural population and possibly affect its abundance (Muir and Howard, 2001, 2002). Moreover, the GM animal, compared with the conventional counterpart, has a lower, equal or higher potential to invade and establish itself as an alien species, particularly when the receiving ecosystem lacks wild or feral relatives (FAO/WHO, 2003). Released fish or fish that have escaped could introduce new diseases or compete for limited resources. Also, if a sterile transgenic male fish mates with a native female, her reproductive effort is wasted (Muir et al., 2001). Muir and Howard (1999) examined the risk to a natural population after the release of transgenic individuals when the transgene trait simultaneously increased transgenic male mating success and lowered the viability of transgenic offspring. The results of their study indicated that a transgene introduced into a natural population by a small number of transgenic fish will spread as a result of enhanced mating advantage, but the reduced viability of offspring will cause eventual local extinction of both populations.

A research carried out by Hedrick (2001) revealed that if such a transgene has a male-mating advantage and a general viability disadvantage, then the conditions for its invasion in a natural population are very broad. Particularly, 66.7% of the probable combinations of the possible mating and viability parameters, the transgene increased in frequency, and for 50% of the combinations, it went to fixation. In addition, by this increase in the frequency of the transgene, the viability of the natural population was reduced, increasing the probability of extinction of the natural population. Devlin et al. (2001) produced transgenic rainbow trout that were on average 1.7% larger than wild-type trout at sexual maturity. With such size differential, transgenic and wild type individuals may not even recognize each other as conspecifics much less be physically capable of crossing. As a result, a successful establishment of transgenic rainbow trout in nature may well represent the instant formation of a new species or evolutionary line, which may produce a significant environmental hazard.

The effects from the release of GE fish into the ecosystem could be uncontrollable, permanent, and irreversible, and by the time they are present it will be too late (http://www.cfr.msstate.edu/courses/wf3141/carpenter.pdf). Adverse genetic and ecological effects of released aquatic GMOs will depend on characteristics such as the nature and degree of change in the physical characters and performance of the GMO; potential for the GMO to disperse, reproduce, and interbreed; and the GMO's potential for adaptive evolution (U.S. Congress, Office of Technology Assessment, 1995). Methods for detection of such GM animals and their transgenes in the environment are likely to involve the application of two well-established bodies of scientific methodologies: (i) diagnostic, DNA-based markers, and (ii) sampling protocols that are adequate (in terms of statistical power) and cost-effective (FAO/WHO, 2003). The greatest environmental risk that a transgenic fish would have is when the gene insert would allow the transgenic genotype to expand its geographic range, essentially becoming equivalent to an exotic species (Dunham, 2003). In this case, natural selection does not regulate transgene frequency, rather forced crossings and other breeding schemes maintain the transgene (Muir and Howard, 2002). About 1% of such releases of exotics result in adverse environmental consequences. Altering temperature or salinity tolerance would be analogous to the development of an exotic species since this would allow the expansion of a species outside its natural range (Dunham, 2003, 2004).

The presence of GE fish does not a priori have a negative effect on native populations. If GE fish are ill-suited to an environment or are physically unable to survive outside the containment, they may pose little risk to the native ecosystems (Van Eenennaam, 2005). New Zealand's Royal Commission on Genetic Modification found that there is no evidence of horizontal gene transfer through consumption of GMOs, but further stated that "(n)evertheless, more investigation into the effects of substances entering the body is required..." (Royal Commission on Genetic Modification, 2001). However, the safest approach to transgenic aquaculture would be to raise the fish in tanks on land. However, the added cost, an increase of about 40%, would make it difficult for producers to compete with conventional marine fish farmed (Stokstad, 2002).

ETHICAL CONCERNS

Ethics is defined as a set of standards by which a particular group of people decides to regulate its behavior to distinguish what is legitimate or acceptable in pursuit of their aims from what is not (Godown, 1987). Ethical concerns about genetic engineering of animals may conveniently be divided into two basic types (Kaiser, 2005). Extrinsic concerns against the technology are related to the consequences of application, and as such are more closely related to perceived risks or outcomes of applications. However, the possible negative consequences of applications of genetic engineering remain uncrystallized and dependent on the course of the technology, and as such are not subjected to extrinsic moral reasoning in the same way as more concrete issues could be (Frewer and Shepherd, 1995). On the other hand, intrinsic concerns are those that have to do with moral concerns about the process of GE; that it is unnatural or against religious views for one or more reasons (http://www.agbioworld.org/biotech-info/articles/agbio-articles/critical.html). Because the DNA change can be precisely designed, an actual targeted genetic change through genetic engineering should be safer than a natural change because it is more under control. Given the results of public opinion surveys that find opposition to cross species gene transfer (Macer, 1994; Macer and Ng, 2000), if the DNA change is made using DNA within the same species entirely, then this concern can be removed. Therefore, there is no new intrinsic ethical dilemma from the modification of DNA structure in genetic engineering as it simply mimics the natural ways organisms use to change genetic structure (Macer, 2003). The following are categories of concern that apply to most situations:

  i) the process of transgenic organism development may produce risks beyond the apparent phenotypic characteristics of the organism;
  ii) unknown human health and environmental impacts, in particular those that are indirect or only realized over the long-term;

   iii) inadequate regulation, oversight, and testing/evaluation;
   iv) inequitable social and economic implications;
   v) the degree of potential for future benefits of transgenic technologies (Vollmer et al., 2007).

Despite strong public support for medical applications of genetic engineering, there is less public support for agricultural biotechnology. Even if GE fish are approved by the FDA in the United States, it will likely be activist, food retailer, and consumer response in the marketplace that will ultimately decide whether GE food fish will sink or swim (Van Eenennaam, 2005; FDA, 2006).

Some people, irrespectively of the application of technology, consider genetic engineering of animals fundamentally unethical. Others, however, hold that the ethical significance of animal biotechnologies must derive from the risk and benefits to people, the animals, and/or the environment (Han and Harrison, 2007). When consumers perceive benefits to themselves and society, they are expected to be more willing to buy GM food, relative to consumers who perceive no benefit. On the other hand, if consumers perceive GM foods as a health risk and risky to the environment, they would be less willing to purchase GM foods (Han and Harrison, 2007). Questions about health risks, social consequences, and moral credentials arise to disturb the usual balance in social interactions. These factors have upgraded biotechnology to a social phenomenon and have brought it to the forefront of public debates in recent years. In a way, biotechnology has become more of a social issue than a technological development (Liakopoulos, 2002). The following factors influenced the overall consumer behavior vis-à-vis GM foods:

   i) awareness, pre-existing beliefs towards food technology and its risks and other general attitudes,
   ii) perceived GM consumption benefit,
   iii) sociodemographic profile,
   iv) social and moral consciousness,
   v) perceived food quality and trust, and
   vi) other secondary influential factors.

This classification assumes a conflict between the beliefs of 'risks' and 'benefits' (Arvanitoyannis and Exadactylos, 2006). Research has consistently revealed an increasing demand for "ethical" choices in the global marketplace. However, very little has been published about the decision-making processes of these "ethical" consumers and the implications for marketing (Shaw and Shiu, 2003). The role of the Ethical Purchasing Index launched in the UK was to ensure that businesses that act ethically are rewarded at the expense of those who do not respond to ethical practices (Doane, 2001).

A study held by Han and Harrison (2007) revealed that when consumers are willing to buy GM foods, food safety was the crucial factor affecting their decision for both GM crops and GM meats. On the other hand, when consumers decide not to buy, ethical issues and concerns about the side effects of GM crops and GM meats on wildlife and the environment were important factors. In addition consumers have stronger risk sensitivity for GM meats than GM crops, which means lower acceptance rates for GM meats compared to GM crops. FAO noted a general trend that respondents in poorer countries tended to show more support than those in wealthier countries, with some exceptions (FAO, 2004). Another survey shows that religious and ethical concerns play a significant role in consumer attitudes towards cloning, and that a significant majority of consumers believe that the government should include ethical and moral considerations when making regulatory decisions about cloning and GM animals (Pew Initiative on Food and Biotechnology, 2004, 2005, 2006).

In general the public are not in favor of GM although the response depends on the socio-cultural background of the customers. In a British survey of customers' views on the acceptability of transgenic fish in fish farming indicated that 36.5% had no opinion, 34.8% thought it was not acceptable and 28.7% thought it was acceptable (Penman et al., 1995). People in Denmark, Ireland, the Netherlands, and the UK are not against modern biotechnology products; they think it is a good technology for the future but that it needs a high degree of consumer education to surmount the initial hurdles of marketing and presentation to the public. French retailers have taken a high moral stance as champions of 'safe wholesome food.' Furthermore, people in Portugal, Spain, Italy and Greece are largely unaware of the transgenic problem; however due to the food adulteration scandals consumers became alert to the quality of food and particularly of processed foods (Moses, 1999). Ethics committees receive most support in the Netherlands (82%), Finland (81%), Greece (75%) and Denmark (70%) whilst the countries most likely to say that "they do not do good work for society" are France (23%), Ireland and the UK (21% each), and Italy (20%) (Eurobarometer, 2000). Consumer acceptance of genetic engineering may be greater for products of similar quality to conventional products but with reduced cost, but not for GE products of improved quality but greater cost. This indicates that consumer acceptance is likely to be highest for products with enhanced qualities but (implicitly) no increase in cost (Hoban and Kendall, 1992). In a survey carried out in Ireland 82% of the respondents stated that they were concerned about the safety of GM foods (39% were very concerned and 43% were somewhat concerned). The remainder, 18%, stated that they were unconcerned. When they were asked about the frequency of concern for the safety of GM foods influencing purchases, the results showed that the concern for the safety of GM foods had a perceived level of influence in purchasing (Morris and Adley, 2001) (Table 9.2). The American public appears far less receptive to the genetic modification of animals. Only 28% of the respondents approve of such practices, while more than two-thirds of the population disapproves of the genetic modification of animals. Many people, more than half (56%) say that the issue of genetic modification causes them great concern, and many (52%) support the fact that the risks involved in GM food are unacceptable (Table 9.2) (Hallman et al., 2002).

Another view focuses on the right of humans to know what they are eating or how their food or pharmaceuticals are being produced and therefore labeling becomes an issue to be addressed (NRC, 2002). Labeling food is the most commonly suggested solution to this problem, and would be in keeping with the principle of the right to informed choice (FAO, 2001). In fact, labeling would allow those who specifically wish to purchase and/or consume GM products the option of doing so (Power, 2003). According to the Eurobarometer survey held in 1997, 74% of the respondents favored labeling of GM foods (CEC, 1997). In Switzerland, most people believed that GE food is not labeled (58.6%). Only 33.5% correctly believed that food containing GE substances has to be labeled (Siegrist, 2003). Clear labeling policies for all fish products would need to be further developed and appropriately introduced if GM technologies were used. In the EU, regulators have explored this issue under the novel food regulations, but further work is needed and this should be carried out in consultation with consumers, producers, and retailers to ensure clear and informative labeling, hence respecting consumer autonomy (Millar and Tomkins, 2007).

RISK ASSESSMENT

Risk assessment is a systematic process used to identify risks posed by certain activities to human health or to the environment (National Research Council, 1989, 1993). That is why most countries have strict regulations concerning a thorough risk assessment before releasing these animals into the environment or introducing them to the market. In fact, it seems that regulation relating to GM animals including GM fish on average is stricter than regulation applying to the introduction of animals that are the result of selective breeding, or the introduction of animals from foreign environments ("exotic species") (Kaiser, 2003).

There are many potential ecological risks associated with GM fish and many of these would have a high consequence, however, the likelihood of these risks occurring is quite low. The overall levels of potential human health risks are relatively low. GM fish would not pose a serious risk to human health mainly because the likelihood of these risks occurring is very low as food products are subjected to intensive food safety assessments (Galli, 2002). However, little is known about the human health impacts from consuming genetically modified aquatic organisms. However, there are several potential concerns, including: allergic reactions from new proteins, exposure to toxins, and long term health effects (http://www.seaweb.org/resources/aquaculturecenter/documents/Aquaculture.GMOD.pdf).

Ecological risk assessment requires an evaluation of the fitness of the GE fish relative to non-GE fish in the receiving population in order to determine the probability that the transgene will spread into the native population (Van Eenennaam, 2005). Potential risk will exist whenever there are large-scale (relative to the wild population) and continuous releases of individuals with laboratory genetic changes (Knibb, 1997). Risk assessment has generally followed two approaches:

- Net Fitness Approach. This approach is based on six net fitness components or critical control points that influence gene spread: juvenile viability, adult viability, age at sexual maturity, female fecundity, male fertility, and mating advantage (Muir and Howard, 2002).
- Observation of behavior in transgenic fish and inference to risk (Muir and Hostetler, 2001).

Risk assessment in cases of deliberate or accidental release of transgenic fish depends on a number of factors, such as:

- the possibility that GM fish could escape from the physical containment under conditions of storm, flood, theft or human interference,
- the species that are released and the biotope they are released into,
- the character of the transgene and the new phenotype,
- the general fitness of GMOs versus wild populations,
- the number of released GM fish,
- whether the transgenic fish are sterile or partially sterile,
- whether the GM fish could become established as a novel pest species if they were to escape or be released, and
- whether the fish survive in cold or warm water, and in salt, brackish or fresh water (Alestrom, 1996; Anonymous, 2003; Maclean, 2003).

Risk assessment of GMOs is given priority both by national and international institutions engaged in biotechnology development (Alestrom, 1995). National policies on oversight of genetically modified aquatic organisms have been developed in a number of countries including Australia, Canada, New Zealand, Norway, the UK and USA. Countries that do not have any policies should be strongly encouraged to develop guidelines for transgenic fish research and the release of genetically modified aquatic organisms into the environment (Sin, 1997). A better documented database of risk assessment results are needed to establish appropriate regulations governing research, use, and release of GMOs that pose risks to the environment and human consumers. Guidelines could be established with involvement from the relevant federal and state agencies as well as representatives of the aquaculture industry, commercial fishing industry, environmental groups, and other stakeholders (U.S. Congress, Office of Technology Assessment, 1995).

Chinese scientists are developing environmental safety assessment strategies for fast-growth GM fish. Since the beginning of 2005 many studies have been taking place to compare the morphology, life history and behavior of fast-growth GM fish with those of wild fish (Fu et al., 2005; Wei et al., 2007). In an experiment carried out by Zhang et al. (2000) test groups of mice were fed with homogenate of transgenics at the dosages of 5 and 10 g/kg body weight for 6 wk, while the control groups were fed with control fish at the same dosages. The results indicated that the test mice showed no significant differences when compared with the two groups of control mice, including in terms of growth performance, biochemical analysis of blood, histochemical assay of twelve organs, or reproductive ability. Furthermore, Chatakondi et al. (1995) compared the body composition of transgenic common carp containing rainbow trout growth hormone gene with non-transgenic full-siblings. The results indicated an increase of 7.5% in protein content and a decrease of 13% in fat content of the transgenic carp muscle. Significant, but relatively subtle, changes occurred in the flesh of transgenic common carp; however they were not detrimental to human health and safety and may have been beneficial.

CONCLUSIONS

In future, specific GM animals may gain approval for widespread production, either with or without approval to enter them in the human food supply. In such situations, it will be important to consider whether or not to apply post-market monitoring for unexpected environmental spread of the GM animals and their transgenes that pose food safety hazards (FAO/WHO, 2003). Whether the European public becomes as accepting of GE and GM foods as the American public will depend on changed perceptions of the risks to human health and the environment. Such changes will hinge on reliable communication of information from scientists, policy makers, industry and the press. It might require that there is more public participation in agricultural research planning in the future (Middendorf and Busch, 1997).

Ethical questions with regard to aquatic GMOs often focus on whether humans have the right to modify natural creations. Are we over-playing our autonomy? However, humans have been modifying plants, animals and the habitats they live in for millennia. Genetic modification allows humankind to modify nature faster and to a greater extent than before (Bartley et al., 2005). As products come "on stream", it is highly likely that the first products to appear will have the greatest impact on consumer attitudes. In the case of food produced with the use of genetic engineering, control by individuals may be linked to labeling of products and without such labeling people may feel they have little opportunity for personal control. However, labeling will have little impact without public understanding of what the labels means. There is a clear need for effective communication strategies to facilitate public understanding of the technology (Frewer et al., 1995, 1996).The success of these technological breakthroughs depends on the dynamic participation of the public in discussions about their development (Leroux et al., 1998).

ABBREVIATIONS

AFP	Antifreeze Protein
FAO	Food and Agriculture Organization of the United Nations
FDA	Food and Drug Administration
FWS	Fish and Wildlife Service
GE	Genetically Engineered
GFP	Green Fluorescent Protein
GH	Growth Hormone
GM	Genetically Modified
GMO	Genetically Modified Organism
GnRH	Antisense Gonadotropin Releasing Hormone
hGH	Human Growth Hormone
ICES	International Council for Exploration of the Seas
IFICF	International Food Information Council Foundation
NFA	Net Fitness Approach
NMFS	National Marine Fisheries Service
OECD	Organization for Economic Cooperation and Development

REFERENCES

Aerni, P. (2004). Risk, regulation and innovation: The case of aquaculture and transgenic fish. Aquat. Sci. 66: 327–341.

Afanassiev, V., Sefton, M., Anantachaiyong, T., Barker, G., Walmsley, R. and Wolfl, S. (2000). Application of yeast cells transformed with GFP expression constructs containing the RAD54 or RNR2 promoter as a test for the genotoxic potential of chemical substances. Mutat. Res.-Genet. Toxicol. Environ. Mutat. 464: 297–308.

Alestrom, P. (1995). Genetic engineering in aquaculture. In: Proceedings of the First International Symposium on Sustainable Fish Farming. Reinertsen and Haaland (eds). Oslo, Norway 28-31 August 1994, pp. 125-130.

Alestrom, P. (1996). Biotechnology Seminar Paper. Genetically modified fish in future aquaculture: technical, environmental and management considerations. Netherlands: International Service for National Agricultural Research [Online: http://www.agbios.com/docroot/articles/02-254-005.pdf].

Anonymous (2003). Genetically modified foods for human health and nutrition: the scientific basis for benefit/risk assessment. Trends Food Sci. Technol. 14: 173-181.

Arvanitoyannis, I.S. (2006). Genetically modified plants: applications and issues. In: Microbial Biotechnology in Agriculture and Aquaculture, Vol. II. R.C. Ray (ed). Science Publishers, New Hampshire, USA, pp. 25-76.

Arvanitoyannis, I.S. and Krystallis, A. (2005). Consumers' beliefs, attitudes and intentions towards genetically modified foods, based on the 'perceived safety vs. benefits' perspective. Int. J. Food Sci. Technol. 40: 343–360.

Arvanitoyannis, I.S. and Exadactylos, A. (2006). Aquaculture biotechnology aiming at enhanced fish production for human consumption. In: Microbial Biotechnology in Agriculture and Aquaculture, Vol. II. R.C. Ray (ed), Science Publishers, New Hampshire, USA. pp. 453-510.

Bartley, D.M., Ryder, J. and Ababouch, L. (2005). Genetically modified organisms in aquaculture. In: Fifth World Fish Inspection and Quality Control Congress, The Hague, Netherlands, 20-22 October, 2003 [Online: www.fao.org/docrep/008/y5929e/y5929e09.htm].

Beardmore, J.A. and Porter, J.S. (2003). Genetically Modified Organisms and Aquaculture. FAO Fisheries Circular No. 989 FIRI/C989(E). FAO, ISSN 0429-9329 [Online: ftp://ftp.fao.org/docrep/fao/006/y4955e/Y4955E00.pdf].

Borgatti, R. and Buck, E.H. (2004). Genetically Engineered Fish and Seafood. CRS Report for Congress. December 7, 2004 [Online: http://www.ncseonline.org/nle/crsreports/04dec/RS21996.pdf].

CEC (1997). The Europeans and modern biotechnology. Eurobarometer 46.1. European Commission, Luxemburg.

Chatakondi, N., Lovell, R.T., Duncan, P.L., Hayat, M., Chen, T.T., Powers, D.A., Weete, J.D., Cummins, K. and Dunham, R.A. (1995). Body composition of transgenic common carp, *Cyprinus carpio*, containing rainbow trout growth hormone gene. Aquaculture 138: 1–4.

Chen, T.T. and Powers, D.A. (1990) Transgenic fish. Trends Biotechnol. 8: 209-214.

Chern W.S. (2002). Consumer Acceptance of GMO: Survey Results from Japan, Norway, Taiwan, and the United States. Department of Agricultural, Environmental and Development Economics, The Ohio State University, Working Paper: AEDE-WP-0026-02, September 2002 [Online: http://aede.osu.edu/resources/docs/pdf/5FE4959A-C1D0-44E7-8DC93811C5418C25.pdf].

Chitminat, C. (1996) Predator avoidance of transgenic channel catfish containing salmonid growth hormone genes. Master of Science Thesis, Auburn University, Auburn, AL, USA.

Cockburn, A. (2002). Assuring the safety of GM food. J. Biotechnol. 98: 79-106.

Cowan, T. and Becker, G.S. (2006). Biotechnology in Animal Agriculture: Status and Current Issues. CRS Report for Congress, March 27, 2006 [Online: http://italy.usembassy.gov/pdf/other/RL33334.pdf].

Davidson, D.J. and Freudenburg, W.R. (1996). Gender and environmental risk concerns: a review and analysis of available research. Environ. Behav. 28: 302-339.

de la Fuente, J., Hernandez, O., Martinez, R., Guillen, I., Estrada, M.P. and Lleonart, R. (1996). Generation, characterization and risk assessment of transgenic Tilapia with accelerated growth. Biotechnol. Applic. 13: 221-230.

Devlin, R.H., Yesaki, T.Y., Biagi, C.A., Donaldson, E.M., Swanson, P. and Chan, W.K. (1994). Extraordinary salmon growth. Nature 371: 209-210.

Devlin, R.H., Johnsson, J.I., Smailus, D.E., Biagi, C.A., Jonsson, E. and Bjornsson, B.T (1999). Increased ability to compete for food by growth hormone-transgenic coho salmon *Oncorhynchus kisutch* (Walbaum). Aquacul. Res. 30: 479-482.

Devlin, R.H., Biagi, C.A., Yesaki, T.Y., Smalius, D.E. and Byatt, J.C. (2001). Growth of domesticated transgenic fish. Nature 409: 781-782.

Devlin, R.H., D'Andrade, M., Uh, M. and Biagi, C.A.(2004). Population effects of growth hormone transgenic coho salmon depend on food availability and genotype by environment interactions. Proc. Natl. Acad. Sci. USA, 101(25): 9303–9308.

Doane, D. (2001). Taking Flight: The Rapid Growth of Ethical Consumerism, New Economics Foundation, London, UK.

Dunham, R.A. (2003). Status of genetically modified (transgenic) fish: research and application. Working paper topic 2. Food and Agriculture Organization/World Health Organization expert hearings on biotechnology and food safety [Online: ftp://ftp.fao.org/es/esn/food/GMtopic2.pdf].

Dunham, R.A. (2004). Aquaculture and Fisheries Biotechnology Genetic Approaches. CABI Publishing, Wallingford, UK.

Eurobarometer (2000). Europeans and biotechnology. Eurobarometer 52.1. Luxemburg, Office for Official Publications of the European Communities. Brussels, European Commission, Research DG, 2000 [Online: http://europa.eu.int/comm/research/pdf/eurobarometer-en.pdf].

FAO (2001). Genetically modified organisms, consumers, food safety and the environment. Food and Agriculture Organization of the United Nations, Rome, 2001.

FAO/WHO (2003). Safety Assessment of Foods Derived from Genetically Modified Animals, including Fish. Rome, 17–21 November 2003 [Online: http://www.who.int/foodsafety/biotech/meetings/en/gmanimal_reportnov03_en.pdf].

Fletcher, G., Shears, M.A., King, M.J. and Goddard, S.V. (2002). Transgenic salmon for culture and consumption. In: International Congress on the Biology of Fish. W. Driedzic, S. McKinley and D. MacKinlay (eds). University of British Columbia: Vancouver, Canada.

Food and Agriculture Organization (FAO) (2004). The State of Food and Agriculture 2003-2004 (Online:http://www.fao.org/docrep/006/Y5160E00.HTM).

Food and Drug Administration (FDA) (2006). Questions and answers about transgenic fish. Center for Veterinary Medicine. (Online: www.fda.gov/cvm/transgen.htm).

Frewer, L.J. and Shepherd, R. (1995). Ethical concerns and risk perceptions associated with different applications of genetic engineering: Interrelationships with the perceived need for regulation of the technology. Agric. Hum. Values 12: 50-57.

Frewer, L.J., Howard, C. and Shepherd, R. (1995). Genetic engineering and food: what determines consumer acceptance? Brit. Food J. 97(8): 31-36.

Frewer, L.J., Howard, C. and Shepherd, R. (1996). The influence of realistic product exposure on attitudes towards genetic engineering of food. Food Qual. Pref. 7(1): 61-67.

Fu, C., Cui, Y., Hung, S.S.O. and Zhu, Z. (1998). Growth and feed utilization by F4 human growth hormone transgenic carp fed diets with different protein levels, J. Fish Biol. 53: 115-129.

Fu, C., Hu, W., Wang, Y. and Zhu Z. (2005). Developments in transgenic fish in the People's Republic of China. Rev. Sci. Tech. Off. Int. Epiz. 24(1): 299-307.

Galli, L. (2002). Genetic Modification in Aquaculture. A review of potential benefits and risks. Bureau of Rural Sciences, Canberra [Online: http://affashop.gov.au/PdfFiles/gm_in_aquaculture.pdf].

Gaskell, G., Allum, N., Bauer, M., Durant, J., Allansdottir, A., Bonfadelli, H., Boy, D., de Cheveigné, S., Fjaestad, B., Gutteling, J.M., Hampel, J., Jelsoe, E., Jesuino, J.C., Kohring, M., Kronberger, N., Midden, C., Nielsen, T.H., Przestalski, A., Rusanen, T., Sakellaris, G., Torgersen, H., Twardowski, T. and Wagner, W. (2000). Biotechnology and the European Public. Nat. Biotechnol. 18(9): 935-938.

Gaskell, G., Allum, N. and Stares, S. (2003). Europeans and biotechnology in 2002: Eurobarometer 58.0. [Online: http://europa.eu.int/comm/research/pdf/eurobarometer-en.pdf].

Godown, R.D. (1987). The Science of Biotechnology. In: Public Perceptions of Biotechnology. L.R. Batra and W. Klassen (eds). Maryland: Agricultural Research Institute, Maryland, USA.

Hallman, W.K., Adelaja, A.O., Schilling, B.J. and Lang, J.T. (2002). Public Perceptions of Genetically Modified Foods. Americans Know Not What They Eat. Food Policy Institute, Rutgers, The State University of New Jersey [http://www.foodpolicyinstitute.org/docs/reports/Public%20Perceptions%20of%20Genetically%20Modified%20Foods.pdf].

Hallman, W.K., Hebden, W.C., Cuite, C.L., Aquino, H.L. and Lang, J.T. (2004). Americans and GM Foods: Knowledge, Opinion and Interests in 2004. Publication Number RR 1104-007, Food Policy Institute, Cook College, Rutgers—The State University of New Jersey, USA.

Han, J.-H. and Harrison, R.W. (2007). Factors Influencing Urban Consumers' Acceptance of Genetically Modified Foods. Rev. Agric. Econ. 29(4): 700-719.

Hedrick, P.W. (2001). Invasion of transgenes from salmon or other genetically modified organisms into natural populations. Can. J. Fish. aquat. Sci. 58(5): 841-844.

Hew, C.L. and Fletcher, G.L. (1997). Transgenic fish for aquaculture. Chem. Ind. April 21: 311-314.

Hoban, T.J. and Kendall, P.A. (1992). Consumer Attitudes about the Use of Biotechnology in Agriculture and Food Production. North Carolina State University, Raleigh, NC, USA.

http://www.agbioworld.org/biotech-info/articles/agbio-articles/critical.html (accessed October 27, 2007).

http://www.cfr.msstate.edu/courses/wf3141/carpenter.pdf (accessed October 27, 2007).

http://www.seaweb.org/resources/aquaculturecenter/documents/Aquaculture.GMOD.pdf (accessed October 29, 2007).

IFIC (2005). IFIC Survey: Food Biotechnology Not a Top-of-Mind Concern for American Consumers. June 2005 [Online: http://www.ific.org/research/upload/2005BiotechSurvey.pdf].

International Council for Exploration of the Sea (ICES) (1996). Report of the Working Group on the Application of Genetics in Fisheries and Mariculture. 1996. ICES C.M. 1996/F:2.

Jönsson, E., Johnsson, J.I. and Bjornsson, B.T. (1998). Growth hormone increases aggressive behavior in juvenile rainbow trout. Hormones and Behavior 33: 9-15.

Kaiser, M. (2003). Ethical issues surrounding the gm-animals/gm-fish production. In: FAO expert meeting 17-21 November 2003 [Online: ftp://ftp.fao.org/es/esn/food/GMtopic6.pdf].

Kaiser, M. (2005). Assessing ethics and animal welfare in animal biotechnology for farm production. Rev. sci. tech. Off. int. Epiz. 24(1): 75-87.

Kapuscinski, A. and Hallerman, E. (1990). Transgenic fish and public policy: anticipating environmental impacts of transgenic fish. Fisheries 15(1): 2-11.

Kapuscinski, A. and Hallerman, E. (1991). Implications of introduction of transgenic fish into natural ecosystems. Can. J. Fish. Aquat. Sci. 48: 99-107.

Kapuscinski, A.R. and Hallerman, E.M. (1994). Benefits, Environmental Risks, Social Concerns, and Policy Implications of Biotechnology in Aquaculture. Contract report prepared for the Office of Technology Assessment, U.S. Congress. Washington, D.C.: National Technical Information Service.

Knibb, W. (1997). Risk from genetically engineered and modified marine fish. Transgenic Res. 6: 59-67.

Lassen, J., Madsen, K.H. and Sandoe, P. (2002). Ethics and genetic engineering — lessons to be learned from GM foods. Bioprocess Biosyst. Eng. 24: 263-271.

Leroux, T., Hirtle, M. and Fortin, L.-N. (1998). An Overview of Public Consultation Mechanisms Developed to Address the Ethical and Social Issues Raised by Biotechnology. J. Consumer Policy 21: 445-481.

Levy, J.A., Marins, L.F. and Sanchez, A. (2000). Gene transfer technology in aquaculture. Hydrobiologia 420: 91-94.

Liakopoulos, M. (2002). Pandora's Box or panacea? Using metaphors to create the public representations of biotechnology. Public Understand. Sci. 11: 5-32.

Macer, D. (2003). Ethical, legal and social issues of genetically modified disease vectors in public health. TDR/STR/SEB/ST/03.1 (World Health Organisation, Geneva, Switzerland, 2003) [Online: http://www.who.int/tdr/cd_publications/pdf/seb_topic1.pdf].

Macer, D.R.J. (1994). Bioethics for the people by the people. Christchurch, Eubios Ethics Institute, 1994. [Online: http://www.biol.tsukuba.ac.jp/~macer/BFP.html].

Macer, D.R.J. and Ng, M.C. (2000). Changing attitudes to biotechnology in Japan. Nat. Biotechnol. 18: 945-947.

Maclean, N. (2003). Genetically Modified Fish and Their Effects on Food Quality and Human Health and Nutrition. Trends Food Sci. Technol. 14: 242-252.

Maclean, N. and Laight, R.J. (2000). Transgenic fish: an evaluation of benefits and risks. Fish 1: 146–172.

Maclean, N. and Talwar, S. (1984). Injection of cloned genes with rainbow trout eggs. Journ. Embryol and Exp. Morphol. 82: 187.

Middendorf, G and L. Busch (1997). Inquiry for the public good: Democratic participation in agricultural research. Agric. Human Values 14: 45-57.

Millar, K. and Tomkins, S. (2007). Ethical analysis of the use of gm fish: emerging issues for aquaculture development. J. Agric. Environ. Ethics 20: 437-453.

Morris, S.H. and Adley, C.C. (2001). Irish public perceptions and attitudes to modern biotechnology: an overview with a focus on GM foods. Trends Biotechnol. 19(2): 43-48.

Moses, V. (1999). Biotechnology products and European consumers. Biotechnol. Adv. 17: 647-678.

Muir, W. (2004). The threats and benefits of GM fish. EMBO Rep. 5(7): 654-659.

Muir, W.M. and Howard, R.D. (1999). Possible ecological risks of transgenic organism release when transgenes affect mating success: sexual selection and the Trojan gene hypothesis. Proc. Natl. Acad. Sci., USA, 96: 13853-13856.

Muir, W.M. and Hostetler, H.A. (2001). Transgenic fish: production, testing, and risk assessment. Biotechnol. Anim. Husband. 5: 261-281.

Muir, W.M. and Howard, R.D. (2001). Fitness components and ecological risk of transgenic release: a model using Japanese medaka (*Oryzias latipes*). American Naturalist 158: 1-16.

Muir, W.M. and Howard, R.D. (2002). Assessment of possible ecological risks and hazards of transgenic fish with implications for other sexually reproductive organisms. Transgenic Res. 11(2): 101-114.

Muir, W.M., Howard, R.D. and Otto, S.P. (2001). Fitness Components and Ecological Risk of Transgenic Release: A Model Using Japanese Medaka (*Oryzias latipes*). American Naturalist 158: 1-16.

Myhr, A.I. and Dalmo, R.A. (2005). Introduction of genetic engineering in aquaculture: Ecological and ethical implications for science and governance. Aquaculture 250: 542-554.

National Research Council, Commission on Life Sciences, Board on Environmental Studies and Toxicology, Committee on Risk Assessment Methodology (1993). Issues in Risk Assessment, National Academy Press, Washington, DC, USA, pp. 247-271.

Needham, E.A. and Lehman, H. (1991). Farming Salmon Ethically. J. Agric. Environ. Ethics 4(1): 78-81.

NRC (National Research Council) (1989). Field Testing Genetically Modified Organisms: Framework for Decisions. National Academy Press, Washington, DC, USA.

NRC (2002). Animal Biotechnology: Science Based Concerns. pp. 7 [Online: http://www.nap.edu/openbook.php?record_id=10418&page=1].

Pardo, R., Midden, C. and Miller, J.D. (2002). Attitudes toward biotechnology in the European Union. J. Biotechnol. 98: 9-24.

Penman, D.J., Woodwark, M. and McAndrew, B.J. (1995). Genetically modified fish populations. In: Environmental Impacts of Aquatic Biotechnolqy. OECD documents, pp. 22-27.

Pew Initiative on Food and Biotechnology (2001). The Gene Is Out of The Bottle: Where To Next?. Pew Initiative on Food and Biotechnology, Washington [Online: http://pewagbiotech.org/events/0523/0523pollfindings.pdf].

Pew Initiative on Food and Biotechnology (2002). Environmental Savior or Saboteur? Debating the Impacts of Genetic Engineering. Pew Initiative on Food and Biotechnology, Washington [Online: http://pewagbiotech.org/research/survey1-02.pdf].

Pew Initiative on Food and Biotechnology (2003). Future fish: issues in science and regulation of transgenic fish. Pew Initiative on Food and Biotechnology, Washington, pp. 3, 10, 20, 21, 34 [Online: pewagbiotech.org/research/fish/fish.pdf].

Pew Initiative on Food and Biotechnology (2004). Americans' Opinions About Genetically Modified Foods Remain Divided, But Majority Want a Strong Regulatory System. Pew Initiative on Food and Biotechnology, Washington [Online: http://pewagbiotech.org/newsroom/releases/112404.php3].

Pew Initiative on Food and Biotechnology (2005). Public Sentiments About Genetically Modified Foods. Pew Initiative on Food and Biotechnology, Washington [Online: http://pewagbiotech.org/research/2005update/].

Pew Initiative on Food and Biotechnology (2006). Pew Initiative Finds Public Opinion About Genetically Modified Foods Remains 'Up For Grabs' Ten Years After Introduction of Ag Biotech. Pew Initiative on Food and Biotechnology, Washington [Online: http://pewagbiotech.org/newsroom/releases/120606.php3].

Power, M.D. (2003). Lots of fish in the sea: Salmon aquaculture, genomics and ethics. Electronic Working Papers Series. W. Maurice Young Centre for Applied Ethics. University of British Columbia [Online: www.ethics.ubc.ca].

Regal, P.J. (1994). Scientific principles for ecologically based risk assessment of transgenic organisms. Mol. Ecol. 3: 5-13.

Royal Commission on Genetic Modification (2001). Report of the Royal Commission on Genetic Modification. Wellington, Royal Commission on Genetic Modification. 465.

Shaw, D. and Shiu, E. (2003). Ethics in consumer choice: a multivariate modeling approach. Eur. J. Market. 37(10): 1485-1498.

Siegrist, M. (2003). Perception of gene technology, and food risks: results of a survey in Switzerland. J. Risk Res. 6(1):45–60.

Sin, F.Y.T. (1997). Transgenic fish. Rev. Fish Biol. 7: 417-441.

Stokstad, E. (2002). Engineered fish: Friend or Foe of the Environment. Science 297: 1797-1798.

U.S. Congress, Office of Technology Assessment (1995). Selected Technology Issues in U.S. Aquaculture. September 1995. Washington, DC [Online: http://www.princeton.edu/~ota/disk1/1995/9555/9555.PDF].

Van Eenennaam, A.L. (2005). Genetic Engineering Fact Sheet 8: Genetic Engineering and Fish. University of California, Division of Agriculture and Natural Resources, Agricultural Biotechnology in California Series, Publication # 8185 [Online: http://anrcatalog.ucdavis.edu/pdf/8185.pdf].

Van Eenennaam, A.L. and Olin, P.G. (2006). Careful risk assessment needed to evaluate transgenic fish. Cal. Agric. 60(3): 126-131.

Volckaert, F. and Ollevier, F. (1998). Transgenic fish. The future of fish with novel genes. Gen. Aquacul. Africa 326: 33-58.

Vollmer, E., Creamer, N. and Mueller, P. (2007). Sustainable Agriculture and Transgenic Crops. In: The Second National Conference on Facilitating Sustainable Agriculture Education, Cornell University, July 11-14.

Wei, H., YaPing, W. and ZuoYan, Z. (2007). Progress in the evaluation of transgenic fish for possible ecological risk and its containment strategies. Sci. China Ser C-Life Sci. 50(5): 573-579.

Winkler, C. and Schartl, M. (1997). Gene transfer in laboratory fish, model organisms for the analysis of gene function. In: Transgenic Animals Generation and Use. L.M. Houdebine (ed). Harwood Academic Publishers, France, pp. 387-395.

Wu, G., Sun, Y. and Zhu, Z. (2003). Growth hormone gene transfer in common carp. Aquat. Living Resour. 16: 416-420.

Zhang, F., Wang, Y., Hu, W., Cui, Z. and Zhu, Z. (2000). Physiological and pathological analysis of the mice fed with "all-fish" gene transferred yellow river carp. High Technol. Lett. 7: 17-19.

Zhu, Z.Y., Li, G., He, L. and Chen, S. (1985). Novel gene transfer into the fertilised eggs of the goldfish *Carassius acuratus* L. 1758. J. Appl. Ichthyol. 1: 31-34.

INDEX

Printed and bound by CPI Group (UK) Ltd, Croydon, CR0 4YY

18/10/2024

01776267-0002

Color Plate Section

Chapter 2

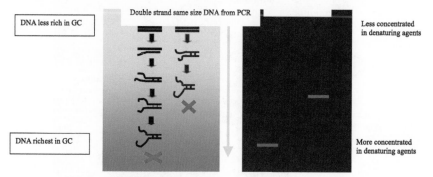

DNA less rich in GC

Double strand same size DNA from PCR

Less concentrated in denaturing agents

DNA richest in GC

More concentrated in denaturing agents

Fig. 2.1 Principle of DGGE migration (Leesing, 2005)

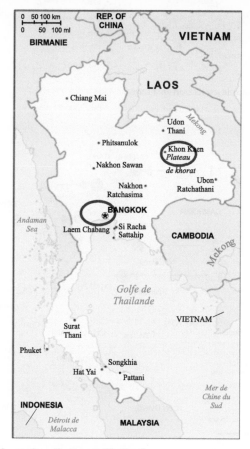

Fig. 2.2 Place of sample collection in Thailand.

Fig. 2.3 Place of sample collection in France.

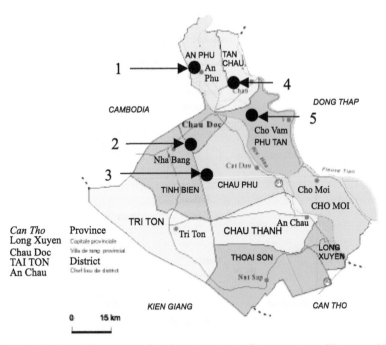

Fig. 2.4 The five different sampling locations in An Giang province, Vietnam : (1) An Phu; (2) Chau Doc; (3) Chau Phu; (4) Tan Chau; (5) Phu Tan.

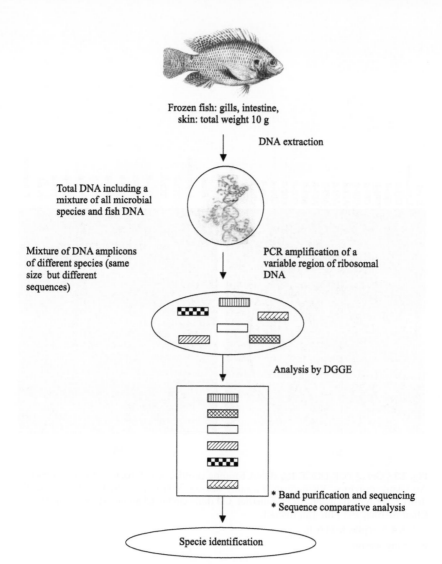

Frozen fish: gills, intestine,
skin: total weight 10 g

DNA extraction

Total DNA including a
mixture of all microbial
species and fish DNA

Mixture of DNA amplicons
of different species (same
size but different
sequences)

PCR amplification of a
variable region of ribosomal
DNA

Analysis by DGGE

* Band purification and sequencing
* Sequence comparative analysis

Specie identification

Fig. 2.5 Protocol for isolating bacterial flora from frozen fish and different steps of PCR-DGGE analysis.

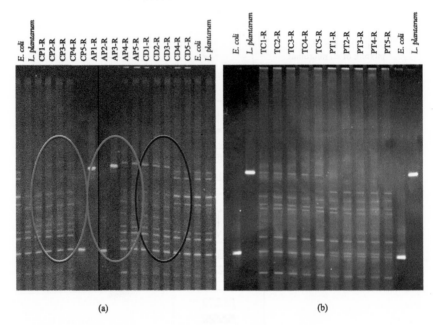

Fig. 2.9 Gel of PCR-DGGE 16S rDNA banding profiles of fish bacteria from 5 districts of An Giang province, Vietnam in the rainy season 2006 (a): CP (red circle): Chau Phu district; AP (green circle): An Phu district; CD (blue circle): Chau Doc district; (b): TC: Tan Chau district, PT: Phu Tan district.

1, 2, 3, 4, 5: replicate of fish.

R = rainy season.

Chapter 6

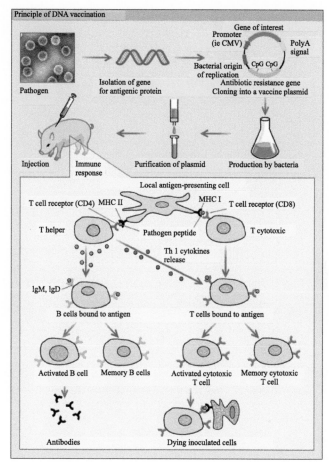

Fig. 6.1 The principle for DNA vaccination, from isolation of the gene encoding the antigenic protein to final immune responses. CD4, cluster of differentiation 4 co-receptor; CD8, cluster of differentiation 8 co-receptor; MHC-I, major histocompatibility complex class I protein; MHC-II, major histocompatibility complex class II protein; IgM, immunoglobulin M; IgD, immunoglobulin D; TCR: T cell receptor; Th1, T helper 1. (From Dafour, 2001)